Pro/ENGINEER Solutions

Advanced Techniques and Workarounds

Bob Townsend and Greg Schmidt

OnWord Press
Thomson Learning™

Africa • Australia • Canada • Denmark • Japan • Mexico • New Zealand
Phillipines • Puerto Rico • Singapore • United Kingdom • United States

Trademarks

Pro/ENGINEER is a registered trademark of Parametric Technology Corporation. Pro/FEATURE, Pro/INTERFACE, Pro/DETAIL, Pro/ASSEMBLY, Pro/SURFACE, and Pro/PLOT are trademarks of Parametric Tehcnology Corporation. Many other products and services are mentioned in this book that are either trademarks or registered trademarks of their respective companies. OnWord Press and the authors make no claim to these marks.

OnWord Press Staff

Publisher: Alar Elken

Executive Editor: Sandy Clark

Managing Editor: Carol Leyba

Development Editor: Daril Bentley

Editorial Assistant: Allyson Powell

Executive Marketing Manager: Maura Theriault

Executive Production Manager: Mary Ellen Black

Production and Art & Design Coordinator: Cynthia Welch

Manufacturing Director: Andrew Crouth

Technology Project Manager: Tom Smith

Cover Design by Lynne Egensteiner and Cammi Noah

Copyright © 2000 by OnWord Press   An imprint of Thomson Learning
SAN 694-0269
First Edition, 2000
10 9 0 7 6 5 4 3 2 1
Printed in Canada

Library of Congress Cataloging-in-Publication Data

Townsend, Bob, 1963–

  Pro/Engineer advanced techniques and workarounds / Bob Townsend and Greg Schmidt.

    p.   cm.

ISBN 1-56690-163-4

1. Engineering design—Data processing. 2. Pro/ENGINEER. 3. Computer-aide design. I. Schmidt, Greg, 1958– .
II. Title.

TA174.T68 1999

620' .0042'028553—dc21   98-55575
                                          CIP

For more information, contact

OnWord Press   An imprint of Thomson Learning

Box 15-015   Albany, New York USA 12212-15015

About the Authors

Bob Townsend is the engineering manager at IMPACT Engineering Solutions, Inc. He has over 15 years of engineering experience with several manufacturing companies in the Milwaukee area. Bob has been a Pro/ENGINEER designer and system administrator since 1989 and helped form the Wisconsin Pro/ENGINEER user group, for which he served as a board member for three years.

Greg Schmidt is a senior product designer with the Harley-Davidson Company. He has over 20 years of engineering experience with several manufacturing companies in the Milwaukee area. Greg has been a CAD designer and administrator since 1983 and has been using Pro/ENGINEER since 1991.

Acknowledgments

I first need to thank my family—my wife, Kathi, and my three children, Andy, Amanda, and Brittany—for giving me the time to write this book. I also need to thank my employer, Harley-Davidson, and my co-workers for sharing some of their modeling techniques with me so I could pass them on to you. A special thanks goes out to Mark Peters, president of Impact Engineering Solutions, for the generous use of their computers and facilities throughout the writing of the book. Last but not least, I need to thank my co-author, and friend, Bob Townsend for giving me the opportunity to work on this book with him. Bob's Pro/ENGINEER skills impressed me back in 1991 when I started using Pro/ENGINEER and they continue to amaze me today. Thank you all.

Greg Schmidt

First off I'd like to thank my wife Barb for her patience and for giving me the time to write this book. I also what to thank my co-author Greg Schmidt for all his hard work in putting this book together. It's been a few years since we've had the opportunity to work together, so this made the time on the book much more enjoyable. In addition, I'd like to thank Mark Peters, president of Impact Engineering Solutions, for the use of the computers and facilities. Last, I'd like to say thanks to Daril Bentley for his work on the editing process. There were times when we were all shaking our heads, wondering what we'd gotten ourselves into.

Bob Townsend

Contents

List of Exercises

Introduction

A Few Concepts Before You Begin

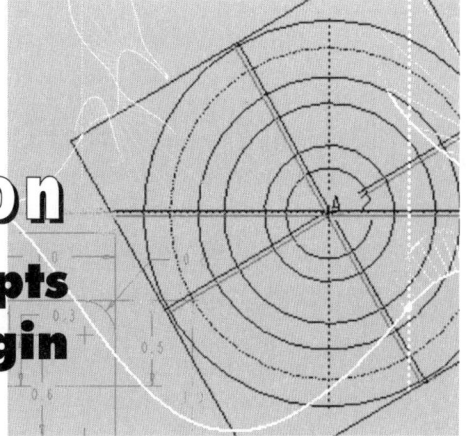

Solid modeling software are tools. They allow us as designers and engineers to create products in virtual space. These virtual products can then be assembled, inspected, analyzed, redesigned, and interrogated to ensure they meet all specifications for form, fit, function, and manufacture. All of this can be accomplished before spending money on tooling, or on potential retooling if you did not catch your design mistakes.

Of course, nothing is free, or is as easy as the previous scenario would seem. Pro/ENGINEER, like any CAD software, expands the abilities of designers and engineers. It does not make a poor designer a good one, but it does have the potential to make a good designer a better one, if the designer understands how to create intelligent, user-friendly solid models.

With Pro/ENGINEER, you have the ability to know important things about your designs: where their strengths and weaknesses are, how they work in regard to manufacture, and how components will work in regard to assembly and how they will interact. All such design constraints can be simulated. Flaws in your designs can be located long before thousands of dollars are spent on tooling changes.

Flexibility and Planning

The key to any design is flexibility. Your product needs to change and upgrade with the market's demands. Your solid models need to be flexible enough to incorporate such rapid changes. To model with flexibility in mind, you must plan your designs prior to creating them on your computer. This does not mean you must know every detail of every part in a product before you start designing it. It does mean that you have a definite idea of the model's fundamental characteristics and overall size, approximately how many components it will consist of, how it might be assembled, and how it might be manufactured. The planning process is what you have some control over in the design process.

Before any design is started, it is critical that you do some planning. Just like the person painting a floor, without a little planning you might find yourself painted into a corner, with the option of waiting until the paint dries before you can continue or redoing a lot of work. A poorly planned design in Pro/ENGINEER can produce the same result. In modeling terms, you end up with nonmodifiable and unusable Frankenstein models with no friends and a million parent/child relations, all needing to be rebuilt.

Fortunately, Pro/ENGINEER offers a great many tools for straightening out a model. Unfortunately, most project schedules do not allow time for reworking a bad design. The ideal is to plan so well at the outset, and make all the right moves along the way, so that you encounter no significant problems. However, finding yourself cornered by your design is inevitable, and is not the end of the world. The point is to learn from such experiences and avoid the same negative situation the next time around, and to be able to correct unavoidable problems.

No one ever learned anything from doing it right the first time, and anyone who has ever learned how to design in Pro/ENGINEER has looked back at their first models and asked "What was I thinking?"

Top Down or Bottom Up

"Top down" and " bottom up" are approaches to modeling, both of which are useful in product design. Top down design is done by looking at your end product and then breaking it down into its subassemblies and component parts, planning how each relates to the others and how they will be assembled, and assessing which are the major components and critical features. You need to keep these things in mind while building related geometry and components.

At the beginning of the planning process, you do not need to know every detail about each component or subassembly. You should, however, have a good understanding of what the end product will be, what its function is, and how it will fit into any existing environment. These are questions you should answer prior to creating any model geometry. Top down design allows you to use areas of Pro/ENGINEER you barely remember and seldom use.

However, techniques such as skeleton model use and master modeling can greatly assist you in capturing your design intent, and serve as communication tools to convey design intent to others that may need to modify your original design. Proper planning, model creation, and the use of some good techniques will allow you to capture and communicate your design intent while maintaining control over the entire product design process.

Bottom up design is a bit of a misnomer. The term basically means starting at the component level and working up toward the master assembly. However, this will not work unless you know at least the fundamentals of what that master assembly is. With this method, you typically are not using the Engineering Notebook, nor are you assigning overall design intent. You may end up with the same result, but you leave more room for interpretation and potential errors.

Most design work today is being done with the bottom up methodology. The main reason for this is that a great many companies have been making the same or similar products for a number of years, employing a "throw-it-over-the-wall" strategy. This has let them develop paradigms in their designs, and because it has always worked that way, there is reluctance to embrace change. However, this way of thinking about the design process will eventually become uncommon as designers and engineers start taking control over their designs.

This paradigm will shift and cause a rapid acceleration in product design cycles, and will allow companies to again capture the knowledge they need to maintain and create new products. In short, the mind-set must change along with the technology and the tools.

Selecting the Right Tool

With Pro/ENGINEER, you have been given more tools than you need to accomplish most design tasks. The key is to select the right tool for each job. You can put screws in with a hammer, but you get better results with a power screwdriver. Out of one hundred ways of doing something, five methods are really good, ten more will work well, and the remaining eighty-five are obnoxious. Your goal should be to consistently select and use the best five methods, resorting to the next ten only if the first five do not work, and to the last eighty-five only because the first fifteen failed somehow.

General Design Rules

The following are general rules to follow when modeling parts and assemblies in Pro/ENGINEER. Most are obvious, but they are all too often overlooked as you work, leading to model modification problems downstream.

1. *Always make models with the understanding that you will not be the last person to work on them.* You have all picked up someone else's models and after going through Regen Info wondered "What were they thinking?" If you keep that in mind as you build models, you will start to build more logically, and impart your design intent. This is like building a model map that can easily be followed and understood. If you do not, you are just leaving your problems for someone else to deal with.

2. *There are two types of relations: the ones you build and the ones you did not know you built.* The second type will kill your design's flexibility every time. You need to plan your work and then work your plan.

3. *Never use a surface where solid geometry will work just fine.* The result of cuts and protrusions made from using quilts may be mathematically correct, but those surfaces are what is needed to make modifications. Because you have used the surface, features may no longer be visible for picking and changing, making the model difficult to work on. Using surfaces can get you out of some tough design problems, but they also have the disadvantage of increasing your feature count.

4. *All circles drawn in Sketch mode consist of two 180-degree arcs.* This is important when creating draft, rounds, and rib features that intersect those circular features. Overlapping geometry, regen failures, and geom-checks can often result from them. It makes for an interesting "gotcha" when working on large models where no geom-checks are allowed.

5. *Rebuilding geometry from scratch is not the worst thing to happen in a design.* As you model a part, you learn how the features interact with each other. The problems that occurred on the first go-around can often be corrected and simplified, resulting in a more user friendly and robust design.

6. *Experiment with new procedures on your own time.* Making critical models into Frankensteins will cause huge modification nightmares during design changes. If you are making great leaps of technique during your Pro/ENGINEER sessions, document what you have done. You may need that trail of bread crumbs two months down the line, when changes come up.

7. *Exercise your designs.* You have planned your design intent into your models. Make sure they do what you have intended.

8. *Keep in mind the KISS principal: Keep it simple, stupid.* There is no need to add complexity to your design. Let your design's complexity develop, as a result of combinations of simple features, rather than begin as one complex feature difficult to modify later.

9. *Start every model with a default datum plane and several named views.* If your design will need geometric tolerancing, set up your dimensional scheme to the default datums, making it easy to follow your design intent. The named views allow for quick model view orientation.

10. Never *fill cuts with protrusions and* never *remove protrusions with cuts.* If you have unwanted protrusions or cuts in a model, delete them and reroute any dependent children.

Planning Your Design

The first step in any design process should be planning. This means gathering all the information you can about the design's fit, form, function, and manufacturing criteria. Then determine how these items will be incorporated into the design to best capture your design intent. Begin with how your part or subassembly will fit into the rest of the product's top-level assembly. Will you be mating model faces or datum planes? Are any axes to be used for alignment? Do you need to represent kinematic motion in your assembly? (If so, you will need to design for that functionality.) Will the master assembly be table driven or possibly a program? What relations should you build into your design? How will your part models be manufactured?

Once these questions are answered, you can figure out how to approach your design. Just as you do in Sketcher mode, you should break the model down into key components or shapes. This is how you will create a complex model through the combination of simple shapes.

In Sketcher mode, you have the opportunity to set up the model's design intent. You determine a dimensional scheme, entity relationships, critical dimensions, and an overall map to your design. The scheme you determine can be used downstream for the creation of detailed drawings. That you have captured your design intent will be determined by whether or not you can use shown (driver) or created (driven) dimensions. The use of shown dimensions in your drawings simplifies the design process. You do not have to worry about forgetting to dimension features, and you maintain complete associativity between the drawings, parts, assemblies, and so on.

A properly used dimensional scheme will lessen the need for model interpretation, by setting up a clear path to the important features of your model. The important model features are the features necessary for assembly, functionality, and inspection. This process may seem overly simplistic, considering the complexity of the designs you are working on, but by always working with this in mind you will create more flexible models.

Pro/ENGINEER and This Book

The standards for your products are high. Your company's product must perform to the level you have said it will, and maintain that performance past the warranty date. This statement speaks to the quality of the goods you are producing. If you maintain this standard, your quality will remain high, but if you allow your standards to diminish, your quality will quickly follow.

The first thing typically overlooked in an engineering department when Pro/ENGINEER is brought in is a set of updated standards for the proper use of this new tool. Typical items such as text height, part numbering conventions, and decimal places can remain intact, but you have opened a Pandora's box of new problems. Every designer has a personal way of building part geometry. Some are good; some are not. How do you maintain some semblance of control or understanding of these new types of models?

The first thing to do is to establish a set of "best modeling practices," such as those stated previously and in the rest of this book. Understanding that each company has unique products and problems means that each company will have unique solutions. The concepts for capturing design intent are absolute methods for avoiding problems downstream, and that is your goal and the goal of the techniques and practices discussed in this book.

Part I

Start Parts and Drawings

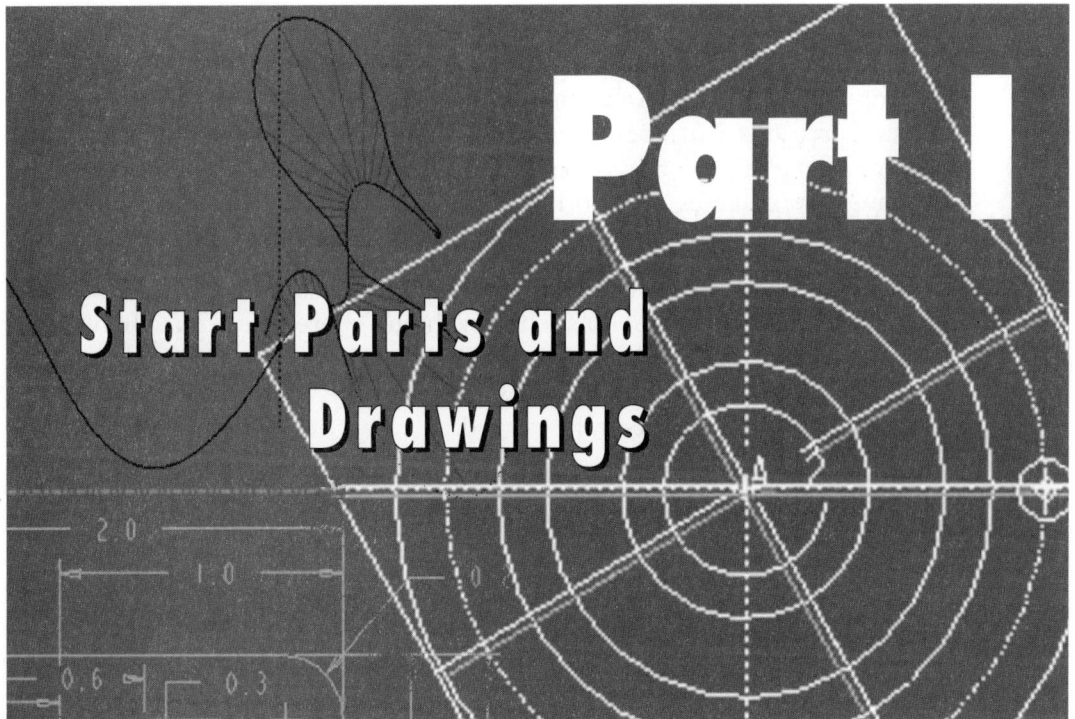

This first section of the book deals with the creation and implementation of start parts and drawings. Among the topics discussed are the types of features that need to be included in a start part, how other items such as parameters can help you become more efficient, and how you can tie these details together to get more consistency in the solid models created by all of your users.

Most companies using Pro/ENGINEER have adopted some form of start part. A start part is considered a method of starting a new part, assembly, or drawing in which some of the basic geometry and/or background information is set up before any work is started on the model or drawing itself. The basic geometry included in a start part is usually the three default datum planes, the default coordinate system, and maybe some datum axes. Background information included typically consists of parameters, view names, and layers.

The advantages of a start part from the administrator's point of view is that the company's modeling standards can be put into effect without the user having to do anything special other than to use the start part. A start part also provides the benefit of consistency of model orientation and nomenclature regardless of who created the model. Users like to use a start part for some of the same reasons, but largely because it saves time. Without a start part, the user must type in a long list of parameters for the drawing title block, and add the same layer names over and over for each part and drawing, set up the view orientation and save each view, and so on through a lengthy process.

None of the components of this process are difficult to do; they just take more time than one wants to spend on them. There is also the chance that each user will do them a little differently than the next. In that consistency is the name of the game when it comes to Pro/ENGINEER, a start part strategy lends a great deal of value in this regard.

Although creating and implementing start parts and drawings is an administrative function and not part of the average user's normal job, using them is, and their input will be invaluable to creating a useful and efficient start part. The more benefit a start part is to the user, the more likely it will actually be used, thereby achieving the greatest consistency among models.

The chapters within Part I discuss what needs to be in start parts and drawings, such as default datum planes, named views, and standard layers. The inclusion of datum axes and parameters for the drawing title block is also explored, as well as whether or not to attach the start drawing to the start part so that both are created at the same time. Part I also presents various ways of implementing start parts and drawings, with a step-by-step example of the most common strategy, mapkeys, included in Chapter 2.

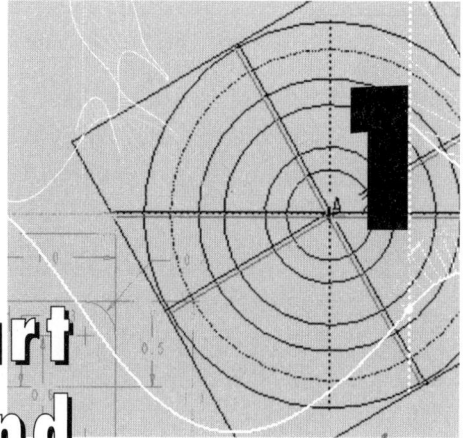

Defining Start Parts and Drawings

This chapter deals with start parts and drawings. Start parts and start drawings are models that contain features that need to be in every model. These models have already been created in Pro/ENGINEER and are available to all users through a standard method such as employing a mapkey. Start parts and drawings offer users and companies a consistency in model creation that greatly enhances ease of use by downstream applications. You will learn in this chapter what features must be included, and what items you may want to include in your start parts. You will also explore items that make your users more efficient, such as parameters and layers, as well as how to embed your company's standards into start parts for maximum consistency.

Company Standards

The two most common reasons companies use start parts and drawings are to save time and to be consistent. Consistency is usually based on some form of company standard, which might be called a drafting, CAD, or modeling standard. By whatever name, the idea behind a standard is the following:

➻ **NOTE:** *Every model must have the identical foundation regardless of what user, or even department, created it.*

Some of the standard elements incorporated into start parts and drawings are parameters for filling in drawing title blocks, named views and orientations, and layer names. These items are explored in detail later in the chapter. The most basic item for any start part is the default datum planes and coordinate system. These four features are the foundation for every Pro/ENGINEER model.

Default Datum Planes and Coordinate Systems

Default datum planes are the very bed-rock of model building, on which the foundation for the rest of the model will be built. They should be three of the first four features created in every model, with the default coordinate system being the fourth. You will be hard pressed to find a start part that does not include default datum planes and a coordinate system. The default datum planes and coordinate system and their default names are shown in the illustration at right.

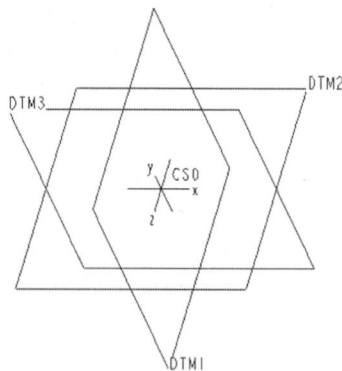

Default datum planes and coordinate system with their default names.

Default datum planes provide an additional function within the start part. Because the start part typically does not have any geometry in it, datum planes are used to orient the model so that saved views can be named in the part and views can be placed in the start drawing.

The default coordinate system also has its place in the start part. It is convenient to have it in place when creating datum point arrays or 3D splines early in the model. A default coordinate system provides a way of assembling the first component into an assembly. With a common coordinate system in place on every assembly and part, you will always have at least one assembly option open to you.

When you have incorporated the three default datum planes and coordinate system, the best practice is to change the name of these features. It is also recommended that you name the assembly datums different from the part datums. Having different names will reduce the chance for confusion when selecting datums for component placement in assemblies. Examples of how to change the name of a feature and naming conventions are covered in depth later in this chapter.

Planning for Assemblies and Drawings

Doing some planning for assembly and drawing creation up front will help you create a more efficient start part. You have been introduced to coordinate systems and datum names and how they aid the user in assembly. There are, however, other features that can be added to a start part to help deal with special cases. If, for example, a company's product and/or its components are mainly cylindrical in shape, a center axis should be added to the start part at the intersection of two of the default datum planes. This axis is a critical foundation not only for feature creation and component placement, but for displaying rotational motion in assembly.

If part rotation is a typical motion in your assemblies, a datum plane should be added through the axis and at an angle to one of the default datum planes. This angled datum plane can then control part or component rotation in assembly. Again, you should change the name of this datum plane so that future users of this part or assembly will have a clear understanding of its function.

Named Views

Standards often include names and orientations for saved views. Users require that these names and orientations be consistent so that the proper orientation will be under the name they expect. The automotive industry is especially particular about view name and orientation because a master coordinate system is typically used for an entire vehicle.

With named views, default datum planes are used to orient the model, and the view saved with the appropriate name. This may seem more than a little obvious, but there is a point to be made. By using default datum planes instead of model geometry to set the model orientation, the saved views are very stable, with little chance of losing their orientation instructions.

If feature geometry is used to orient a model view, the view's orientation instructions are dependent on that feature. If the feature is later removed, the saved view loses its orientation. This is particularly important when you use a saved view to set up the first view in drawing mode. If the main view loses its orientation, all views projected from it lose their orientation as well. By using default datum planes to set up saved views, you do not need to worry about when the drawings of these models are being called up, or that the views will lose their references and be displayed as a series of isometric views. Many of the dimensions that were on the drawing will be gone, and those that are still displayed will be highlighted in red, which indicates to you that they too will be deleted if the view is reoriented.

Standard Layers

Having a set of standard layers set in the start part, drawing, and assembly can aid substantially in the use of layers in Pro/ENGINEER. Try to keep layer lists in start parts from getting too long. The following basic lists tend to work very well.

Start Part Layer List

Part Mode	Drawing Mode	Assembly Mode	Items To Be Placed on Layer
Axis	Axis	Aaxis	Datum axes
Curves	Curves	Acurves	Datum curves
Def_cs	Def_cs	Adef_cs	Default coordinate systems
Datums_All	Datums_All	Adatums_All	All datum planes other than default
Def_datums	Def_datums	Adef_datums	Default datum planes
Geomtol	Geomtol	Ageomtol	Datum planes & axes used for geometric tolerancing
Pnts	Pnts	Apnts	Datum points
Surfs	Surfs	Asurfs	Surfaces
—	Drw_text	—	Text such as axis labels and section names
Extra_Dims	Extra_dims	Aextra_dims	"Show" dimensions not needed on the drawing

Because of the way layers with the same name interact with each other between parts and drawings (parts and assemblies, assemblies and drawings, and so on), you may want to duplicate these layer names into the downstream start parts. This can, however, lead to some long layer lists, in which case you might want to create separate mapkeys that will add the required layers to parts, assemblies, or drawings as needed. Layers can be set up to automatically add a specific type of geometry or feature to them as they are created, but this is often too general and does not work out as well as it would first seem. For example, you may not want every datum plane on only one layer. This can limit the amount of control you have over how the datums are displayed. The list of these automatic layers, however, is quite extensive and should not be dismissed out of hand. There may be one or several of these layers that will work for your particular situation. The table that follows presents a partial list of these layers. These entries may have more general applications than others.

Automatic Layers

Type Option	Description
layer_assem_member	Assembly members
layer_axis	Features with axes and cosmetic threads
layer_surface	Surface features
layer_datum	Datum planes
layer_curves	Datum curves
layer_dim	All dimensions
layer_driven_dim	Driven dimensions

➠ **NOTE:** *For the complete Default Layers list, see "Managing Layer Information" in Chapter 12 of the manufacturer's* Introduction to Pro/ENGINEER *guide.*

To get a good idea of the type of geometry being put on layers, as well as the layer names used, you need look no further than your existing database. Chances are you will see a pattern in the naming of the layers or the type of geometry placed on those layers. These layers are obvious choices to be added to the start parts, but not all users are employing some of them. Do not throw these layers out. Go to these users and find out why and how they are using these layers; they just may be on to something the others have overlooked. Again, get your users' input. If they feel they have been part of the process, you will almost certainly get their buy-in, and they will use the start part layers more effectively.

There are several methods of incorporating layers into your company's standard modeling practices. The two discussed in the following sections are:

- Manually entering each layer into the start part, start drawing, and start assembly, then selecting each entity for each layer

- Automatic assignment of entities to preset layer names

Manual Layering Method

As the name implies, this is a manual entry technique for creating layers in your start files. Whether working with the start part, assembly, or drawing, the picks used for the following example of the manual method are the same. In this example, you will use part mode and create layers for a start part. Select Layer ➡ Setup Layer ➡ Create to select from the Part menu at any point in the start part creation process. In the message window, enter the layer names you want to use. The following illustration is a representative screen shot of the LAYER menu and message window data required for the creation of a layer.

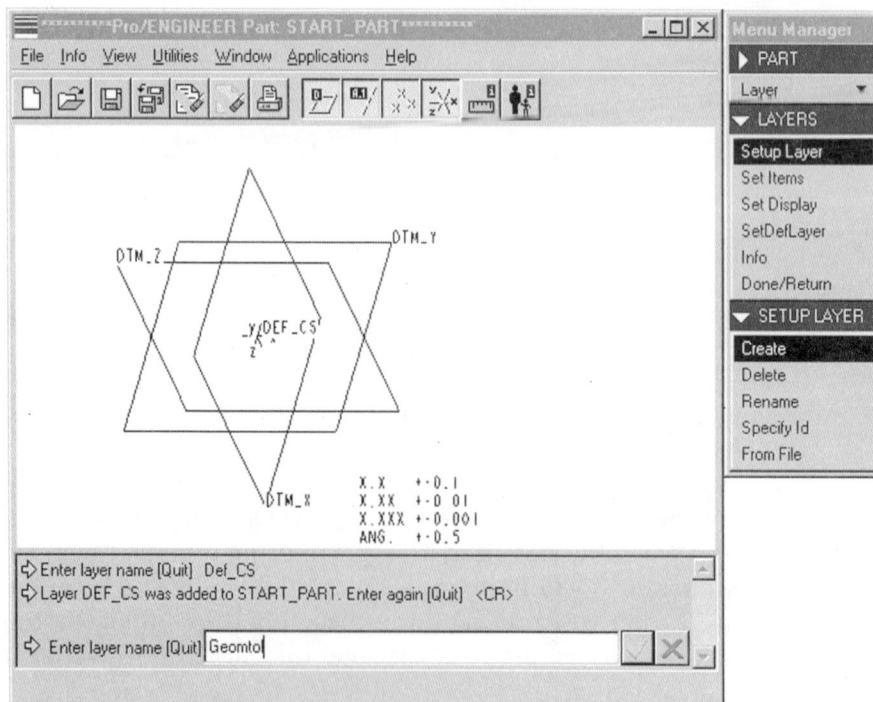

Menu structure and message window when creating new layers.

After you enter the layer names, go to Layer ➥ Set Item ➥ Add Item. Select the layer you want to add the items to (as shown in the following illustration at left), then select Done Sel.

Menu structure for items added to a layer.

Select the objects/types you want to add to the selected layer. When finished, select Done/Return. Use the layer display dialog window to control the display setting for each layer, as indicated in the center illustration.

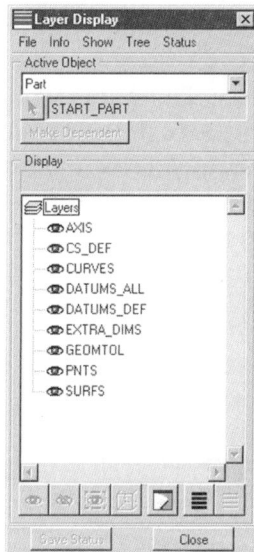

The three display setting icons are located along the bottom of the display dialog window, as shown in the right-hand illustration.

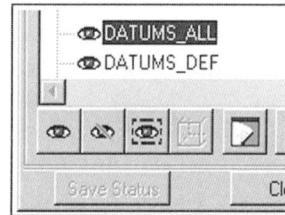

The new "eye" icons for layer display.

What a typical start part layer list might look like.

The first two "eye" icons are largely self-explanatory, but the dashed box eye is a little unclear. When layering entities in Pro/ENGINEER, there are two approaches available. One method is global, meaning that you assign all items of a specific type to a layer. That is, for example, you would assign all datum planes to the Datums_All layer. When you turn off that layer, all datum planes are blanked. However, when you want to see only the default datum plane, you can either remove those default datums from the Datums_All layer or leave them set, but add them to the Datums_Def layer as well. This is where the "dashed eye" icon comes in. It treats the selected layers as if they were isolated, meaning that if you assign the default datums to the Datum_Def layer as well as the Datums_All layer, you can use the "dashed eye" to display them even though they may be on a blanked layer.

Automatic Layering Method

The automatic method is a little more system administrator aimed. You will need to have access to the Pro/ENGINEER load point directory and files located there, as well as some knowledge of file editing to modify some of the Pro/ENGINEER layer setup files. In addition, you will need to make a quick edit to your *config.pro* file.

First, modify your *config.pro* file. To do this, select Utilities ➡ Preferences ➡ Edit Config from the pull-down menus across the top of the Pro/ENGINEER window. In the *config.pro* file, you will want to set the option "create_ numbered_layers" to Yes. Next, set up some default layers, such as the one mentioned earlier in this chapter. To do that, you added def_layer callouts to the *config.pro* file. The format for these default layers is shown in the illustration that follows. On the left is the "def_layer" command, followed by the type option (the portion of the right-hand box in capital letters; e.g., LAYER_AXIS), as well as the name you want to assign those options to in the models (the last word in the right-hand box; i.e., Axis).

> ➡ **NOTE:** *The type option descriptions are located in the "Managing Layer Information" section of Chapter 12 in the* Introduction to Pro/ ENGINEER *guide.*

Default layer assignment list from a modified config.pro *file.*

```
Pro/TABLE TM  Release 20.0  (c) 1988-95 by Parametric Technology Corporation ...  _ □ ✕
 File  Edit  View  Format  Help

DEF_LAYER

!=================================================================
! LAYER CONTROL SECTION
!=================================================================

CREATE_NUMBERED_LAYERS                          Yes
DEF_LAYER                                       LAYER_AXIS Axis
DEF_LAYER                                       LAYER_DATUM Datums
DEF_LAYER                                       LAYER_CSYS Coords
DEF_LAYER                                       LAYER_POINT Pnts
DEF_LAYER                                       LAYER_SURFACE Surfs
DEF_LAYER                                       LAYER_GTOL GeomTol
DEF_LAYER                                       LAYER_CURVE Curves
DEF_LAYER
```

Once you reload your *config.pro* file and start a new model, you will have layers numbered 1 to 32 and named Axis, Datums, Coords, and so on. Layers 1 through 32 will have nothing assigned to them, but the named def_layers will, once you create the specified geometry. You then need to replace those numbered layers with some meaningful names.

✗ **WARNING:** *Because these file modifications are being done at the system level, you must not make any of these changes without your system administrator's knowledge and approval.*

Edit/create a *layersetup.pro* file. The easiest method of doing this is to copy the existing *layersetup* file from the text directory under your Pro/ENGINEER loadpoint and place it in the directory in which you start up Pro/ENGINEER. This is where some administration knowledge comes in very handy. The following shows the information in the default *layersetup.pro* file. This is where the numbers come from.

```
!LAYER SETUP 518
!
! Comment lines should begin with a "!"
!
! 1. The first column must always contain the layer name.
! 2. The display status may optionally be specified by entering one of
!  the following: NORMAL, BLANK, DISPLAY or HIDDEN.
! 3. Interface id may optionally be specified by preceding the id with #
! 4. You may specify as many sub-layers as you wish.
!
! eg.
!
! DATUMS  NORMAL #100 DTM_PLANES AXES POINTS COORD_SYSTEMS
!
! NAME | STATUS | #ID | SUB-LAYER NAMES
!
  1
  2
```

```
3
4
5
6
7
8
9
10
11
12
13
14
15
16
17
18
19
20
21
22
23
24
25
26
27
28
29
30
31
32
```

You will need to replace those numbers with the names you entered in the *config.pro* file. Therefore, in whichever file editor you are familiar with, delete numbers 1 through 32 and type in the names you want. When finished, your file should look like the following.

```
!LAYER SETUP 518
!
! Comment lines should begin with a "!"
!
! 1. The first column must always contain the layer name.
! 2. The display status may optionally be specified by entering one of
!  the following: NORMAL, BLANK, DISPLAY or HIDDEN.
! 3. Interface id may optionally be specified by preceding the id with #
! 4. You may specify as many sub-layers as you wish.
!
! eg.
!
! DATUMS  NORMAL #100 DTM_PLANES AXES POINTS COORD_SYSTEMS
!
! NAME | STATUS | #ID | SUB-LAYER NAMES
!
  Axis

  Coords

  Curves

  Datums_All

  Geomtols

  Pnts

  Surfs
```

Once you have saved this layer setup file, when you start Pro/ENGINEER again and look at the layers, the seven previously listed will already exist. As you create entities in the model, they will automatically be assigned to those layers. The automatic assignment is due to the "layer_def" command lines in the "config.pro" layer control section.

At first glance it would appear that you lose individual entity control by using automatic layer assignment. In the case previously discussed, every datum plane you create will go on the Datums_All layer. If you want control of the default datums only, you will need to tie them to the Def_Datums layer. The layer itself can be created using a *layersetup.pro* file, but assigning default datum planes to that layer will be done manually. You can then use the "dashed eye" icon to display the default layers only.

Standard Layers in Summary

Model and drawing layers are created to control the display of all types of entities. By turning layers on and off you control the amount of clutter visible on the screen. This makes it easier to select the entities you want without going through a long series of Next commands during query selection. When setting up layering in standard and start parts, you need to determine whether you want to use the manual or the automatic layering method.

Entering the layer names using the manual method needs to be done only once, during the start part creation. By editing your *config.pro* file, you can automatically assign specific entities during model creation, thus combining the manual and automatic methods. If you use the automatic method only, you do not need to enter layers into the start parts because the *config.pro* and *layersetup.pro* files will assign entities and layers every time you start a new model.

Parameters and Start Parts

There are two modes in which parameters can be created: Model and Drawing. Model parameters can be used in the drawing, but drawing parameters can be accessed only in Drawing mode. In addition to these two types of parameters there are system parameters. System parameters preexist within Pro/ENGINEER and cannot be changed. They are used, for example, to automate filling in data in a drawing format.

Model Parameters

Pro/ENGINEER includes the following model parameters.

Parameter	Value
Integer	Its value must be a number.
Real Number	Its value is a decimal number.
String	Its value is a text string.
Yes/No	Must have a value of Yes or No.

◆ **NOTE:** *The Yes/No parameter's value is the ID of a model note.*

Of these four types of model parameters, the one most commonly used in start parts is the String parameter. String parameters are used for items such as "detailer," "designer," "checker," "title_1," "title_2," "material_1," "material_2," and so on. The drawing format will then call out these parameters and their values will appear in the title block. Creating a drawing format is covered in Part V.

Drawing Parameters

As previously mentioned, drawing parameters can be accessed only in Drawing mode. In most cases you will want the parameter values to be in the model. However, the drawing revision is one that may best reside in the drawing only. The reason for this is that if a drawing-only change is required, the drawing revision parameter can be modified without having to go into the model to force a modification. Depending on how your company's product development workflow is set up, this may be the best way to handle drawing revisions.

System Parameters

The parameters used in the previous examples are user-defined parameters. This means that the user is the one that defines the parameter name, and then assigns a value to it. There are, however, system parameters that have specific names and values associated with them. Some of these system parameters and their values and listed in the table that follows.

System Parameters

Parameter Name	Description
&todays_date	Enters date of note creation
&model_name	Enters model name used for drawing
&dwg_name	Enters name of drawing
&scale	Enters scale of drawing
&type	Enters drawing model type (asm, prt, and so on)
&format	Enters format size
&linear_tol_0_0 thru &linear_tol_0_000000	Enters linear tolerance values for one to six decimal places
&angular_tol_0_0 thru &angular_tol_0_000000	Enters angular tolerance values for one to six decimal places
&current_sheet	Enters the current sheet number
&total_sheets	Enters total number of sheets

The system parameter can be used in the same ways as the user-defined parameter. It can be used to fill in areas of the title block automatically or used in notes on a drawing. You will be using them together with repeat regions in tables to automatically create a Bill of Material or BOM table of assemblies in Part V.

Summary

This chapter explored start parts and the basic geometry that needs to be in them, how to look at your situation for additional features that can be a great benefit to your users, the use of named views and standard layers created by the manual and automatic methods, and how the use of parameters, models, drawings, and systems can increase the efficiency and consistency of your users. Start parts are an effective means of getting users to consistently adhere to company standards. They also save users a lot of time by eliminating the time-consuming chores of typing in long lists of layer and view names. Chapter 2 explores step-by-step procedures for implementing start models.

Creating Start Models

There are basically three phases involved in creating and implementing a start part. The first phase involves creating the start part itself. This phase begins with a standard Pro/ENGINEER part that contains the geometry, layers, parameters, and anything else you have determined should be in the start part. The second phase consists of placing the start part in a central directory or folder where users will have access to it, but will not be able to modify the start part. The third phase consists of creating mapkeys that will select the start part, prompt the user for any required input, and then save the start part to the correct name at the correct directory or folder.

Exercises

In the following exercises, you will create a start part, a start drawing, a start assembly, and a start assembly drawing. You will then ensure start part file safety, implement the start part, and save a mapkey.

Exercise 2-1: Creating a Start Part

1. Create the start part as you would create any Pro/ENGINEER part. The first four features should be the three default datum planes and the default coordinate system. The order in which these features are created is not important, but *how* they are created is.

2. Create the default datum planes by selecting **Feature** ➥ **Create** ➥ **Datum** ➥ **Plane** ➥ **Default**. This command sequence option is available only when creating the datums as the first feature in the model, or as the second feature only if preceded by the default coordinate system. If the datums are created using the option, offset from a coordinate system, the offset can be modified later on, defeating the purpose of having the datums in the first place. To create the default coordinate system, select **Feature** ➥ **Create** ➥ **Datum** ➥ **Coord Sys** ➥ **Default**.

3. Now that you have some geometry to work with, change the names of these features so that their purpose will be understood by anyone using them. To change the name of a feature, select **SetUp** ➥ **Name** ➥ **Feature**. Then select the feature to be renamed. Pro/ENGINEER will prompt you for the new name. Select **Enter**. The illustration at right shows the default datum planes and coordinate system with their default names.

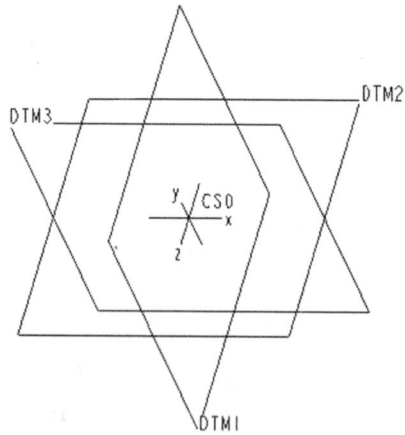

Default datum planes and coordinate system with their default names.

4. With the default datum planes in place with the proper names, the view orientations can be set and saved under the desired names. To orient the model, select **View** ➥ **Orientation** ➥ **Front**, select the datum DTM_Z, then select **Top** and pick the datum DTM_Y. The model is now oriented looking toward the yellow side of DTM_Z, and the yellow side of DTM_Y is pointed up. Save this view by selecting **Saved Views**, type in the name *Front*, and select the **Save** button. Continue this process using various combinations of front, back, top, and bottom datums for the other five standard views. The following illustration shows the new style of dialog box for view orientation and storing saved views.

→ **NOTE:** *If you have decided that the standard layers are going to be created automatically, they will already be in the start part. Manually created layers can be created now. For advantages and disadvantages, as well as step-by-step instructions, for both methods, see the section "Standard Layers" in Chapter 1.*

View Orientation and Saved Views dialog box style menu.

5. Regardless of the method used to create the layers, you will need to add the default datum planes to the layer Datum_Def, and will need to add the coordinate system to the layer Cs_Def. This ensures that these items will be preset to the proper layers in all parts.

6. The last step is to create the user-defined parameters. Select **SetUp** from the PART menu, then **Parameters** → **Create** → **String**, and enter the parameter name *Designer* and select Enter. Pro/ENGINEER will now prompt you for a value for this parameter. Select **Enter**, and continue. After all parameters shown in the adjacent list have been created, they need to be designated so that other applications such as Pro/PDM can read and display them. To designate the parameters, select **SetUp** → **Parameters** → **Designate** → **Parameters** → **Select All** → **Done Sel** and **Done/Return**. The illustration that follows shows a typical list of parameters found in a start part. These parameters would then be used to automatically fill out the title block of a drawing.

Typical start part parameters.

```
INFORMATION WINDOW (param.inf)                              _ □ ✕
File  Edit  View                                                  ▲


Symbolic constant              X-refs      Current value
------------------             ------      ------- -----


PART_NO                        Local        " "
NAME_1                         Local        " "
NAME_2                         Local        " "
MATERIAL_1                     Local        " "
MATERIAL_2                     Local        " "
FINISH_1                       Local        " "
FINISH_2                       Local        " "
DETAILER                       Local        " "
DETAILER_DATE                  Local        " "
CHECKER                        Local        " "
CHECKER_DATE                   Local        " "
DESIGNER                       Local        " "
DESIGNER_DATE                  Local        " "
MFG_SIGN_OFF                   Local        " "
MSO_DATE                       Local        " "
MANAGER                        Local        " "
MANAGER_DATE                   Local        " "
                                                                  ▼
◄                                                              ►

                            Close
```

✓ **TIP:** *It is a good idea to make mapkeys for creating both standard layers and parameters, as well as for designating the parameters. This allows standard layers and parameters to be easily added to models or drawings not started with the start parts. These mapkeys are usually considered system mapkeys and often have the forward slash ("/") in front of their name. Example: /al for add layers, /ap for add parameters, and /dp to designate the parameters.*

7. The start part is now complete. Return the model to the default view orientation and save the part by selecting **File ➦ Save** and **Enter**.

Exercise 2-2: Creating a Start Drawing

1. Create a new Pro/ENGINEER drawing, and enter the name of the start drawing. Select a size for the drawing (C or D is the most common). For this example, use the D-size drawing. Next, add a drawing format to the drawing. If you do not already have a format created, follow the example in Chapter 20. When the format is complete, finish creating the start drawing. To add the drawing format, select **Sheets** from the DRAWING menu, select **Format** ➥ **Add/Replace**, select the format name "D-Size," and select **Open**. If there are no values for the parameters, Pro/ENGINEER will prompt you for them. Select **Enter** for each parameter prompt. Your drawing should look similar to that shown in the following illustration.

Start drawing with the D-size format in place.

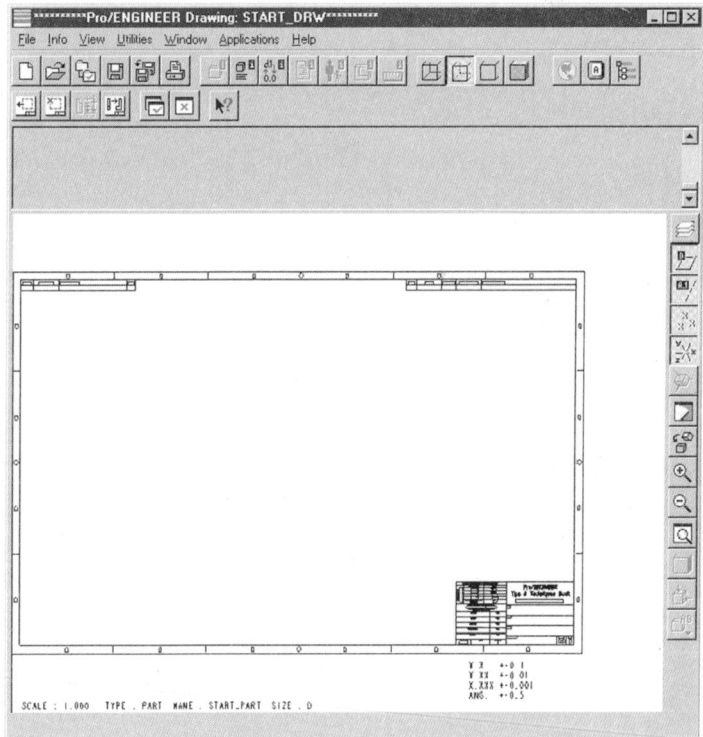

2. Add the standard views of the start part to the start drawing. Select **Views**. The model list window will pop up. Select the model to be linked to the drawing, select "start_part" and **Open**. The first view must be a general view and will be oriented once in the drawing. When setting the orientation for this first view, use the **Saved Views** option (select the blue bar that says SAVED VIEWS), and select Front from the list of saved views. Then select **Set** and **OK**. The rest of the views will be created as projection views from the previous views. However, for this to occur, there needs to be a piece of solid geometry in the start part. To do this, create a simple block feature in the start part. The view can be placed anywhere on the drawing at this point and moved to the desired location later. In the illustration that follows, the first view has been oriented by using the Saved View option and has been placed in the approximate proper location on the drawing.

✗ **WARNING:** *Do not save the start part or drawing once the block has been added. The block will be deleted later.*

➥ **NOTE:** *If it looks as if the block feature does not appear in the drawing, check the scale to be sure it is 1.000. The scale will automatically adjust to a very small number when the first view is placed without geometry.*

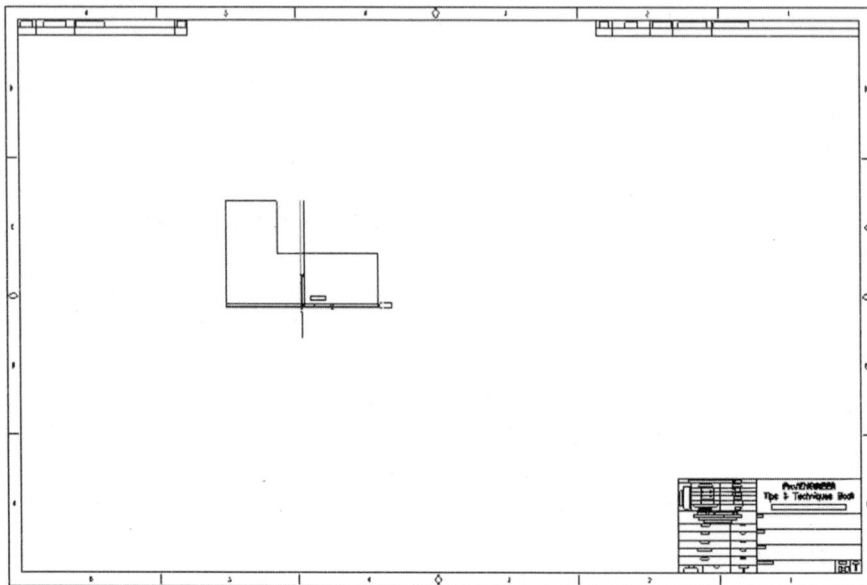

Oriented and placed first view.

3. Add the remaining views to the drawing using **Projection** as the view type. In this example, there will be six standard views, as shown in the following illustration. In the illustration, block geometry has yet to be deleted from the start part. After you create the views, reopen the start part and delete the block feature. Return to the start drawing. All six of the standard views will remain on the drawing.

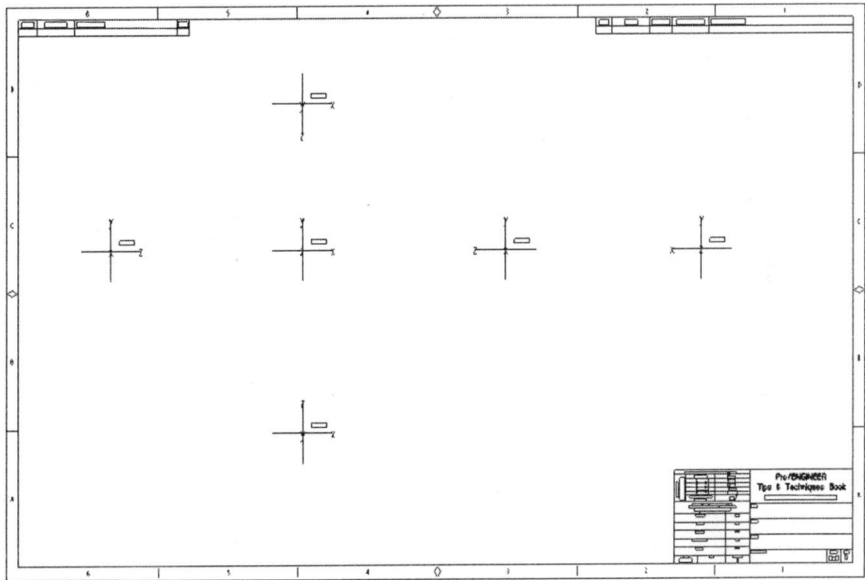

Start drawing with all six standard views in place.

4. Creating layers in the drawing with the same name as layers in the start part will help control the appearance of such items as datums and points in the drawing without having to change the environment settings. To create a drawing layer, select **Layer** from the DRAWING menu, then select **Drawing ➥ Setup Layer ➥ Create**, and enter the layer names just as you did for the start part. To get a better understanding of how layers can be used to set the appearance of the drawing, select **Layer ➥ Drawing**, select the Def_Datums layer name, select the Do Not Display icon (the eye with the red slash through it), select the Repaint icon, and select **Close**. The drawing should look as it does in the following illustration. The datums are not displayed. The coordinate system marks that indicate where each view is located are displayed.

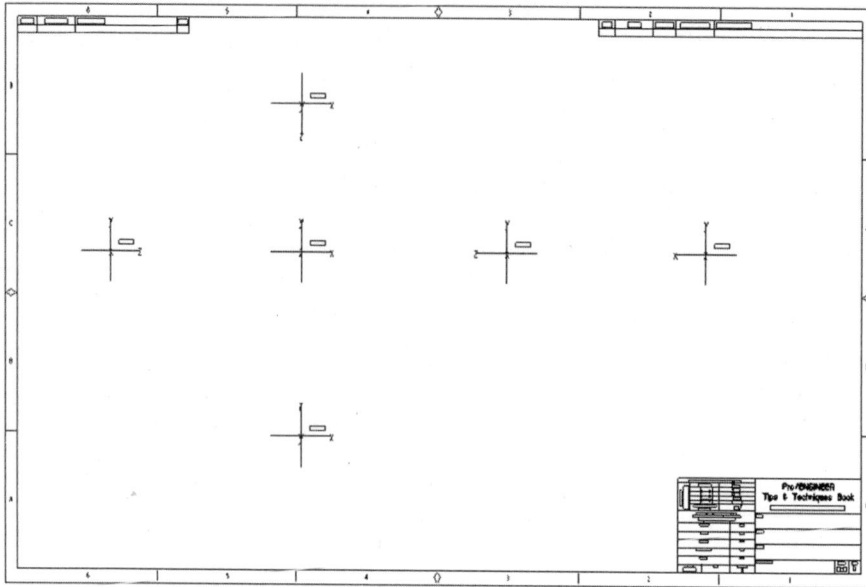

Start drawing with block geometry removed and datums blanked by layer.

5. The final step before filing the start drawing is to create the drawing parameter for the drawing revision. Select **Advanced** ➥ **Parameters** ➥ **Create** ➥ **String,** enter the parameter name *drw_rev,* and press Enter. When prompted for a value, enter a "-" (a hyphen, without the quotes) and press Enter. Once the parameter is created, it must be designated. To designate the parameter, select **Advanced** ➥ **Parameters** ➥ **Designate** ➥ **Parameters**. At this point you can select the Drw_Rev parameter or **Select All**, then select **Done Sel**. The drawing is finished. Make sure the block feature has been deleted from the start part; then save the drawing.

Exercise 2-3: Creating a Start Assembly

1. Many of the steps used to create a start part are also used to create a start assembly, with a few subtle differences. These are discussed in a step-by-step manner in the material that follows. Create the start assembly as you would create any assembly, and enter *start_assy* for the name. The first four features should be the three default datum planes and the default coordinate system.

Change the names of these features so that their purpose will be understood by anyone using them. To change the name of a feature, select **Setup ➥ Name ➥ Feature**, then select the feature to be renamed. Pro/ENGINEER will prompt you for the new name. Press Enter. The start assembly will appear as shown in the first illustration at right. Note in the illustration that the names for the assembly are different from the names used in the start part.

✓ **TIP:** *The best practice to follow when naming assembly datums is to use a different name than the one used for parts. This allows for easier differentiation between an assembly datum and a part datum when making selections for component placement.*

2. With the default datum planes in place and properly named, the view orientations can be set and saved under the desired names. Using the Orientation dialog box style menu, shown in the second illustration at right, orient the assembly. Select **View** from the top menu bar, select **Orientation ➥ Front**, select the datum ADTM_Z, and select **Top** and pick the datum ADTM_Y. The model is now oriented looking at the yellow side of ADTM_Z, with the yellow side of ADTM_Y pointed up. Save this view by selecting **Saved Views**, type in the name *Front*, and select the **Save** and **OK** buttons. Continue this process, using various combinations of front, back, top, and bottom datums for the other five standard views.

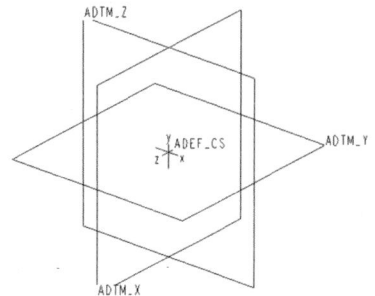

Default datum planes and coordinate system with the modified names.

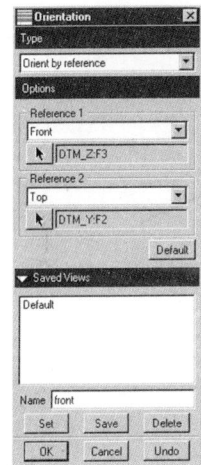

View Orientation and Saved View dialog box style menu.

➥ **NOTE:** *This would be a good time to set up your assembly layering if you opted to use the manual layering method discussed earlier in this chapter.*

3. Create the user-defined parameters. Select **Setup** from the ASSEMBLY menu, select **Parameters** ➥ **Create** ➥ **String**, enter the parameter name *Designer*, and press Enter. Pro/ENGINEER will prompt you for a value for this parameter. Press Enter, and continue. After you have created all parameters, designate them. To designate the parameters, select **Setup** ➥ **Parameters** ➥ **Designate** ➥ **Parameters** ➥ **Select All** ➥ **Done Sel**. The start assembly is now complete. Return the model to the default view orientation and save the assembly by selecting **File** ➥ **Save** ➥ **Enter**.

Exercise 2-4: Creating a Start Assembly Drawing

To create a start drawing for assemblies, follow the same procedure as a start drawing, except for the differences discussed in the following material. As with creating views in a start drawing for a part, the drawing for an assembly needs some solid geometry for creating projection views.

1. Use the start assembly Start_Assy as the model the drawing is linked to. When adding standard layers to the start drawing for assemblies, you need to add the standard layer names from the start part, start assembly, and start drawing. Having common layer names among these three items will help with drawing cleanup and simplifies layer operations. Add the layer names.

Exercise 2-5: Ensuring Start Part File Safety

The next phase in the start part process is to place the files in a central directory or folder that users can access but not use to modify the start part files in any way.

1. Access to a directory in UNIX can be set up as a "search path" in the global *config.pro* file. The full path name to a folder can be entered in the Pro/ENGINEER dialog box for access on an NT operating system. Be sure that these directories and files are set for READ ONLY access.

2. It is also a good idea to store the start parts and drawings in your Pro/ENGINEER database system, whether it is Pro/PDM, INTRALINK, or some other database manager. This will ensure that should anything happen to the original start files they can be easily restored.

Exercise 2-6: Implementing Start Parts

The most common method of implementing start parts involves using a set of special mapkeys. The procedure described in the following phases was written using Pro/ENGINEER version 20; therefore, the actual picks will differ slightly from earlier versions. However, the steps are valid regardless of the version of Pro/ENGINEER you are working with.

Set Up the Mapkey

1. Log in to a Pro/ENGINEER session, and select **Utilities ➠ Mapkeys** from the top menu bar. A Mapkeys dialog box will pop up in the lower right portion of the screen. Select the **New** button. The Record Mapkey dialog box will pop up on top of the first dialog box. These two dialog boxes are shown in the following illustrations. Enter a key sequence for the mapkey. The key sequence is the actual keystrokes required to run the mapkey. It is common practice to start system-type mapkeys with the forward slash character ("/") to help distinguish them from user-defined or other mapkeys. This practice also helps prevent the mapkey from being accidentally run. Enter */sp* for the key sequence; then enter */SP* for the mapkey name.

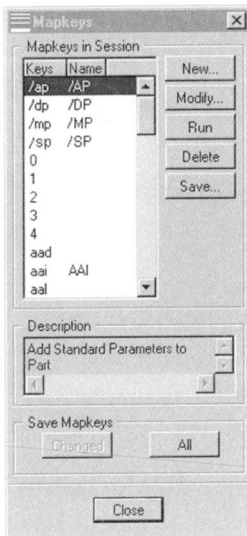

Mapkeys dialog box.   *Record Mapkeys dialog box.*

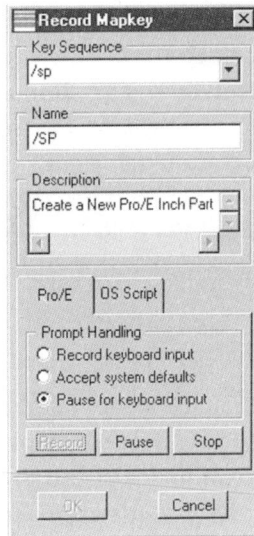

2. Below the mapkey name box is a box for entering a description of the mapkey. This description is also what appears when the cursor is placed on the mapkey if the mapkey is displayed on a toolbar. Enter *Create a New Pro/E Inch Part.* You are now ready to begin recording the mapkey. Select the **Record** button. Any further keystrokes or menu picks will be recorded to the mapkey.

Start Part Mapkey Keystrokes

1. Retrieve the start part from the secure folder. To do this, select **File ➡ New ➡ Copy From**. In the Name portion of the dialog box, type in the full path name to the start part folder. Be sure to include the start part name. After the full path name has been typed in, select the **Open** button. The "Choose template" dialog box for an NT operating system is shown in the first illustration at right.

"Choose template" dialog box with the full path name to the start part entered in the Name entry box.

2. Rename the start part to the desired name. To do this, select **File ➡ Rename**. The Rename dialog box, shown in the second illustration at right, will open. Do not enter a new name at this point. Every keystroke is still being recorded to the mapkey.

3. This is all that is required to create a start part mapkey. Return to the Record Mapkey dialog box and select the **Stop** button. This ends the keystroke recording to the mapkey. The next step is to save the mapkey to the *config.pro* file.

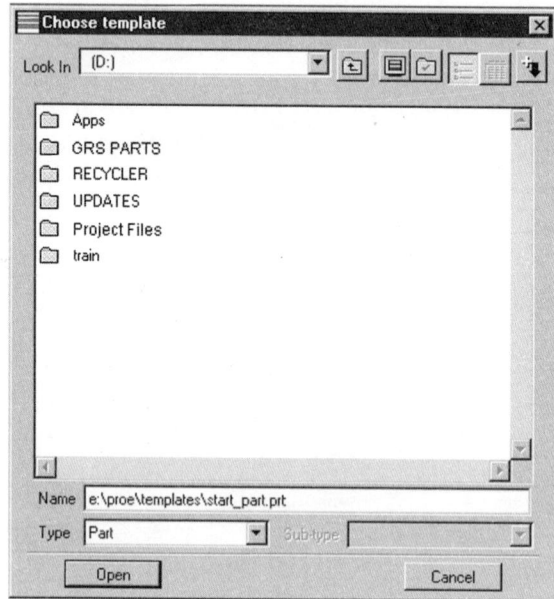

Rename dialog box.

Exercise 2-7: Saving a Mapkey to a *config.pro* File

When the Stop button is selected, the Record Mapkey dialog box will close, but the Mapkeys dialog box will remain open. To save the new mapkey to the *config.pro* file, select the **Changed** button, type in the full path name to the global *config.pro* file, and press Enter. The *config.pro* file should now contain the new mapkey named /sp. All that is left now is to log out of Pro/ENGINEER and then log back in and run the /sp mapkey to verify that it is working correctly. When the Mapkeys dialog is selected, the new mapkey /SP will appear in the list, as shown in the illustration at right.

> ✓ *TIP: Select* **Utilities** ➥ **Preferences** ➥ **Edit Config** *to edit the* config.pro *file. The advantage of editing within Pro/ENGINEER and using Pro/TAB is that each keystroke is a new line, making it easy to figure out where to make the desired corrections. Editing the* config.pro *file outside Pro/ENGINEER allows you to use more powerful text editing tools, but reading the mapkey is more difficult.*

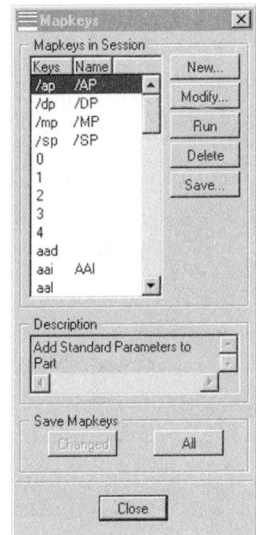

Mapkeys dialog box with the new start part mapkey /sp listed.

Summary

Start part models and drawings run the gauntlet from very simple to extremely complex. Some set up the part and drawing at the same time, whereas others call up a blank part and ask you to rename it. The factors in determining what should be in your start parts and drawings are a matter of what works for you and your company. The layers and parameters included in the exercises in this chapter are intended to serve as guidelines and to help give you some idea of what you should consider including in start parts.

Try creating start parts and drawings on your own. Let a pilot group of users try them out for a week or so to make sure the parts and drawings function correctly. Get the users' input for any changes that might be required. Make the requested changes and test the parts and drawings again. When the pilot group is satisfied, release the start parts and drawings to the entire user community.

There are several ways of implementing start parts, each having its own set of advantages and disadvantages. The method described in this chapter is only one of them, but it is the most common. Find the one that works best for your company and users; then make sure to train users in how the start parts and drawings should be used. You will find that users will like the idea of start parts and will use them regularly. Once users have gotten in to the habit of using start parts, the consistency of models and adherence to company standards will increase dramatically. After all, the consistency of Pro/ENGINEER models and drawings among users is why you should take the time to create start parts and drawings in the first place.

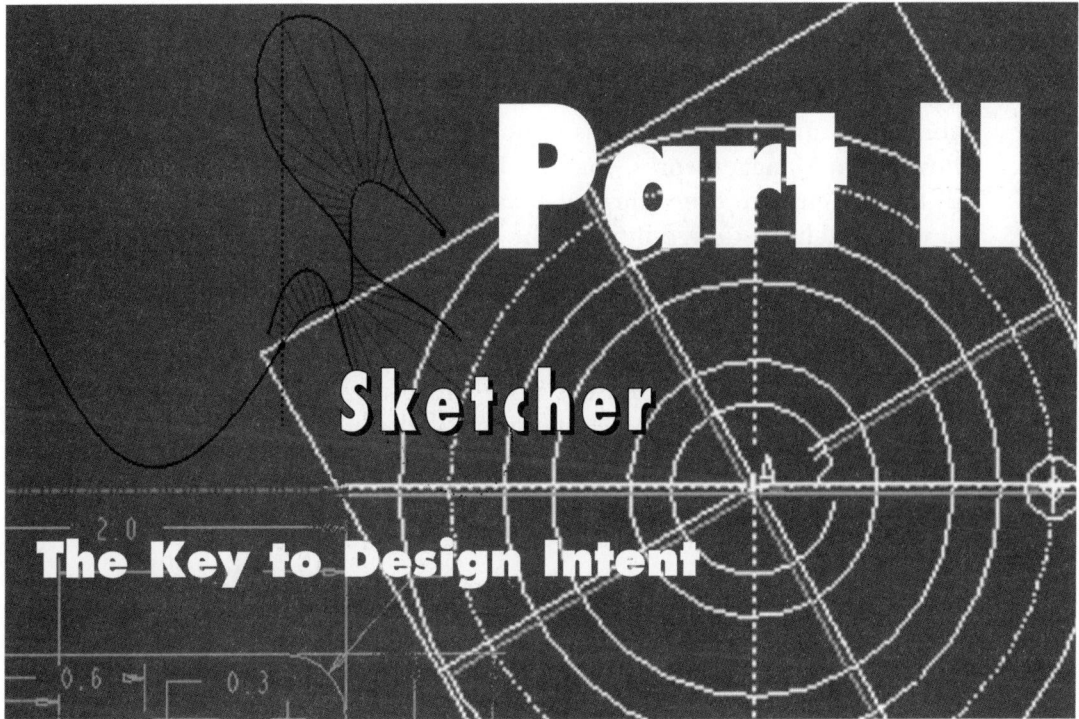

Part II

Sketcher

The Key to Design Intent

Just as you need a language—be it verbal, written, or sign—to communicate your thoughts and ideas to others, you need a language to communicate your design intent to anyone requiring that information. In today's engineering world, that language is the solid model. The model contains all of the information required for fit, form, function, analysis, inspection, and manufacture. How does all of this information get into the model? The answer is that such data is built into the model with individual features; that is, protrusions, cuts, draft, and the like. However, if you reduce it to essentials, you end up at the sketcher.

In Pro/ENGINEER, the language of design intent is represented in Sketcher. You must have command of Sketcher to convey your design intent effectively in your model. In Part II you will explore sketching and dimensioning sections, constraints and assumptions, and parent/child relationships.

The Sketcher mode in Release 20/2000i of Pro/ENGINEER has been updated considerably, with the biggest change being the addition of Intent Manager, which has dynamically created dimensions and highlights sketcher constraints and assumptions as entities are sketched. When you understand how Intent Manager works, you will find that it greatly speeds up the creation of the sketch. However, the real advantage will be that the sketch will more closely match the design intent—the goal of any sketch.

Part II also explores Intent Manager and how to use its new functionality to create more accurate sections faster than ever before. Along with providing some basic sketcher techniques, the step-by-step examples show how Intent Manager works and how to use it more effectively. A large portion of Part II is devoted to explaining the least understood of all the sketcher geometry: splines and conics. The text, examples, and exercises in Part II cover, among other topics, creating and controlling splines and conics, dimensioning techniques, and creating tangency as well as continuity between entities.

Sketcher Basics

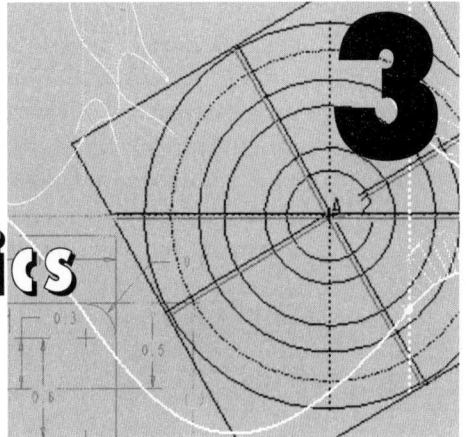

In Pro/ENGINEER, the sketch is a 2D section of the feature being created. Therefore, before you can create sketch geometry, you need to select and orient a sketching plane. The sketching plane can be any flat entity, such as a face of existing solid geometry or a datum plane. You can create section geometry once the sketch plane has been oriented.

Starting a Sketch

When you select the sketching plane, an arrow pops up showing the direction in which the feature will be created. Different features are created in different directions. For example, the cut feature's default direction is into the screen, and protrusions are created out from the screen. You can of course flip this arrow to change the direction of the feature creation, but remember that the viewing direction will always be set up so that the feature is created in its default direction. An example is shown in the following illustration, in which a protrusion is created from the default datum planes. The datum DTM_Z is selected as the sketching plane and the feature creation arrow is pointing out toward the screen.

If an orientation plane were selected, the sketch view will be oriented to show the front (yellow) side of DTM_Z. Flip the direction arrow so that it is now pointing into the screen. Select DTM_Y as the top plane for view orientation. When you flip the feature creation arrow Pro/ENGINEER will rotate the sketch view orientation 180 degrees so that a protrusion feature will be created coming out from the screen. You can easily be fooled into thinking that your sketching plane will still be oriented to show the "yellow" side of DTM_Z. This is especially common when there is little or no geometry to provide you with some visual clues to help determine model orientation.

The sketch view is oriented so that the protrusion will always be created coming out of the screen, toward you. Therefore, Pro/ENGINEER will orient the sketch view to show the "red" side of DTM_Z, and the "yellow" side of DTM_Y will be up, as shown in the illustration at right. This view can, however, be reoriented 180 degrees about the screen's vertical axis and Sketcher will function as if this were the original orientation.

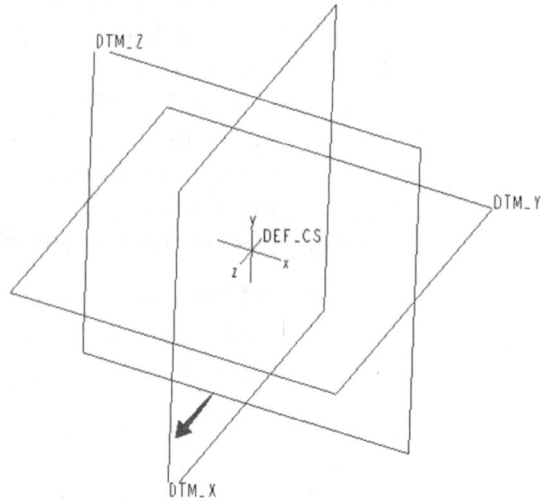

Direction of feature creation for a protrusion created from default datum planes.

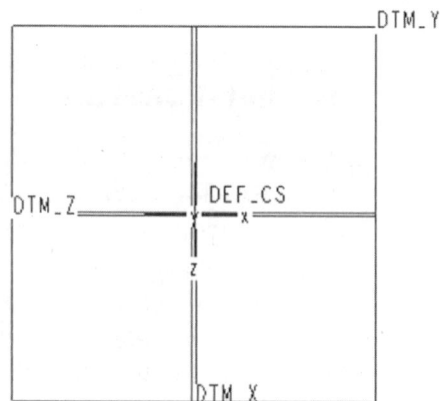

Sketch view will be oriented to show "red" side of DTM_Z, and the "yellow" side of DTM_Y is up when the feature creation arrow is flipped.

There is one other thing to take into consideration when selecting the sketching and orientation features for the sketch view. Parent/child relationships will be created between the sketching and orientation features and the feature being created. It is all right to pick solid geometry faces as the sketching plane, but try to select one of the three default datum planes as the orientation plane as often as possible. The default datum planes are stable geometry. As a result, there is less of a chance your sketching plane will lose its orientation references and fail regeneration.

> ◦ **NOTE:** *Creating a model with design intent in mind is good design practice, and creating features of a model with parent/child relationships in mind is good modeling practice.*

New in Release 20/2000i of Pro/ENGINEER is the automatic selection of a sketching plane orientation reference called the "Default" orientation. After you select the sketching plane and set the feature creation direction, instead of picking Top, Bottom, Left, or Right and a second reference, select the Default button, shown in the illustration at right.

Pro/ENGINEER will now automatically select an appropriate orientation reference, reorient the model, and place the model in Sketcher mode. Long-time users of Pro/ENGINEER will have difficulty getting used to the idea of parent/child relationships being automatically selected by Pro/ENGINEER. However, the use of the default orientation will not create any additional parent/child relationships in the feature. If an orientation plane cannot be selected without adding a parent/child relationship, the default option will be grayed out.

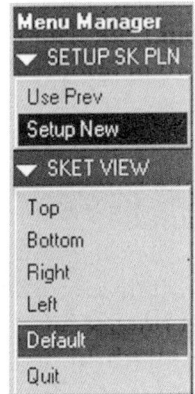

New sketch view orientation option Default.

Intent Manager

One of the most significant changes in Release 20/2000i of Pro/ENGINEER was the introduction of Intent Manager in Sketcher mode. Sketcher had always remained pretty much unchanged until about Release 14, which added basic productivity enhancements such as sketching rectangles and constraint symbols and, later, auto dimensioning.

These enhancements were helpful, but sketching sections was still as much art as it was science. You still had to know just how much to exaggerate a sketch so that it would regenerate and not make assumptions such as to snap lines vertical, or produce regeneration failure messages such as "Line segment too small" or the dreaded "Extra dimensions found." How to make things work was often a matter of knowledge passed on from one user to another via experience. Intent Manager, however, does a great job of removing almost all of the guesswork involved in working in Sketcher. The following are the major features of Intent Manager, which are explored in depth in sections that follow.

- Fully automatic dimensioning created as the section is created

- Fully dynamic creation of constraints

- Sketch geometry "snaps" to help with alignment and understanding constraints

- Addition of Undo and Redo buttons

- Better access to the most common SKETCHER menu picks using the right mouse button

- Complete control of Sketcher constraints and a more complete list of constraints

- The section will update after each modification if desired

- Improved Move command

References for Dynamic Dimensioning

Before Intent Manager there was AutoDim, which would dimension a completed section after some model references were selected. Intent Manager's way of performing the same function is completely different. Intent Manager asks for the model references first. You can select as many model references as you need to help lock in section geometry. Sketcher references are represented in Sketcher by orange lines that have a phantom font. The first of the following illustrations shows the entities that were selected as the sketch plane, and the sketcher references. The second of the following illustrations shows how the section reference entities will look in Sketcher mode.

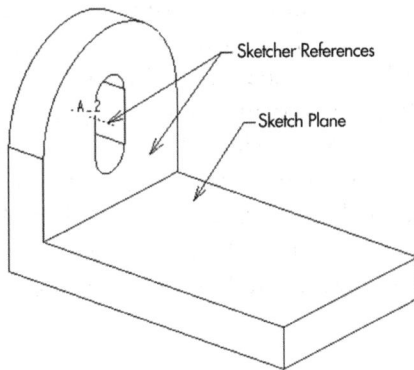

View showing the surface selected as the sketch plane and the two entities selected as sketcher references.

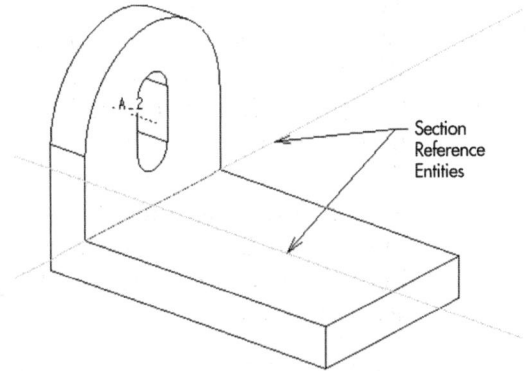

How section reference entities are shown in Sketcher. They are projected onto the sketch plane and displayed by an orange line in phantom font.

If you have not selected enough references, however, Pro/ENGINEER will let you know that additional references need to be selected. Pro/ENGINEER will then ask if you want to go back and select additional references or proceed with the sketch. Once the references have been selected, press the middle mouse button to exit the SPECIFY REFS menu and enter the Sketch ➡ Line menu. This allows you to begin sketching line entities for the section. Additional reference entities can be selected during and after the section has been sketched, and used in much the same way the Alignment command was used in previous releases.

Dimensions are created from the selected references to the sketch geometry at the end of each sketch operation. This means that the section is always regenerated. In fact, there is not even a regeneration button at this point. All that needs to be done is sketch the section, then modify the dimensional values and move on to the next feature, right? Not exactly, there is that little thing called design intent.

> ➥ **NOTE:** *When you use Intent Manager, you must satisfy design intent before moving on to the next feature. Just because the section regenerates and Pro/ENGINEER will allow the feature to be created, it does not mean the feature is correct.*

Selecting Sketcher References

Selecting Sketcher references creates parent/child relationships between the feature being created and the features that contain the reference entities. The following are some things to look for and tips to use when selecting Sketcher references.

- Use Query Select when picking geometry for Sketcher references. Information on what entity is actually highlighted will appear in the query bin and the message window.

- Use datum planes as references as much as possible (default datums are better yet). Datum planes tend not to be deleted or changed as easily as solid features.

- Use base features of the model instead of secondary features such as edges of rounds or chamfers. Again, there is less likelihood you will lose this reference, causing the child to fail regeneration.

Following these general guidelines will result in a more robust model that will regenerate the way you expect it to when modified.

> ✓ **TIP:** *When selecting Sketcher references at the beginning of the sketch, the Delete pick is not available to remove an unwanted reference. To remove an unwanted reference before the section has been started, use the Undo button.*

Geometry Creation and Dynamic Constraints

After you have selected the references that will link the section to the model, you can begin to sketch the section. There are two things about the Intent Manager sketcher you will notice right away. The first is that sketch entities such as horizontal and vertical lines "snap" into position. The second is that any constraints Sketcher is using will be displayed as the entity is being placed, as indicated in the following illustration. The combination of these two features makes for the fastest and most accurate Sketcher Pro/ENGI-NEER has developed to date.

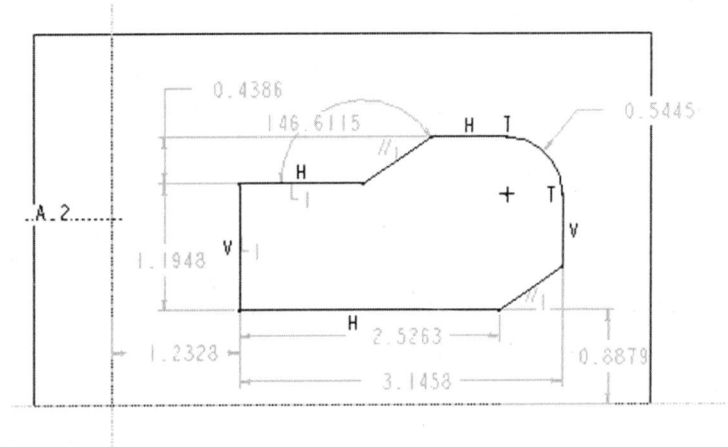

Simple sketch showing constraints used by Pro/ENGINEER to determine the section.

Geometry Creation

Now that sketch entities "snap" into place, many of the geometry creation menu picks have been eliminated because they are no longer needed. Horizontal and Vertical line menu picks have been eliminated, which can be made more quickly using the 2-Point option. When a sketch entity "snaps" to a position it is because Pro/ENGINEER is making some type of assumption or alignment. With Intent Manager, these assumption and alignments are indicated by symbols placed next to the section entity at the time it is created.

What makes this symbology such a desirable feature is that you now have a visual indicator as to when an assumption or alignment is being made. If the "snap" is expected, you know you are on the right track and can continue with the section. If the "snap" is unexpected, you can stop there until you understand exactly what assumption or alignment Pro/ENGINEER is making, and then make an informed decision as to whether or not to accept the constraint and continue sketching.

There will be times when this constraint is undesirable, requiring that you alter the section in some way before moving on. This method of sketching makes it very easy; however, to completely understand every assumption and alignment your section contains. This can only lead to more robust models that more closely follow your design intent.

In addition to making the section easier to understand, Pro/ENGINEER has also included some sketching short-cuts. The first of these shortcuts involves clicking and holding the right mouse button anywhere in the main window. This opens a pop-up menu box that contains the most commonly used Sketcher options. While continuing to hold the right mouse button down, move the mouse to highlight the option to be selected, then release the right mouse button. The option is selected. The majority of sketch options needed for the section can be selected through this pop-up menu, shown in the following illustration.

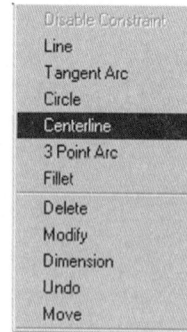

Disable Constraint
Line
Tangent Arc
Circle
Centerline
3 Point Arc
Fillet
Delete
Modify
Dimension
Undo
Move

Pop-up shortcut menu of most commonly used Sketcher options.

The second shortcut is more specialized. It involves the creation of tangent arcs and lines. You sketch line entities as you would normally, until you reach the point where a tangent arc is required. Press the middle mouse button to complete the line creation operation of the section, then press the middle mouse button a second time. This activates the Tangent Arc option. Use the left mouse button to create the tangent arc. Once the tangent arc is sketched, press the middle mouse button again to return to the Line creation option and continue with the sketch.

> ➥ **NOTE:** *A step-by-step exercise of the creation of a slot using this technique is presented in Chapter 4.*

Dynamic Constraints

Constraint symbols are now displayed as section geometry is created. This helps keep you informed on how Pro/ENGINEER is going about solving the section. Without knowing all of the constraints in use, the section can easily become tied down in ways that make the model difficult to modify later on. In earlier releases of Pro/ENGINEER you could show all constraints after the section was regenerated, but it was difficult to go through each of the constraints to make sure the section was constrained the way you wanted.

Intent Manager Constraints

With Intent Manager, however, constraints are displayed as each section entity is created. Therefore, you can look at each constraint more closely and can make an immediate decision whether the constraint is to be accepted or avoided. Intent Manager provides you with useful tools for constraint management. The following table lists the constraints Pro/ENGINEER can create while sketching.

Sketcher Constraints and Symbols

Constraint	Symbol
Horizontal entities	H
Vertical entities	V
Line segments with equal lengths	L, with a subscript index number*
Perpendicular entities	∧
Parallel entities	//, with subscript index number*
Equal coordinates	Small, thick dashes between points
Tangent entities	T
Symmetry	Arrow pointing toward centerline of symmetry from each vertex
Equal radii	R, with subscript index number*
Same points	"ϒ" drawn around points
Entities that line up horizontally or vertically	"- -" horizontally " ≠ " vertically
Collinear entities	" = " straddles collinear line
Point on entity	"–ϒ∠" "– –"

* Index numbers in subscript indicate corresponding entities for the constraint.

Constraint Tools

Now that you are able to make an informed decision about each constraint as it is created, you can act on these decisions using the mouse buttons as well as the Shift and Tab keys. As you sketch geometry, you will notice constraint symbols in two different colors. All of the constraints will be gray in color except for the latest one, which will be red. The gray color indicates that Pro/ENGINEER considers the constraint "weak." A "weak" constraint simply means that Pro/ENGINEER can remove the constraint at any time without notification should it become unnecessary.

The red constraint is the "current" constraint, which can be modified at the time of creation using the right mouse button, the Shift key, and the Tab key. If while sketching a section you want to disable the current constraint, simply press the right mouse button. The international symbol for "No" will appear over the constraint symbol. As long as no new section geometry has been created and the disabled constraint is still current, selecting the right mouse button a second time will enable the constraint.

To lock or strengthen a constraint, press and hold the Shift key and press the right mouse button. Repeating the process toggles the constraint between locked and unlocked. You will often have more than one active constraint. To step through the active constraints, press the Tab key until the desired constraint becomes current (red).

Modifying the Dimensioning Scheme and Constraints

Intent Manager dimensions and constrains the section as it is sketched. Therefore, when you finish sketching a section, it is a completely regenerated and ready-to-use section. However, the dimensioning scheme created by Pro/ENGINEER using weak dimensions (gray in color) and the references selected before the sketch was started most likely do not meet all needs of the design intent. Additional dimensions will need to be created to complete a functional section that satisfies design intent. As the new dimensions are created, weak dimensions are removed. It is also possible to strengthen a weak dimension by selecting the DIMENSION menu and then the Strengthen option.

Created and Strengthened Dimensions

Explicitly created dimensions and strengthened dimensions are displayed in the familiar yellow color and cannot be deleted by Pro/ENGINEER automatically. The same holds true of constraints. They can now be explicitly added, and Pro/ENGINEER will automatically remove any unnecessary weak constraints. To explicitly create a constraint, select Constrain from the SKETCHER menu and then select Create, as indicated in the illustration at right.

The following table lists constraints you can explicitly create. The table also provides information on how each option constrains the geometry of a section.

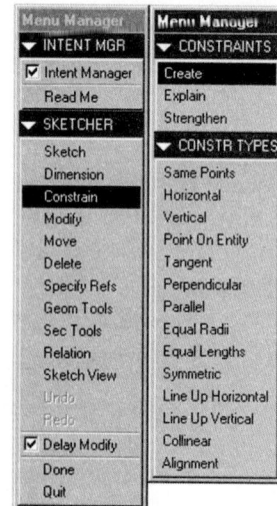

Constraints and Constraint Types option menus available in Sketcher mode.

Created Constraints and Effects on Geometry

Constraint Name	Effect on Geometry
Same Points	Makes two points coincident
Horizontal	Makes a line horizontal
Vertical	Makes a line vertical
Point on Entity	Places a point on an entity (Pick the point, then pick the entity the point is to be on)
Tangent	Makes two entities tangent
Perpendicular	Makes two entities perpendicular
Parallel	Makes two entities parallel
Equal Radii	Makes the selected arcs/circles have equal radii
Equal Lengths	Selected entities will be of equal length
Symmetric	Makes entities symmetric about a selected centerline (Pick a centerline, then pick the two entities to be symmetric)
Line Up Horizontal	Lines up two vertices horizontally
Line Up Vertical	Lines up two vertices vertically
Collinear	Makes two lines collinear
Alignment	Aligns two entities (Same as Alignment from previous releases of Pro/ENGINEER)

Constraints can also be strengthened, just as with dimensions, by selecting Constraint from the SKETCHER menu, selecting Strengthen, and then selecting the constraint. When a dimension or constraint is added that conflicts with other strengthened dimensions or constraints, Pro/ENGINEER highlights the conflicting dimensions and constraints in red and asks you to select which of the restrictions are to be deleted. Create dimensions and strengthen constraints until the section dimensioning scheme matches design intent. The section will still be fully regenerated and ready to go.

Modifying Dimensions

At this point the section is fully regenerated and the dimensioning scheme meets all design intent requirements. All that is required now is to enter the correct dimensional values and the section will be complete. You have two options: (1) selecting Done, without modifying any of the dimensional values, and continuing with the feature creation, modifying the values later, or

(2) modifying the dimensional values in Sketcher and finishing the creation of the feature.

Unless the dimensional values are quite close to the final values, the feature may not be able to be created correctly in the model. The feature geometry may have an overlap condition that will not exist when the proper values are set. Therefore, the best practice is to modify the values and then complete the feature creation.

There are some special circumstances to take note of when modifying dimensions. First, modifying a weak dimension will strengthen it. It is best to have the dimensioning scheme that matches design intent before modifying too many dimensions. Second, a section will automatically update after the dimensional value is changed. The reason behind this is that Intent Manager will always maintain a fully regenerated section, and will therefore regenerate the section after any modification to guarantee that the new dimensional value is valid before moving on.

Pro/ENGINEER does, however, give you the option of turning off the automatic update by selecting Delay Modify from the SKETCHER menu. When the Delay Modify box is checked, a Regenerate option will appear on the SKETCHER menu. The section can now be regenerated at any time, just as in previous releases.

New Sketcher Tools

A new option in the MODIFY menu for Release 20/2000i is Scale. This option allows you to modify the scale of an entire section by modifying one dimension. This is most useful when creating the first feature of a model. Because there is no geometry in the model to base the scale on, the first section always has very large dimensional values.

By selecting Modify ➡ Scale and then selecting any dimension and entering a new value, Pro/ENGINEER will scale each of the other dimensions in the section by the ratio determined by the original and new values of the modified dimension. In previous releases, each dimension had to be selected and modified with a new value by the user, and often the dimensional values caused problems regenerating the section. The menus shown in the illustration at right indicate the menu structure for the new Scale option.

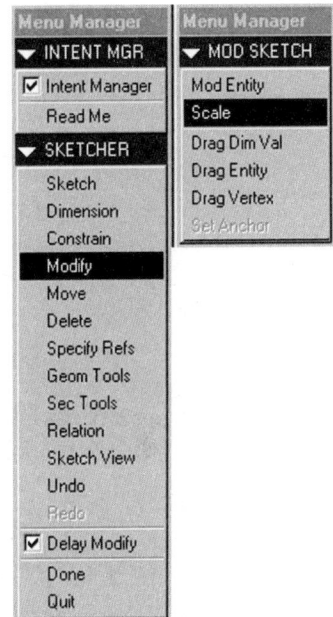

Menu structure of the new Scale option in Sketcher.

Technically not a new feature in Sketcher, the Move option has been enhanced enough that it can be included in this section. When the Move option is selected from the SKETCHER menu (shown in the following illustration), the MOVE SKETCH menu opens, containing the following options.

- **Drag:** A section entity, vertex, center point, or dimension can be selected to be moved to a new location.

- **Drag Chain:** Works by selecting the first and last section entities in the chain, selecting a vertex that belongs to the chain, and then dragging the chain to a new location. To select a complete chain, simply select one of the section entities and then select Done Sel. The complete chain will be highlighted, which can then be dragged to a new location.

- **Lock/Unlock:** A section entity, vertex, or dimension can be selected to be locked or unlocked. Locking an item prevents Pro/ENGINEER from modifying it in any way. Pro/ENGINEER displays a locked entity by placing a triangle over the entity or, in the case of dimensions, an L in front of the value. Locking an angular line locks its angle. Locking an arc or circle locks its radius. Locking a point locks its location from being moved. Locking a dimension locks in its value.

- **Lock All Dims:** Locks all dimensional values of a section while it is being moved, shown in the illustration at right.

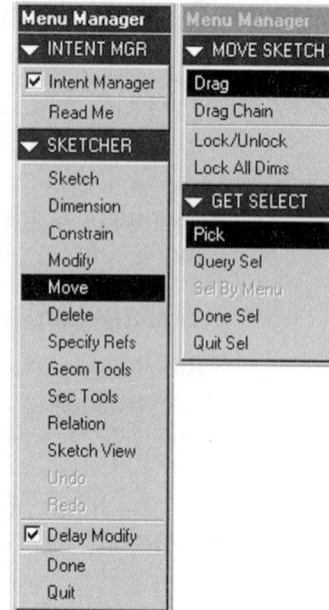

Menus showing the new items located under the MOVE SKETCH menu.

Undo and Redo are two new options in Release 20/2000i. They are available only in Sketcher mode. The Undo feature works by removing the last operation performed in Sketcher. Redo will replace operations removed by the Undo command. You can step back through the Sketcher operations by repeatedly selecting the Undo button. Repeatedly selecting the Redo button will replace each of the operations. The Undo button is found in the SKETCHER menu and replaces the Undo Last, Unregenerate, and Untrim Last commands.

There is a limit of 500 steps to both the Undo and Redo commands, with 200 being the default. Use the configuration option sketcher_undo_stack_limit to set the number of steps to something other than the default.

You can also now within Sketcher create an axis normal to a sketch. You place a point in the sketch by selecting Sketch ➡ Adv Geometry ➡ Axis Point. An axis is created that runs through this point and is normal to the sketching plane when the feature is completed. This allows you to create an axis for non-cylindrical features such as slots, where a center axis is desirable. It eliminates the need to create several additional features just to create the reference axis after the non-cylindrical feature has been created.

Basic Sketcher Rules

The following are basic rules to keep in mind when working in Sketcher.

- *Keep it simple.*

The first rule of thumb to sketching in Pro/ENGINEER is to keep the sections as simple as possible. Remember, a simple sketch is not only flexible but robust. This becomes a great asset when design modifications are required.

- *Exaggerate the sketch.*

The second key to creating sections in Sketcher is to exaggerate the sketch. By exaggerating the differences between entities and angles, Sketcher is prevented from making any assumptions about the section during regeneration. Intent Manager does an excellent job of giving you information and control over all assumptions the software is making, but exaggerating the sketch is still the best way of preventing unwanted assumptions from occurring in the first place.

- *Be aware of reference entity and alignment parent/child relationships.*

Alignments as you knew them in releases previous to Release 20/2000i and its Intent Manager are no longer needed, having been replaced with reference entities. (However, the term alignment is still present in Intent Manager.) In many ways, alignments and reference entities are very similar. The most important similarity to the designer is that both create parent/child relationships. You need to understand the parent/child relationships formed when reference entities are used.

The existing geometry selected as a Sketcher reference entity or that a section entity aligns to becomes a reference and parent of the feature you are creating. Make use of Query Select and the query bin to verify that the geometry selected is in fact the geometry you want to reference. Using an isometric view rather than a planar orientation prevents several model edges or surfaces from lining up with one another. Therefore, geometry selection is clearer. The following two illustrations indicate the advantages of using an isometric view and the query bin (shown in the illustration at left) to help determine entity selection.

Query Bin
Surf:F5(PROTRUSION)
Surf:F5(PROTRUSION)
Surf:F6(CUT)
Surf:F7(CUT)
Surf:F8(SLOT)
Edge:F8(SLOT)
Surf:F6(CUT)
Surf:F8(SLOT)
Edge:F8(SLOT)
Surf:F7(CUT)
Edge:F7(CUT)
Surf:F5(PROTRUSION)
Edge:F5(PROTRUSION)
Edge:F5(PROTRUSION)
Edge:F6(CUT)
Edge:F6(CUT)
Edge:F7(CUT)

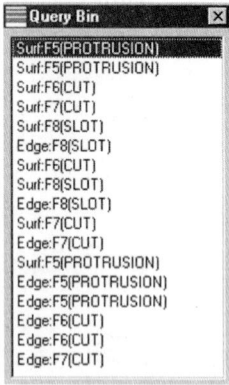

Actual query bin list of entities when the top edge of the block from the sketch view in the previous graphic is selected.

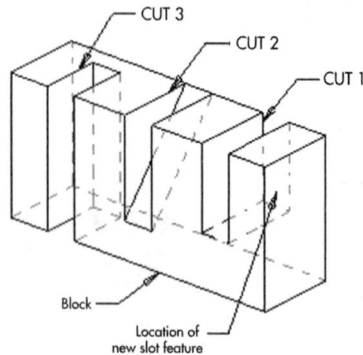

Selecting entities using an ISO view

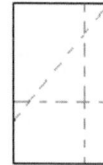

Selecting entities using standard sketch view

An example of how using an isometric view clarifies entity selection.

- *Make use of construction geometry.*

Construction geometry is another tool not used as much as it probably should be, but that can go a long way to help you dimension your sketch to convey design intent. Construction geometry is sketched entities not used in the creation of the solid feature. Any method used to create a line or circle can also be used to create a centerline or construction circle. Points are the other forms of construction geometry.

The most common uses of centerlines are for lines of symmetry and axes of rotation, but they can also be used for tangent lines, intersection lines, perpendicular lines, and collinear lines. Construction circles can be used in much the same ways as centerlines. Points will snap to vertices (including intersections), where they can be aligned to existing geometry or dimensioned to any geometry.

- *Use Edge, Offset Edge, and parent/child relationships.*

A word of caution: when you employ Use Edge, Offset Edge, or any other geometry creation command that requires you to select existing model geometry to create section entities, it will create a parent/child relationship with every feature those edges belong to. For example, in the following illustration, unless the part is rotated to an orientation where you can see what edge is selected for the Use Edge picks, cut 3 could be a child of block 1, cut 1, or cut 2. Likewise, depending on which edge is selected for dimensioning, the slot can be a child of block1, cut 1, cut 2, or cut 3.

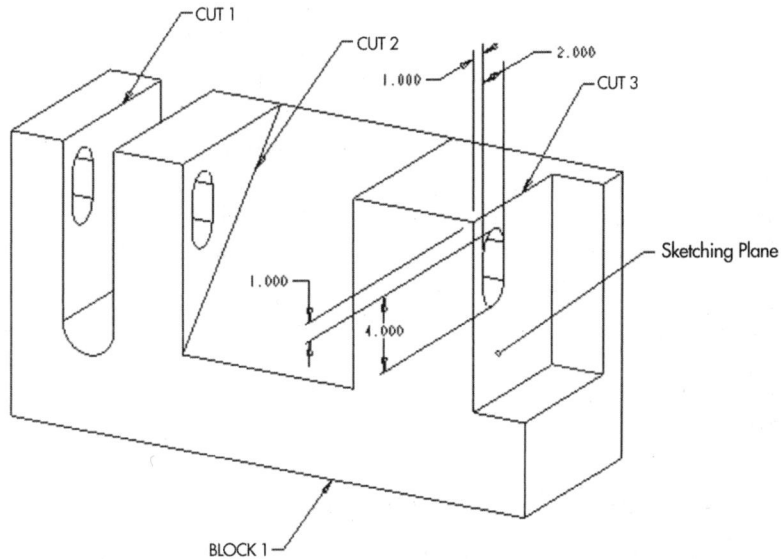

Depending on what entities are selected to create the dimensions that locate the slot, the slot could be a child of the block 1, cut 1, cut 2, or cut 3.

One Last Word on Dimensioning a Section

To define a section in Pro/ENGINEER, you need to dimension the sketch to itself and/or the model, or align the section entities to existing geometry, or a combination of both. When dimensioning a section, always dimension with design intent in mind. This cannot be overstated. Dimension the section for design intent first, and then add only the dimensions required to regenerate the section.

> ✓ **TIP:** *Dimensioning the sketch to show your design intent first and regenerating the section second will help downstream users of your model understand what is important not only to the model but to the design.*

An additional benefit of this technique of dimensioning is that it greatly reduces the time required to create a drawing (if you are forced to create one). It becomes a simple matter of showing and moving dimensions rather than figuring out a dimensioning scheme and creating a series of draft dimensions in Drawing mode. The section entities not constrained by dimensions must be constrained so that the sketch can be regenerated.

Summary

In this chapter you explored the mechanics of Sketcher, including what a sketch is and how to set up a sketch view. The chapter also reviewed the numerous improvements and new features included in the new sketch tool, Intent Manager. Among these improvements are automatic dimensioning, geometry creation shortcuts, sketcher constraint management, and the new look and feel of geometry creation in Sketcher now that the geometry "snaps" into place.

This chapter also introduced the theory of using Sketcher as the language of design intent. Design intent is the underlying theme of this book, and you will see it reiterated throughout. The reasoning behind this is that conveying design intent is the entire purpose of building a Pro/ENGINEER model in the first place. Therefore, most of everything you do in Pro/ENGINEER should be to further that goal.

Remember that dimensioning the sketch to show your design intent first and regenerating the section second will help downstream users of your model understand what is important not only to the model but to the design. In addition, you should understand parent/child relationships and their significance. They are the keys to building modifiable and robust models that will have a long and useful life.

Sketcher in Action

Examples of Sketcher Application

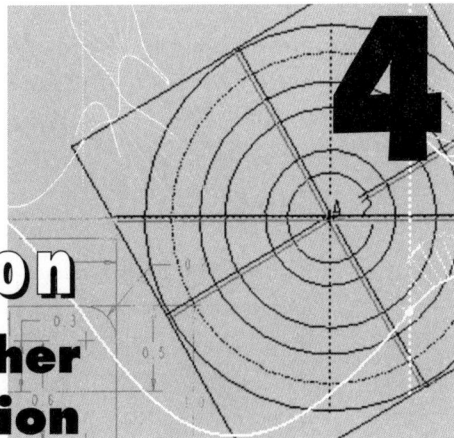

In this chapter you will be going through several example exercises using Sketcher mode and Intent Manager. Each example serves two purposes. First, each example covers a specific topic. Second, each example highlights various aspects of its respective topic using Intent Manager. After completing these simple exercises, you should have a better understanding of how Pro/ENGINEER Intent Manager "thinks" and how it controls your section with dimensions and constraints. The chapter also covers Sketcher shortcuts and includes tips and techniques for better and faster sketching.

Exercises

In the following exercises, you will create a V-block, create a slot, create a bolt circle, and dimension to a theoretical sharp corner.

Exercise 4-1: Creating a V-block

In this first exercise you will create a simple V-block. The V-block will be a single feature using a standard start part. The first step is to select the sketching plane and orientation planes.

    1. Select one of the vertical default datum planes as the sketcher plane and the horizontal datum plane as the sketcher orientation plane.

2. Intent Manager is not activated by default. Therefore, the first time you enter Sketcher mode, a window will open to give you more information about Intent Manager. This window has two options, one for opening a readme file that covers many of the Intent Manager features, and one for closing the window. It is recommended that you read the readme file at least once. The file contains a lot of very good information. If you have read the readme file, go ahead and close the window. Check the Intent Manager box in the SKETCHER menu.

✓ **TIP:** *There are two new* config.pro *options involving Intent Manager activation and the announcement window you need to consider. To have Intent Manager on as the default sketcher environment, add the following config option to your* config.pro *file.*

```
sketcher_intent_manageryes
```

3. To prevent the announcement window from opening whenever Sketcher mode is entered the first time, add the following config option to your *config.pro* file.

```
sketcher_readme_alertno
```

With Intent Manager activated, you next need to select the sketch references. The SELECT REFS menu pick is already highlighted.

⟿ **NOTE:** *Sketch references do not need to be selected at this time, but Intent Manager will not be able to dimension the section on the fly as if references had been selected. Additional sketch references can be selected at any time in Sketcher mode. Simply select the Select Refs option and pick the new references. To delete a sketch reference, select Delete from the SKETCH menu and select the unwanted reference.*

4. Select the two crossed default datums. An orange line with a phantom font will mark any sketch reference. After you have selected the references, press the middle mouse button to exit the SELECT REFS menu and enter the SKETCH menu. Before you start to sketch any section geometry, press and hold the right mouse button to pop open the sketcher shortcut menu and select the **Centerline** option. Sketch a vertical center-line on the vertical default datum. You will notice that the centerline "snaps" to the datum

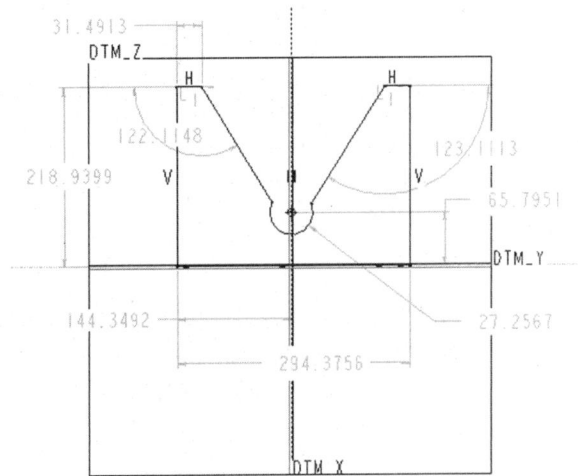

Section to be sketched.

but also into a vertical position. This centerline will be used to create symmetry constraints later on. Press and hold the right mouse button and pick the **Line** option to begin sketching the rest of the section. Create the section shown in the illustration at right, starting with the vertical line on the left side.

Depending how you sketched the section, Intent Manager may have dimensioned the section differently, but it will not matter because you will dimension the section for design intent later on. There are two important constraints to notice, and both are at the top of the block. First is the Lined Up Horizontally constraint. This constraint makes sure that the height of the block will be the same on both sides of the V. The second is the Equal Length constraint L1. This constraint makes sure that the length of the top segment of the block on each side of the V will remain equal, thus centering the V in the block.

Modify Scale

A new feature in Sketcher allows you to modify the size of one dimension and rescale the entire section by that same ratio.

5. To do this, select **Modify ➡ Scale** and then pick the overall width of the V-block. Modify the dimensional value to *3.000* and select **Regenerate**.

All of the dimensions of the section will be rescaled, but the section itself will look unchanged, as shown in the illustration at right. It is also worth noting that the 3.000 dimension is now a strengthened dimension and is displayed in yellow.

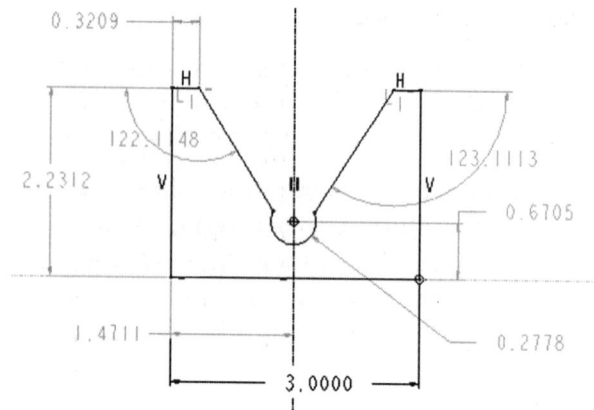

Regenerated V-block.

Adding a Constraint

6. To add a constraint to the section, select **Constrain ➡ Create** and then select the type of constraint you want to create. In this case, you will be adding a symmetry constraint. Select **Symmetry**. The symmetry constraint requires you to select a centerline and then the two vertices that are to be made symmetrical. Select the vertical centerline created at the beginning of the sketch and then the two vertices at the relief radius.

You are also going to create a second symmetry constraint to guarantee symmetry of the V-block width about the vertical centerline.

7. Select the vertical centerline and the two vertices at the top of the outside of the V-block. Verify that the symmetry constraint symbols are placed at the vertices selected. You should also notice that the dimensioning scheme has changed with the addition of the two constraints. Compare the previous illustration with the following.

Further modified V-block.

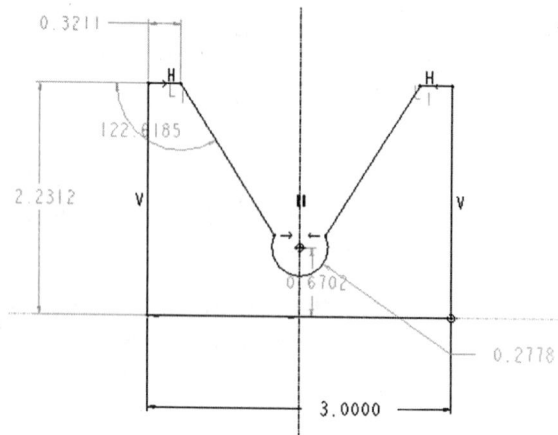

Dimension for Design Intent

The section is fully constrained and regenerated. However, the dimensioning scheme still does not meet the needs of the design intent. To modify the scheme, all you need do is create the specific dimensions required to fulfill the needs of design intent.

8. Create the dimensions shown in the illustration at right. Again, notice how the dimensioning scheme created in Intent Manager is modified with each new dimension.

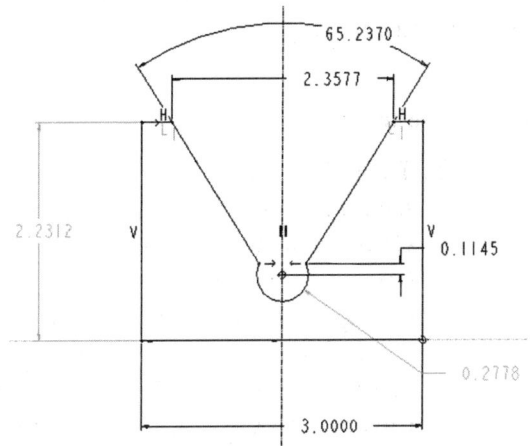

Modify Dimensional Values

Now that the dimensioning scheme matches the design intent, all you need to do is modify the values of the dimensions to the proper values.

9. Select **Modify** from the SKETCHER menu or press the right mouse button and select Modify there. Select the dimension and enter the desired value. Modify the dimensions according to the following illustration.

Dimensions to be created to meet design intent.

V-block dimension values modified to match the design values.

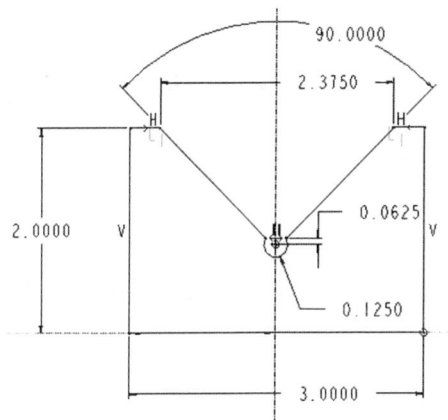

10. The section is now complete. The dimensioning scheme matches design intent, and the dimensional values are correct. Finish the feature creation. The V-block will look as it does in the illustration at right.

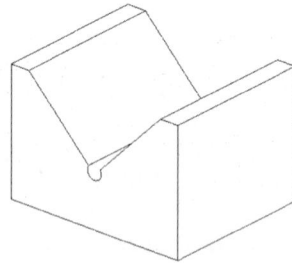

Finished V-block part.

Exercise 4-2: Creating a Slot

This example is used to show a number of new features and shortcuts implemented in Pro/ENGINEER's Release 20/2000i Sketcher. For this exercise, you will need some existing geometry.

1. Create an "L" block with a full round radius at the top of the vertical leg, as shown in the illustration at right.

Design Intent

The design intent of the slot is to be a vertically oriented slot with its center on the center of the full round. This is so that if the height of the vertical leg should change, the slot will move with it.

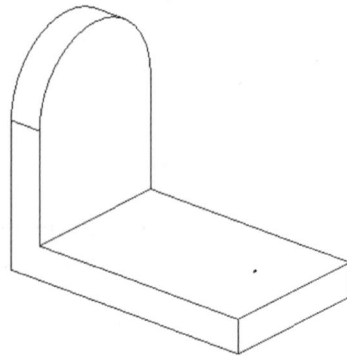

Default Sketcher Orientation

2. Select **Feature** ➥ **Create** ➥ **Slot** ➥ **Done** ➥ **Done** and pick the inside surface of the vertical leg of the block. The SKET VIEW (Sketch View Orientation) menu will now be available. It is the same menu as in previous releases; however, the Default option has been added. Select the **Default** option.

The part will reorient itself and place you in Sketch mode. Remember that in selecting the default option Pro/ENGINEER will not create an additional parent/child relationship when selecting the feature to orient the sketcher plane. If it needed to create an additional reference, the default option would be grayed out.

Now that the sketching plane has been selected and Intent Manager is checked and active, Pro/ENGINEER will be asking for Sketcher references. For this example, you want to tie the slot geometry to the full round at the top.

3. Place the model in an ISO view, which makes the round easy to pick, and select the full round. The full round will be the only Sketcher reference required for this example. See the section in the first illustration at right.

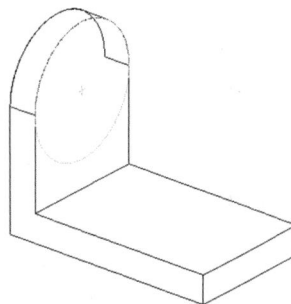

Full round Sketcher reference.

4. Press the right mouse button to exit the SELECT REFS menu and enter the SKETCH menu at the Line option. Select the **Sketch View** option from the SKETCHER menu to begin sketching. Press and hold the right mouse button to open the shortcut menu. Select the **Centerline** option. Place a vertical and horizontal centerline through the center point of the full round, as shown in the second illustration at right. At this point you are ready to create the slot geometry.

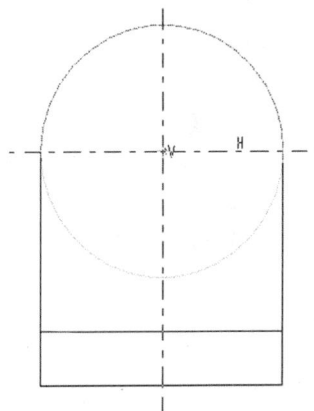

Vertical and horizontal centerlines placed in the section.

Slot Creation Shortcut

5. Press and hold the right mouse button and select the **Line** option. Start the slot section with the vertical line on the left side. As the line is being sketched, it will "snap" to a point where it is symmetrical about the horizontal centerline. Look for the symmetry constraint symbol to be displayed at that point. Pick this end point for the line and click the middle mouse button. Intent Manager will dimension the line as shown in the third illustration at right.

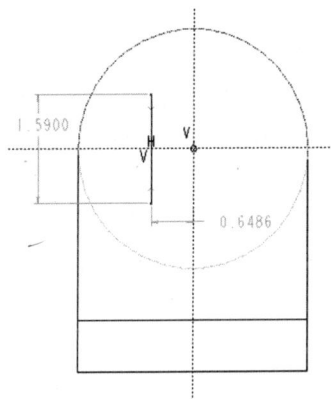

Dimensioned line.

6. At this point you want to create a tangent arc. You could select Sketch ➡ Arc ➡ Tangent Arc and then create the arc, or simply click the middle mouse button a second time. This ends the line entity creation and takes you to the TANGENT ARC menu. As you sketch the tangent arc, it will "snap" at 90 degrees and at 180 degrees. The Line Up Horizontal constraint symbol will also be displayed. Select this point as the tangent arc end point. Your section should look similar to that shown in the first illustration at right.

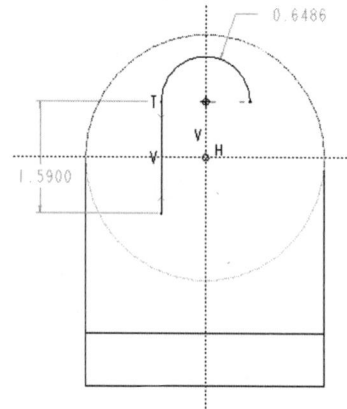

Tangent arc with sketcher constraints displayed.

7. After you select the endpoint for your tangent arc, a second tangent arc will rubberband from the end of the first. Click the middle mouse button once to exit the TANGENT ARC menu and return to the Sketch ➡ Line menu, ready to start sketching line geometry. This allows you to sketch the vertical line on the right-hand side of the slot.

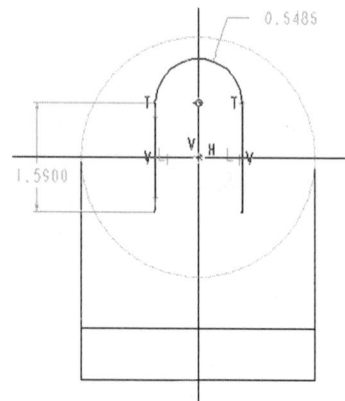

8. Sketch the line. Look for the L1 constraint symbol and the line to "snap" where its length will be equal to that of the first line. Select this as the end point and double click the middle mouse button to enter the TANGENT ARC menu to create the final arc. Your sketch should look like that shown in the second illustration at right.

Slot section ready for the last tangent arc.

9. Sketch the last tangent arc to complete the section. It will "snap" nicely into place. The completed slot should look as it does in the following illustration.

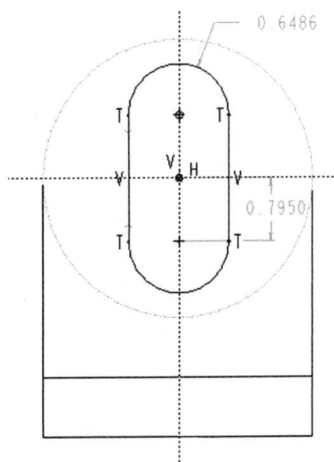

Completed slot section.

Create an Axis Point

The next phase in completing this exercise is to create an axis at the center of the slot. To do this you will be using a new option in Release 20/2000i called Axis Point. An axis point will create an axis through the sketched point and normal to the sketch plane when the feature is created.

> **10.** Select **Sketch** ➥ **Adv Geometry** ➥ **Axis Point**. Place this point at the intersection of the horizontal and vertical centerlines of the slot. It will want to "snap" to this location.

Dimension for Design Intent

> **11.** Add dimensions to the section for the overall length and width of the slot to match the design intent. Modify the values of the dimensions to the desired size, and the section will be complete, as shown in the illustration at right.

After completing the feature creation, the slot will look as it does in the following illustration. The slot will move with the height of the vertical leg of the block, and the size of the slot will be controlled by its overall length and width. It will also have an axis placed at the center of the slot to aid in the assembly of a fastener later on.

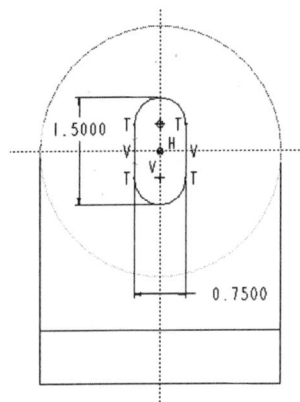

Completed section of slot with axis point and dimensioned for design intent.

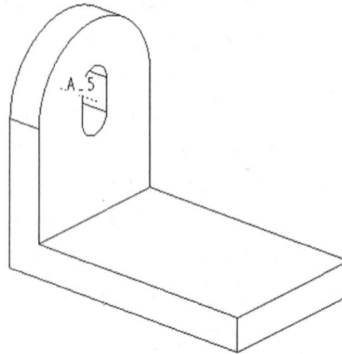

Slot after feature creation.

Exercise 4-3: Creating a Bolt Circle

In this example of creating a bolt circle pattern of holes, you will be using some construction geometry to create the diameter on which the pattern of holes will lie. This exercise illustrates how you can develop a dimensioning scheme that matches design intent through the use of construction geometry. You will take the process one step further, to the drawing phase. This will show you that by using construction geometry the dimensions you needed for the bolt circle diameter and design intent can be shown on the drawing without the need to create driven dimensions. As with the previous exercise, you need to create some background geometry for this example.

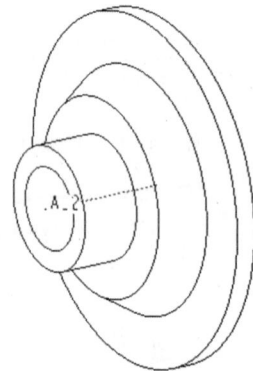

1. To begin, create a flange part similar to the part in the illustration at right.

Flange part.

Set Up the Sketching View

The sketch plane will be the face of the flange. Because this feature will need to be patterned about the center axis, the sketcher plane orientation will have to be a Make Datum.

2. After selecting the flange face as the sketching plane, select the **Make Datum** option for the orientation plane.

3. Select the **Through** option from the DATUM PLANE menu and pick the center axis. Select the **Angle** option from the DATUM PLANE menu and pick the horizontal default datum. Select **Done**. The OFFSET menu will appear, asking how to specify the new datum orientation. Pick Point is the default; this option creates the datum through the center axis and through a point you pick on the screen. The other option is Enter Angle, which prompts you to enter an angle of rotation in the direction of a red arrow displayed on the screen. The option creates the datum at that angle. Select the **Enter Angle** option and enter an angle of rotation, or accept the default value of 45°.

✓ **TIP:** *The Enter Angle option works better when working in the default or some other isometric view. It can be difficult at times to pick a point on the screen at an angle close to the required angle when the model is in an odd orientation.*

Sketcher References

The sketch view has now been set and the model is oriented to begin sketching. With Intent Manager active, the next step is to select the sketcher references. For this section, the only sketcher reference required is the center axis.

4. Select the center axis and click the middle mouse button to enter the SKETCH menu.

Bolt Circle Diameter

Sketcher is now ready to sketch some line geometry. The first thing you want to sketch is the diameter of the bolt circle.

5. Press and hold the right mouse button and select the **Circle** option. Select the **Construction** option from the CIRCLE TYPE menu located at the right-hand side of the graphics display window. The other option is to select the Circle option from the GEOMETRY menu and then select Construction ➡ Center/ Point from the CIRCLE TYPE menu.

6. Sketch the circle with its center on the center axis of the flange and place its diameter to be about in the center of the flange area. This circle should have a centerline font so that you can easily identify the geometry as construction geometry, as shown in the following illustration.

Bolt Hole

7. Sketch the circle for the bolt hole diameter. The center of this circle is at the intersection of the construction circle and the datum created to orient the sketch view (DTM1 in the first illustration at right). Sketcher will ask if you want to align the center of the bolt hole to the horizontal datum. Respond **Yes** and align the center to the datum. A sketcher reference entity, running through the datum, will be added to the sketch, as shown in the second illustration at right.

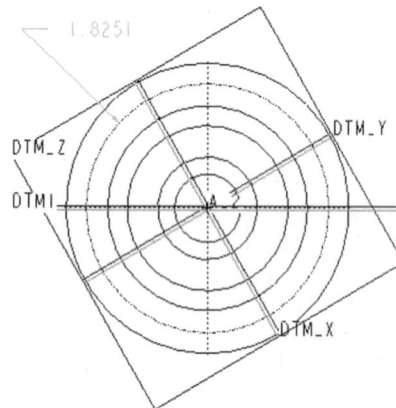

Bolt circle sketched as construction geometry.

Dimension for Design Intent

8. Create the dimensions required for design intent. In the last illustration at right, the construction circle has a radius dimension that needs to be a diameter.

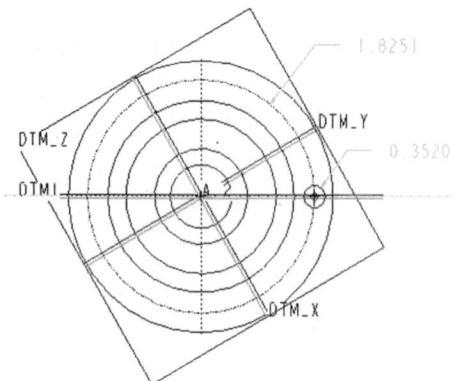

Sketcher reference entity added to sketch.

Modify the Dimensions

9. Modify the dimensions to the desired values. Remember to modify the angle of the orientation plane to 0° and **Regenerate** the sketch if the Delay Modify box is checked. Even with Intent Manager on, the section will not automatically regenerate if the Delay Modify option is checked. The bolt hole feature, ready to be patterened, is shown in the following illustration.

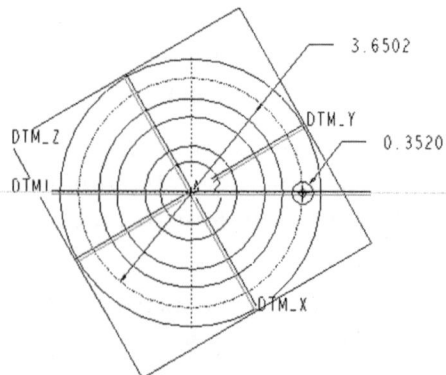

Dimensions fulfilling design intent.

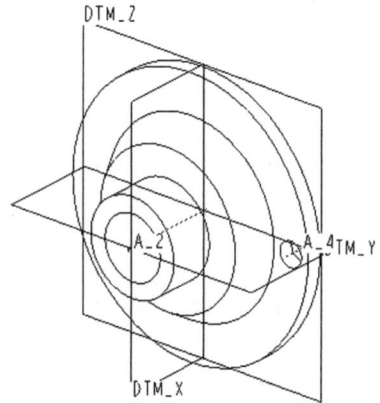

Completed bolt hole feature ready to be patterned.

Pattern the Hole

10. To create the bolt hole pattern on the bolt diameter, select **Feature ➡ Pattern**, pick the bolt hole to be patterned, and select **Done**. Pro/ENGINEER will prompt you for the dimensional increment. Enter *30* and press the Enter key. The next step allows you to vary additional dimensions. However, for this pattern, the angle is all you need. Therefore, select **Done**. Pro/ENGINEER prompts you again, this time for the total number of instances in the pattern. This pattern consists of 12 instances, including the original (360°/30° = 12). Enter *12* and press the Enter key; then select **Done** and the pattern will be created, as shown in the illustration at right.

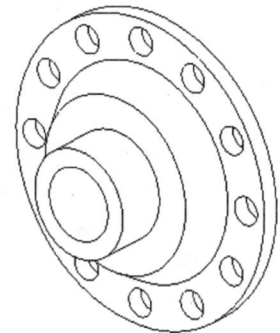

Bolt hole pattern completed.

Advantages in the Drawing

11. Using the construction circle to create the bolt circle diameter in the sketch, display that dimension in the drawing per the design intent.

The same geometry could have been just as easily created without using the construction circle. However, additional steps would have to be taken in Drawing mode to create the dimensions needed to convey design intent. Worse, the proper dimensions would have never been created and design intent lost. The example shown in the following illustration indicates the design intent.

Drawing of the flange part displaying bolt circle pattern and dimensions matching design intent.

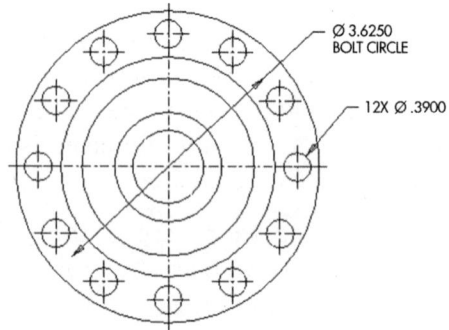

Ø 3.6250
BOLT CIRCLE

12X Ø .3900

Exercise 4-4: Dimensioning to a Theoretical Sharp Corner

The last example for this section is using point construction geometry to drive a dimensioning scheme to a theoretical point where no model geometry exits.

Sketcher References

1. Using a standard start part, create a protrusion feature, selecting one of the vertical default datum planes as the sketching plane and the horizontal default datum plane as the sketch plane orientation. With Intent Manager on, select the two intersecting default datum planes as the sketcher references. Using the middle mouse button to exit the SKETCH REF menu and enter the LINE menu, sketch the section as shown in the first illustration at right.

151.0696
74.9965
230.9366
V
127.7549

Section to be sketched.

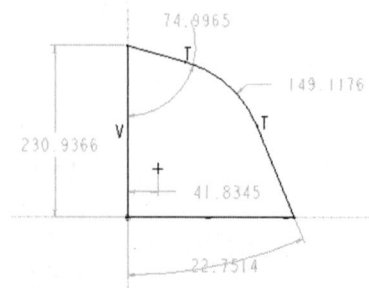

2. Sketch a fillet at the vertex of the two angled lines, as shown in the second illustration at right. Press and hold the right mouse button to open the sketcher shortcut menu. Select the **Fillet** option and pick the two lines.

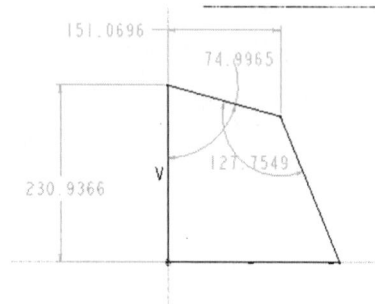

74.9965
230.9366
V
149.1176
41.8345
22.7514

Sketched fillet.

Sketching the Point

The sharp corner where the two lines met is now gone, as well as the dimension to that vertex. To replace them, you need to sketch a point at that same location so that the dimensions required by design intent can be created.

3. Select **Point** from the GEOMETRY menu and place the point at the vertex of the two lines. The sketcher point will "snap" to this location, so placing it is not difficult. Notice the Line Up constraint symbols displayed beside the sketcher point in the first illustration at right. These symbols will show you exactly what section geometry is being used to place the point, as in the illustration at right.

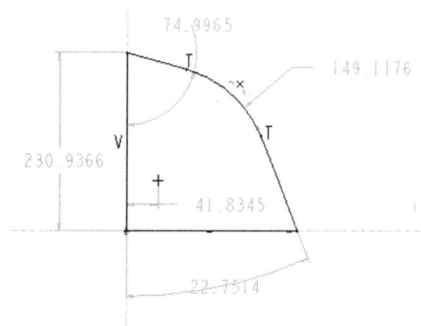

Dimension for Design Intent

With the sketcher point in place, you can now dimension for design intent.

4. Create dimensions to the sketcher point from the horizontal and vertical lines, as shown in the second illustration at right. Notice how the dimensions Intent Manager created change with each new dimension added.

Modify the Dimensions

5. With the dimensioning scheme created to satisfy design intent, modify the dimensional values and finish the feature creation. The sketch incorporating the final feature creation is shown in the third illustration at right.

Sketched point with Line Up constraint symbols displayed.

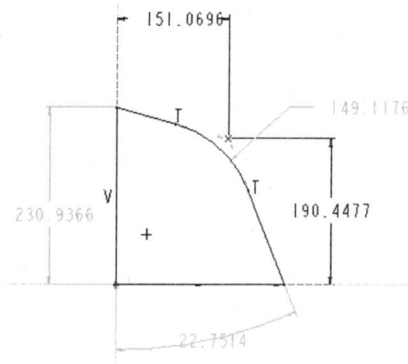

Section dimensioned for design intent.

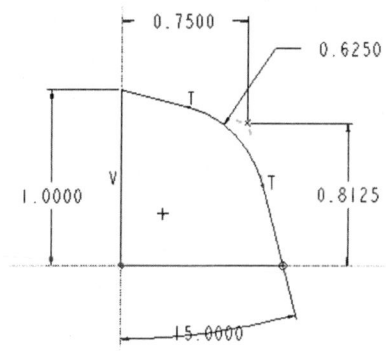

Section to be used for the feature creation.

The location of the fillet is now controlled by the location of the theoretical sharp corner, as indicated in the following illustration. This same geometry, and even dimensioning scheme, could also have been achieved by creating two features: the block without the fillet dimensioned to the sharp corner, and the block with the fillet added as a round feature. Often there are many ways to get to the same point in Pro/ENGINEER. Few are always right or always wrong. Most are just a little better than the others.

Fillet located and dimensioned to a theoretical sharp corner.

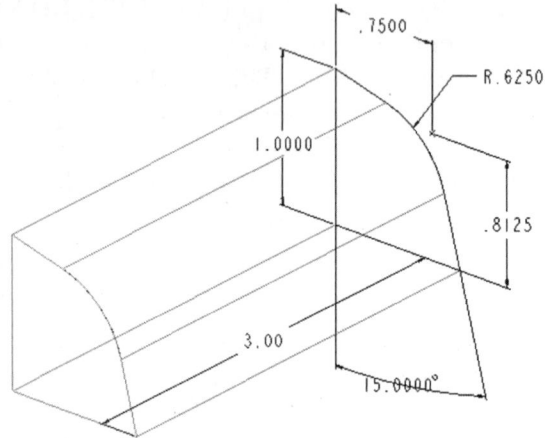

Summary

The examples used in the sketcher section were set up to demonstrate some of the new features of Pro/ENGINEER Sketcher, and to familiarize you with Intent Manager's features, shortcuts, and constraint tools. Intent Manager is a great productivity tool to make sketching faster and easier once you have taken the time to understand how it works. It is not going to finish your design work for you; that is still going to be your job. It does, however, eliminate the work of trying to get a section to regenerate, and gives you a much better understanding of how the section is being constrained, as well as control over those constraints. Intent Manager allows you to concentrate on getting the design intent right—and that is what is important in today's engineering world.

Advanced Geometry

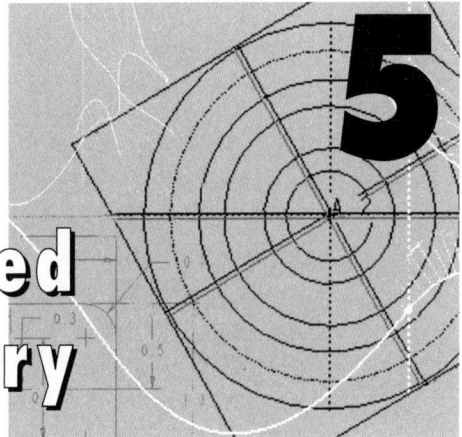

This chapter explores the creation and uses of spline and conic entities in Sketcher mode. In a lot of cases, these entities are used as the basis of complex geometry such as stylized surfaces and organic shapes. Users of Pro/ENGINEER will create splines and conics in Sketch mode toward becoming "advanced users." However, many never really grasp how and when they should use these features. Often, models produced without such knowledge do not capture the intended design and become very difficult to modify.

Most designers and engineers are very comfortable with geometry based solely on arcs, circles, and lines. In all of our solid part modeling lives we have heard the phrase "keep it simple." Simple refers to an angle, radii, or length value that can be changed; that is, straightforward geometry, with no rho values, smoothing factors, or other "black magic" needed to resolve design constraints.

In some cases, using spline and conic entities gets uncomfortably close to the world of industrial design, where phrases such as "make it sweeter" and "I don't like how that feels" rule the creation of model geometry. This chapter is intended to give you enough information to feel comfortable creating and modifying spline and conic entities in Sketcher—getting the results you anticipated in your part geometry.

Splines

A spline is created by stretching a mathematically smoothed curve between selected points on your screen. The curve itself acts very much like a thin piece of wire would if bent around pegs in a board. Each additional bend or flex in the curve affects the curvature around the previously selected couple of points. A sketched spline is shown in the illustration at right.

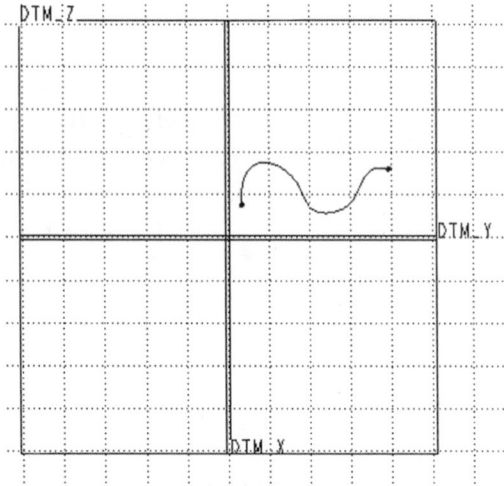

Sketched spline.

Splines allow designers to control the blending of geometry in part models. The designer can set up tangency between the ends of sketched entities to add stylistic control over features. Most models and feature geometry can and should be created without using spline geometry. However, when non-radius blends are specified, the flexibility of splines is necessary.

As with all part and feature modeling, the issue of modifiability comes into question. Splines are modifiable, but not everyone is familiar with how splines work. This can cause concern about who has the ability to work with a given model. Because of this, the designer should be careful when employing splines. When using spline entities in a sketch, the designer has the ability to use varying amounts of defining information. In its simplest form, such definition is limited to the X and Y dimensions of spline end points. In its most complex form, definition includes end and internal point positions, tangency, curvature, and relationship to coordinate systems. The sum of this defining information makes splines very powerful entities.

In Pro/ENGINEER's Sketch mode, the SPLINE MODE menu is located under the ADV GEOMETRY menu, shown in the illustration at right. One of the first things you will notice about the SPLINE MODE menu is its additional options. You have the ability to sketch the spline internal points or select sketched points on the screen. Additionally, you can assign tangency and curvature control conditions for your sketched spline. In the following exercises you will create a few spline curves, which will demonstrate how some of the menu options affect your sketch.

Menu Options and Controls

To set up the exercises that follow, go into Sketch mode and select Sketch Points from the SPLINE MODE menu under ADV GEOMETRY. Using the left mouse button, start picking a few points on the screen. Notice the curve and how it bends as you select each new point. Each point you select on the screen affects the spline curvature around the previous point. After you have selected about five points, click the middle mouse button to end the spline. Your sketch should look like that shown in the illustration at right.

ADV GEOMETRY menu.

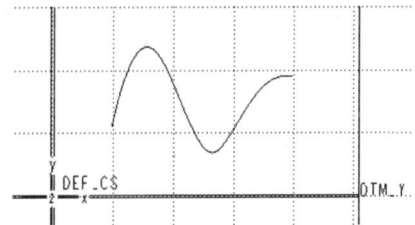

Sketched spline.

If you had imported a point file, or had previously sketched points on the screen, you could have created a similar spline by using the Select Points option from the SPLINE MODE menu.

> ➬ **NOTE:** *Point files and their creation and uses are discussed later in the chapter.*

Exercise 5-1: Creating a Four-point Spline

The second set of options under the TANGENCY menu lets you control the curvature and tangency conditions of a spline. Tangency conditions are applied to the end points of splines, allowing you to set an angular control to either end or both ends of the curve.

> ⊷ **NOTE:** *You do not have to add tangency conditions when you first sketch a spline. Tangency can be added to the spline end points later, during modification.*

1. To see how tangency conditions work, in Sketch mode, draw a square with one side missing. Next, select Both from the Tangency menu and sketch a spline across the open end of the square. The sketch should look similar to that shown in the first illustration at right.

2. Dimension the sketch and regenerate it.

At this point, Pro/ENGINEER will make some assumptions regarding the spline tangency as it meets the vertical sides of the box. With constraints turned on and displayed, as previously discussed, the result should be similar to that shown in the second illustration at right. Notice that both ends of the spline have a tangency constraint displayed and have been made vertical as they join the sides of the box.

If the design intent is to have a specific angle at the top of this sketch, you will need to add an angular dimension to each end of the spline as it relates to the side walls. In this case, dimensioning the spline end angles is very simple.

3. Picks 1 and 2 are on the spline. (After pick 1, you will notice that the spline's internal points become visible.) Pick 3 is on the vertical side, and pick 4 is on the spline's end point. Last, place the dimensions using the middle mouse button. The result should look similar to that shown in the following illustration.

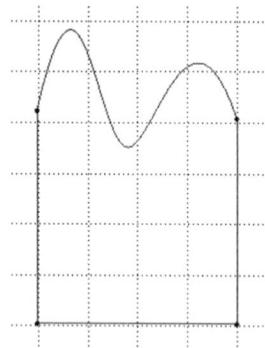

Spline sketched across open end.

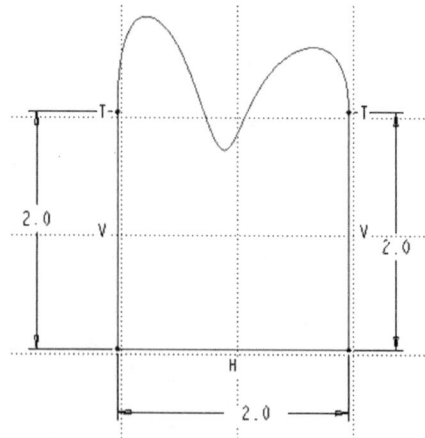

Default tangency constraints.

Dimensioning spline tangency.

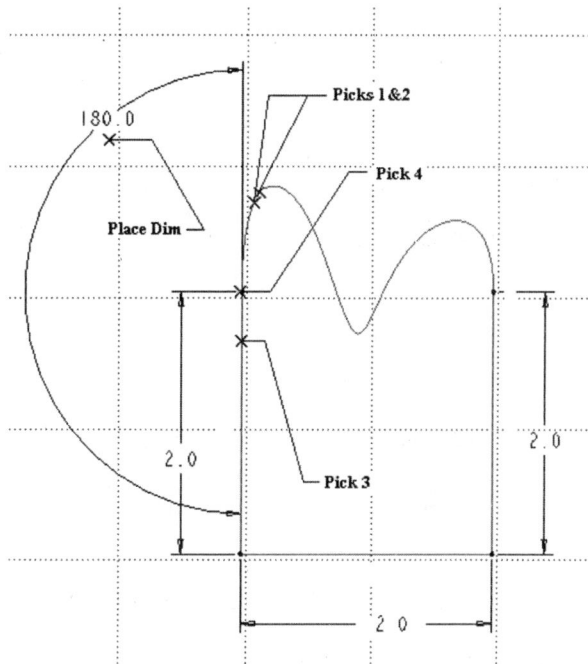

Picks 1 & 2

Pick 4

180.0

Place Dim

Pick 3

2.0

2.0

2 0

4. Repeat the previously described process to place a dimension on the other end point of the spline. Upon regeneration, both angular dimensions will be 180 degrees, tangent to the two vertical sides. To see the effects of the angular control at the end points of the spline, modify one of the two angular dimensions. Notice how the spline flexes around its control points, like a thin wire bending around pegs on a board, as indicated in the illustration at right.

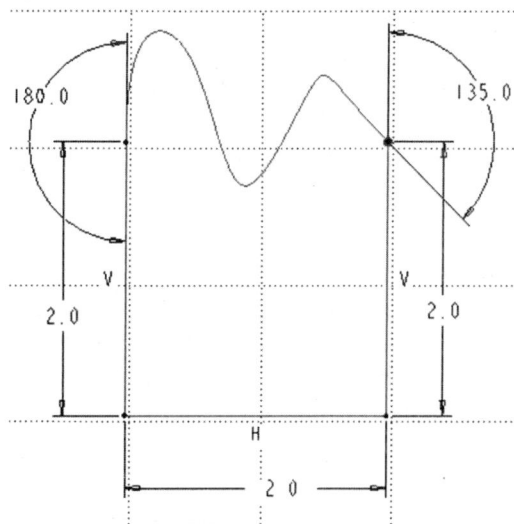

180.0

135.0

V

V

2.0

2.0

H

2 0

Flexing the spline.

This example is quite simple, but it does demonstrate some important characteristics about how spline geometry acts. In most cases, however, you will not be controlling spline tangency to line entities. Instead, you will be trying to work out smooth blends between splines, arcs, and conics. The following exercise takes you through an example of this concept.

Exercise 5-2: Blending Between Curves

In most cases, when you start having to consider tangency conditions and complex blends between curves, you have probably stepped outside your comfort zone. In this exercise, you will encounter a few techniques that clarify design intent, which will increase your comfort level when working with this type of geometry. The first step is to agree on and understand a definition for tangency. The definition used here is: Tangency is achieved when two curves meet at one point and the resulting angle between those curves at that point is 180 degrees. In this exercise you will be creating a spline that blends tangentially between the left ends of two arcs.

1. Draw two 3 Point arcs like those shown in the illustration below left.

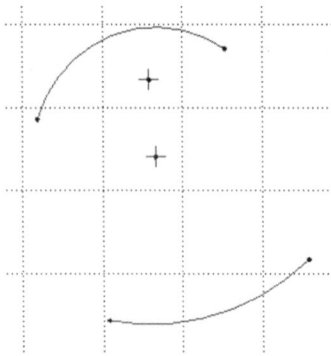

Sample three-point arcs.   Drawing the two-point spline.

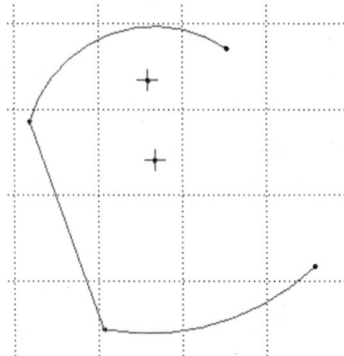

If you simply wanted to have a tangent curve between the ends of each arc, you could create a fillet between them. Although a fillet would give you the tangent end condition, it would also change the end point position of at least one of the arcs. Therefore, here you will use a spline.

2. Draw a **2 Point** spline with **Both** ends tangent between the upper end points of the two arcs. The resulting sketch should look similar to that shown in the illustration above right.

As you can see, a two-point spline looks like a straight line. This, however, will change when you add the angular dimensions to the spline end points. This brings up an interesting problem. In Sketch mode you have no vertical or horizontal references. Therefore, in this case you need to draw a couple of centerlines to dimension to. Because you are interested in controlling the tangency between the arcs and splines, you need to determine the angle of the arcs' end points.

3. Draw a centerline with **2 Points** from the center of each arc through each arc's end points. In addition, create a vertical or horizontal centerline through each arc center. The first illustration at right is an approximation of your centerline layout.

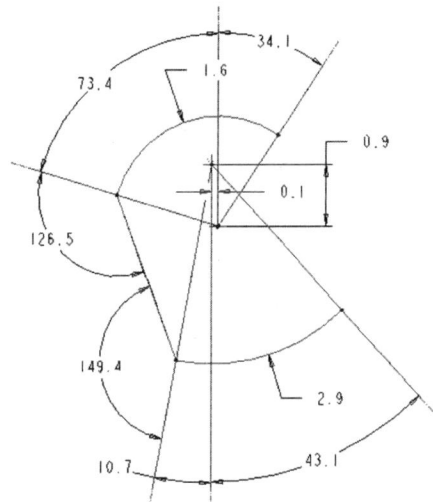

Centerline layout.

4. Dimension the sketch.

The end points of the arcs will be controlled by radii and angular dimensions, and the end points of the spline will be dimensioned angularly to the centerline through each arc. The dimensional scheme should be similar to that shown in the second illustration at right.

5. Modify one of the spline angular dimensions to 90 degrees and regenerate the sketch.

This will make the spline tangent at that end point and will change the shape of the spline between its two end points. In addition, per the rules of trigonometry, regardless of the angular

Spline dimensional scheme.

value controlling the arc end points, the spline will always remain tangent at that point. By modifying both spline ends' angular dimensions to 90 degrees, you have made the spline tangent to both ends of the sketch arcs, as indicated in the following illustration.

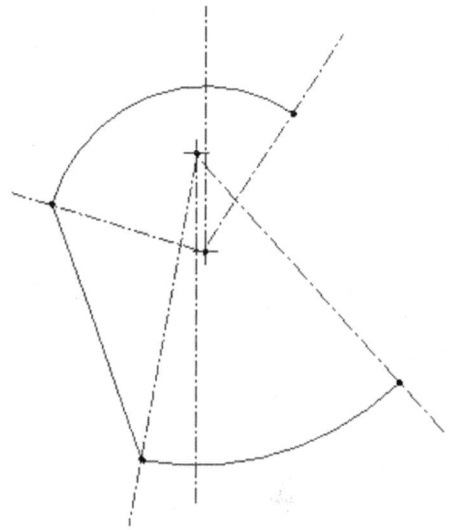

Tangency between arc and spline.

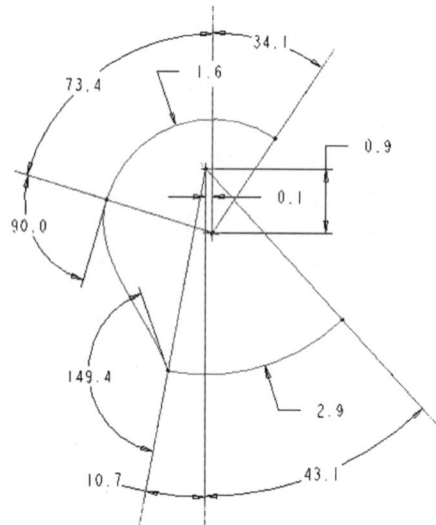

Exercise 5-3: Establishing Spline Tangency to Existing Geometry

This exercise involving spline tangency control demonstrates how to make a sketched spline curve tangent to an existing curve. The existing curve in this example could be any existing feature edge geometry. You will be creating parent/child relationships; therefore, make sure you understand this: Any subsequent features based on this spline are related, so modifications can become messy.

1. In Part mode, create a spline datum curve. The menu picks are **Feature ➡ Create ➡ Datum ➡ Curve ➡ Sketch**. Sketch a four-point spline and dimension the end points to your default datum planes. After you regenerate the sketch and select **Done** and **OK**, you should have a part that consists of a datum curve in space, as shown in the illustration at right.

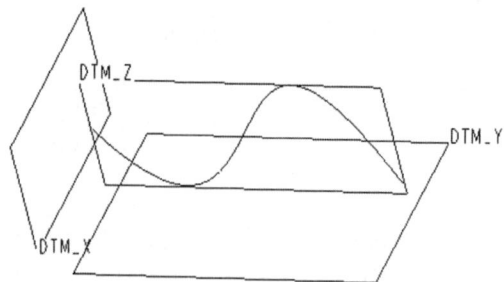

Spline curve part.

This curve is what you will make the next spline curve tangent to, and relate its tangency angle to. You do this so that as the first curve is modified the new curve will remain tangent as well. You want to create the new curve on the same plane as the existing geometry, which in this case is DTM-Z. Once you get into the sketch, you want to create a spline curve with tangency at its Start.

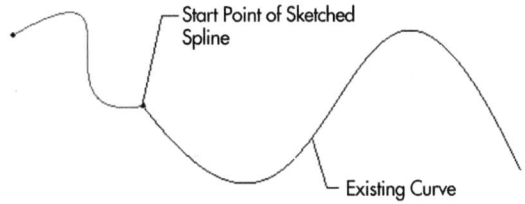

Four-point spline sketch.

2. Sketch a four-point spline with its start point being the left end of the existing curve. Your sketch should look similar to that shown in the first illustration at right.

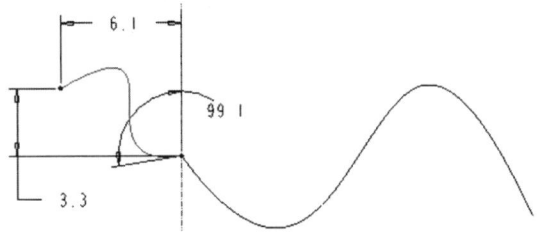

Regenerated sketch.

3. Draw a Vertical Centerline through your spline's start point and dimension the sketch. Make sure you align the start point of your spline to the existing curve, and do not forget the angular dimension to the centerline you sketched. Your sketch should look similar to the regenerated sketch shown in the second illustration at right.

The start point of the sketched curve is now related to the end of the existing curve such that wherever the end of the parent curve moves the sketched curve will follow. Now you need to get tangency under control. Toward this end, you are going to create some additional geometry by using the existing curve edge.

4. From the SKETCHER menu, select **Geom Tools** ➡ **Use Edge** ➡ **Sel Edge** and pick the curve to the right of the centerline. This will add a sketch entity over the existing datum curve. The sketch entity is a child of the datum curve and will move and flex when its parent curve does. Now add an angular dimension between the Use Edge curve and the centerline. This new dimension, shown in the illustration at right, will highlight as an extra dimension, but do not delete it because you will be using it to drive the tangency relation.

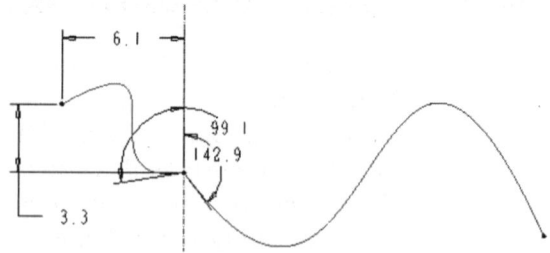

Extra angle dimension.

You now need to add a relation between the sketched spline and the aligned curve's tangency angles.

5. From the SKETCHER menu, select **Relations** ➡ **Add**. In the message window, enter *sd2=180-sd5*, where sd2 is the symbolic dimension for the sketched spline's angular dimension and sd5 is the symbolic dimension for the aligned curve's angular dimension (the highlighted extra dimension). The illustration at right shows the result of the regeneration after the relation has been entered. Now as you modify the original datum curve the newly sketched spline will stay attached and tangent.

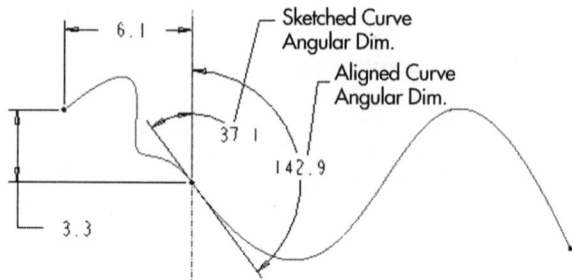

Two tangent splines.

Spline Modification Menu

Through the previous exercises you have seen how splines are drawn and how a spline's internal points help shape it. You have also learned how tangency conditions are assigned to either end of a spline curve and how the angular dimensions flex the curve. In this section, you will explore some of the tools under the spline modification menu, as well as some techniques for snapping splines into shape.

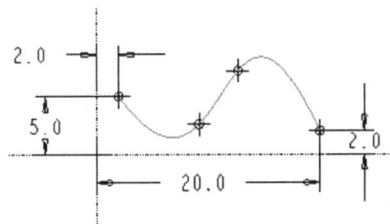

Internal spline points.

1. Splines can be modified in Sketch mode by simply selecting Modify and picking on a sketched spline. Once you have picked the spline, you will notice that the spline's internal points are now visible and that the spline modification menu opens. Select **Modify** and pick the spline. The illustration above left shows a spline's internal points. The illustration at right shows the MOD SPLINE menu and its options.

Spline modification menu.

The first thing you may notice is that the MOD SPLINE menu contains a few grayed-out options: Coords, Read Pnts, Save Pnts, and Info Pnts. Each of these options comes into play once you have associated the spline with a local coordinate system.

2. Draw a coordinate system in the sketch. A local coordinate system is a sketched coordinate system. The coordinate system from the start part file will not work as the local, but you can dimension or align the local coordinate system to it.

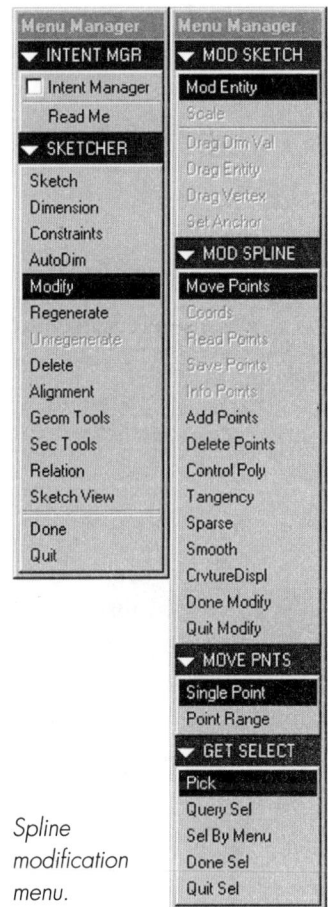

3. To create a local coordinate system for the sketch, select **Sketch ➥ Adv Geometry ➥ Coord Sys** and pick a spot on the screen at which the coordinate system will be created. If you choose to not align your local coordinate system to the default datum planes or any existing geometry, you will need to dimension it to the existing part. Once you have created the local coordinate system, you can associate a spline to it by selecting **Dimension** from the SKETCHER menu, picking twice on the spline, picking on the local coordinate system, and clicking the middle mouse button. The following illustration shows the picks involved in the association procedure.

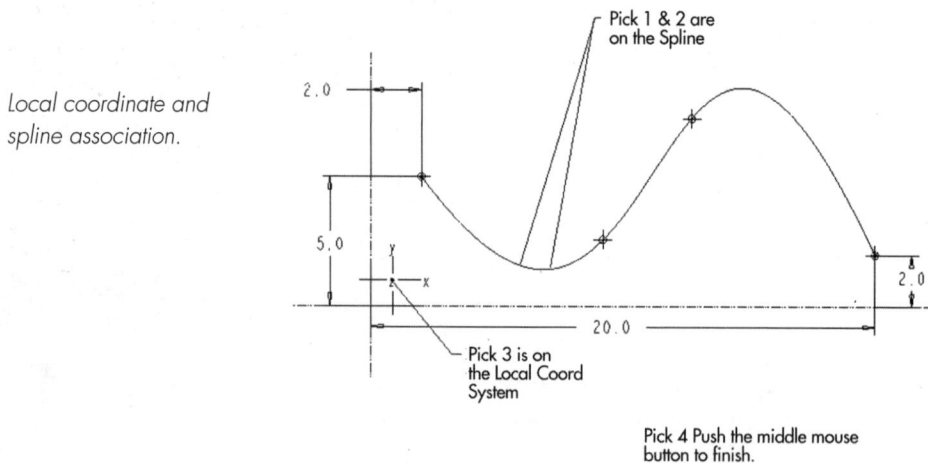

Local coordinate and spline association.

The note in the message window will read "Spline is dimensioned to local coordinate system." Now when you modify your spline, all menu options will be available. The following is a brief description of each of the MOD SPLINE menu options.

- **Move Points** allows you to select and drag any internal point to a new position.

- **Coords** allows you to modify the X an Y coordinates of each selected internal spline point.

- **Read Pnts** allows you to set spline points to those of a specified text file.

- **Save Pnts** saves the current spline points to a text file, which you can edit and then read back in using the Read Pnts option.

- **Info Pnts** displays the coordinates of all spline points in an Info window. You must select the coordinate type you want your values displayed in: Cartesian, Cylindrical, or Spherical.

- **Add Points** adds internal points to the spline.

- **Delete Points** deletes internal spline points.

- **Control Poly** displays a set of polylines you can use to modify the shape of the spline curve. The display will look similar to that shown in the following illustration.

Control Poly display.

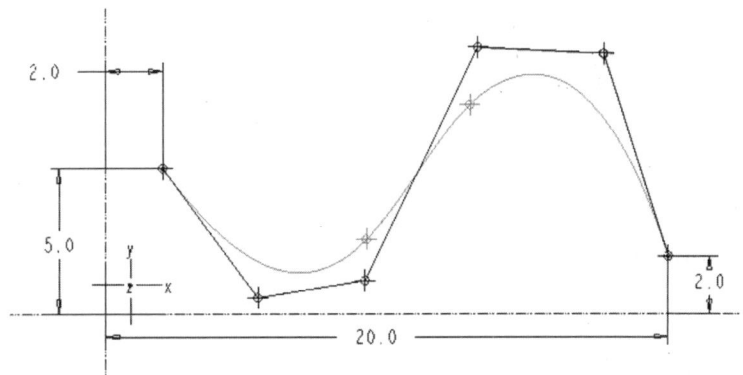

- **Tangency** allows you to add or remove tangency control to the end points of your spline.

- **Sparse** allows you to enter a deviation value that will eliminate points from the spline. The resulting spline will be displayed in green and you will have the option to Accept or Reject the resulting curve. Entering different values for deviation will result in different curves.

- **Smooth** allows you to enter an odd number of points the system will use to average the spline's curvature between. Again, you will have the option to Accept or Reject the resulting green curve.

  ➠ **NOTE:** *The options Sparse and Smooth are most commonly used when importing datum curve, IGES, and point text files. Many of these files can contain excessive numbers of points, which will need to be reduced to more easily control their shape. Importing data into Pro/ENGINEER is discussed in Part III.*

- **Crvture Displ** displays the curvature of spline curves. You have the options to set Scale (length of display line) and Density (number of display lines). In the following illustration, you can see how the curvature display changes as the spline goes from concave to convex.

Curvature display graph.

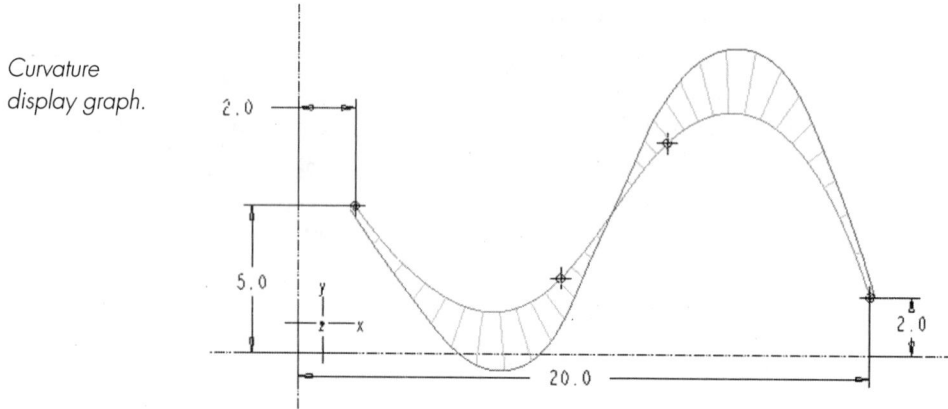

- **Done Modify** accepts changes and closes the MOD SPLINE menu.

- **Quit Modify** cancels changes made during current spline modification and closes the MOD SPLINE menu.

Exercise 5-4: Dimensioning a Spline

As you have seen in the previous exercises, a spline must be dimensioned to the existing part by locating the X and Y positions of its start and end points. Additionally, you can assign tangency conditions to either or both end points. In this exercise, you will be examining how the dimensional scheme can help control a spline's shape. You will also be using a local coordinate system like that created in the last exercise to generate a point table. You will then edit that point file and read it back in to control a spline.

1. Draw a four-point spline and a local coordinate system in Sketcher mode. Then dimension the start and end points to the local coordinate system. The result should look similar to that shown in the following illustration.

2. To dimension the interior points of the spline: pick the spline to display the internal points (pick 1), pick the point you wish to dimension (pick 2), pick the reference geometry (pick 3), and place the dimension (pick 4). The pick sequence is shown in the second illustration at right.

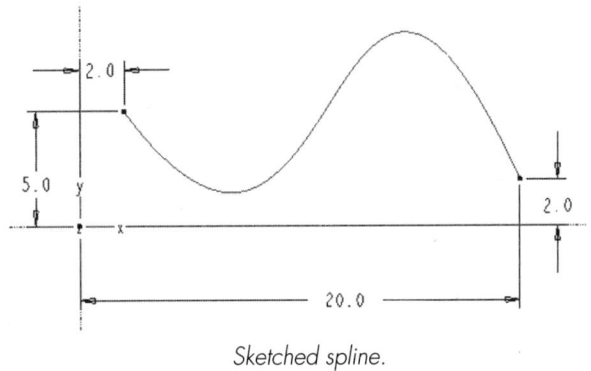

Sketched spline.

•◦ **NOTE:** *When dimensioning the internal points of a spline, you must dimension both the X and Y directions.*

3. Finish dimensioning the internal spline points to the local coordinate system. When finished, your sketch should look like that shown in the third illustration at right.

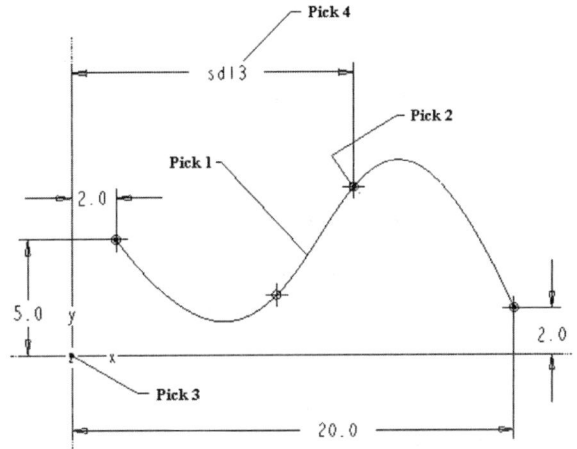

Dimension pick sequence.

4. Once the spline is dimensioned, modify the end point dimensions.

As before, the curvature of the spline is changed across the next point. However, because you have locked the internal points with dimensions, they do not move. Here, you are controlling the effects of dimensional modification on the spline. The dimensional scheme is associated with the position of the local coordinate system. This scheme is the same as the scheme

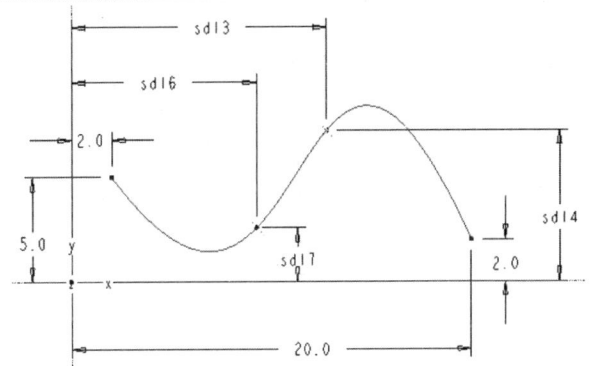

Internal point dimensions.

developed by reading in a point file. However, a point file will not anchor the internal point positions the way actual hard dimensions do.

Exercise 5-5: Creating a Point File

In this exercise, you will generate a point file for the spline you have just sketched. You should know the following about reading in a point file.

- Point files specify the X, Y, and Z coordinates for spline points, sketched points, datum points, and datum curves.

- Point files can be used to control the internal point locations of a spline. However, to modify a spline's point positions, you have to either dimension the point positions or modify the point file and re-read it in.

- The start and end points of the spline will still need to be dimensioned to the part before the sketch will regenerate.

- You do not need to dimension the spline before saving its points.

- If you have dimensioned the internal point positions of a spline you will be prompted to delete all internal dimensions before you can successfully read in a point file.

- A point file that is read in will not override the existing dimensions of a spline's start or end points.

1. Associate the spline with the local coordinate system. (Refer to the foregoing "Spline Modification Menu" section for the picking sequence.) Next, select **Modify** from the SKETCHER menu, select the spline to be modified, and then pick **Save Points** from the MOD SPLINE menu. You will be prompted for a file name. Enter *Curve*, and select the Cartesian Coord Type. This will save a file named *curve.pts* containing the current spline point positions. The created point file will look similar to the following.

```
         Coordinates of spline points:
(they may be edited using available editor; changes in X and Y
coordinates of the first and the last points will be ignored)
CARTESIAN COORDINATES:
X          Y          Z
20         2          0
12.70639   7.259887   0
9.196275   2.572891   0
2          5          0
```

↝ **NOTE:** *The statement in the previously shown file that states that changes to the first and last points will be ignored is not correct. If you have not dimensioned the first and last points in Sketcher, the file values can be used to set them.*

2. On your computer, modify the *curve.pts* file values for the second and third entries and save your changes. Select **Read Points** from the MOD SPLINE menu, and enter in the modified point file's name. You will next need to select the Coord Type, **Cartesian**, and the system will read in the point dimensions. If you are reading points in for a spline with its internal points dimensioned, you will be prompted to delete the internal dimensions. After selecting **Yes**, the new point values will be entered and the spline will be modified. You can use the point file to change the number of points in your spline by adding or deleting values from lines in the file.

A typical use for the generation of a point file in Sketch mode would be when you have 2D curves or geometry profiles. These could be several sets of points that define curves (which will be used for surface blends) or actual geometry edges.

↝ **NOTE:** *If the curves you need to build are 3D, you will need to generate datum points in Part mode and then drive a datum curve through those points. This process is described in Part III.*

Exercise 5-6: Using a Dimensional Scheme

1. Sketch and dimension a six-point spline. Use a dimensional scheme similar to the one from exercise 5-4.

Look at how you can control the spline when it reacts to a different dimensional scheme. As designer, you need to determine what the design intent is and incorporate that in sketched geometry. In the first illustration at right, the intent is to keep the positions of the two left-most internal points associated with the left end point of the spline.

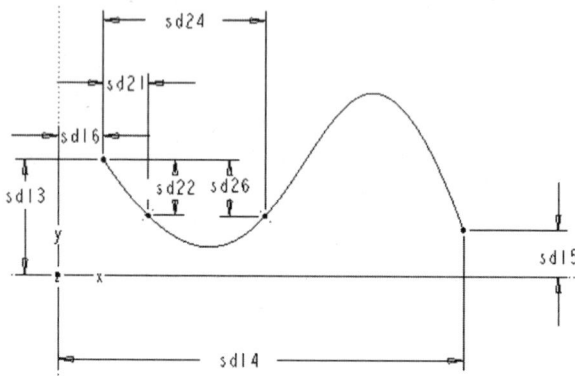

End point scheme.

With this type of dimensional scheme, as you modify the location of the left end point, the curvature and shape of the spline between the next two points will be maintained.

2. To control the shape at both ends of the spline relative to each end point, dimension from each end point to the next two internal points. This causes all spline flexing to occur between the third and forth spline

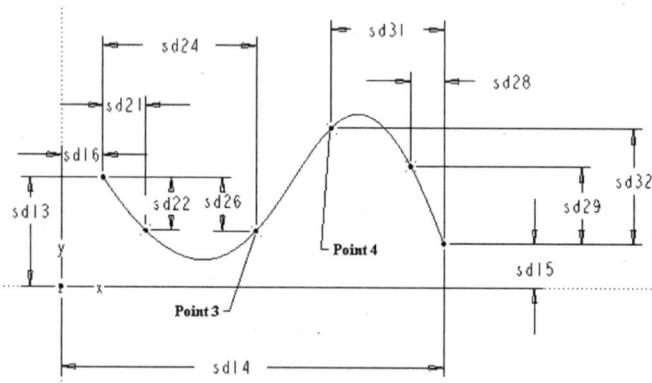

Internal dimensions to each end.

point. The second illustration at right shows the dimensional scheme.

This type of dimensional scheme is useful when you have existing geometry you need to align to and maintain blending between. As in exercise 5-3, you can control tangency at the end points. However, here you can also control the curve's shape away from the ends.

Exercise 5-7: Assigning a Radius of Curvature Dimension

Curvature allows you to assign a dimension to the spline's end points to help control and adjust blending between adjacent features. With the use of the Crvture Disp option from the MOD SPLINE menu, you can visually see how closely the curvature between adjoining curve sections match, allowing you to achieve as smooth a blend area as possible.

> ➥ **NOTE:** *The spline end point you want to dimension with a radius of curvature must have tangency control assigned to it or the curvature dimension will be considered extra upon regeneration.*

1. Assign curvature dimensions to the end points of a spline. To get a feel for how this type of dimension is used, you will first need to create a part consisting of one datum curve, similar to the operation performed in exercise 5-3. Once you have created that datum curve, switch to Sketch mode to create another datum curve.

Using this new curve, you will see the dimensioning of curvature as well as the Crvture Disp option in practical use. As in exercise 5-3, you have an existing curve and you will be sketching your new curve as a spline attached to one end of it, similar to that shown in the illustration at right.

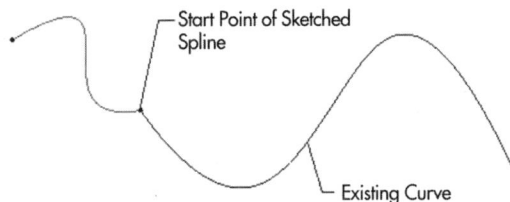

Spline attached to end of existing curve.

2. Assign tangency to the start point of your new spline. Select **Modify**, pick the spline, select **Tangency** ➥**Add**, pick the start point of the spline, and select **Done**. This will make the selected end point tangent. You then need to dimension it. When finished dimensioning, and after regeneration, the regenerated sketch should look similar to that shown in the illustration above.

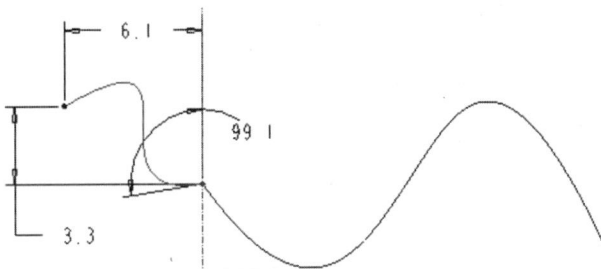

Regenerated sketch.

3. After this regeneration is complete,
go to **Geom Tools**
and **Use Edge**,
select the existing
curve, and, as you
did in exercise 5-3,
set up a relation-
ship to control
tangency between
the two curves.
Next, add the
radius of curva-
ture dimension to
the start point of

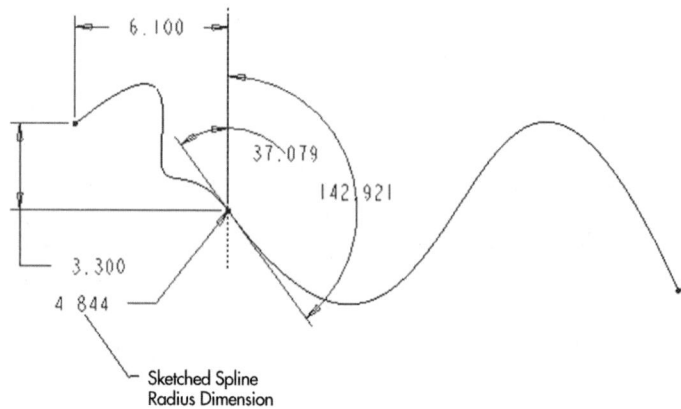

Curvature dimensions.

the spline. To add the radius of curvature dimension, select **Dimension** from
the SKETCHER menu, select the start point of the spline, and use the middle
mouse button to place the dimension. Regenerate the sketch. The resulting
dimensions should look like those shown in the illustration above.

4. Use the **Crvture Disp** option to look at a curvature graph for the sketched
spline. You may need to adjust the scale of the graph to get a better look at the
end curvature. To look at the curvature of the aligned spline, first unalign the
spline, then apply **Modify** and **Crvture Disp** to see the spline's curvature
graph. Because you want to compare the graphs, make sure you set the graph
scale factor to the same value.

Unfortunately, there is no way to see both curvature graphs displayed at once in
Sketcher. Therefore, you will need to remember what the two displays look like. The following illustrations are two images of the curvature graphs. The illustration at left is the
graph for the sketched spline, and the illustration at right is the graph for the aligned
spline.

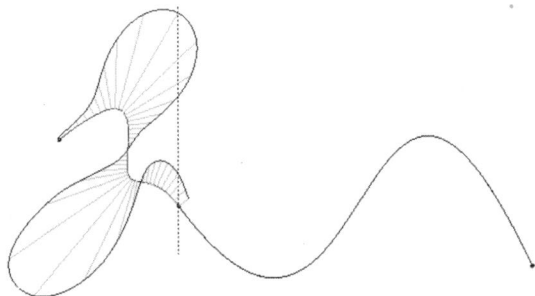

Sketched spline curvature graph.   *Aligned spline curvature graph.*

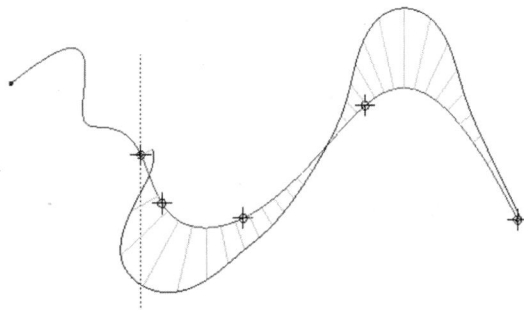

As evident from the previous illustrations, the curvature displayed by the graphs is not the same, even though where they both share a point they are close. You can also see that the graph normals are both above and below the curves, showing positive and negative curvature. Keep in mind that you are using short, multipoint spline sections and that due to their interior point locations the curves will have both concave and convex areas. Most curve and surface blending you will work with will not be this extreme. In this exercise, you are exploring a technique for curvature control, your main goal being to get the radius of curvature values to be the same.

At this point, you could eyeball the curvature between the two curves and tweak the curvature dimension until the Crvture Disp graphs appear closer. This could take some time. Alternatively, you could do what you did to control the tangency, assigning another curvature dimension to the aligned curve entity and using that to drive the unaligned spline entity's curvature. This second method will be quicker and take some of the guesswork out of the job.

5. Assign a curvature dimension to the end point of the Use Edge spline you created. Use **Query Sel** to select the end point. Place the dimension. Regenerate the sketch and the resulting dimensions should look like those shown in the illustration at right. Add a relation between the two curvature dimensions. The relation will be sd10 =sd9 for the sketch shown in the following illustration

6. After the relationship is written and
the sketch regener-
ated, use **Crvture
Disp** again to check
your results.

Congratulations, you have
just made two spline curves
with C2 continuity between
them. Somewhere down the
road you may get to make
Class A surfaces using them.
The term *Class A* comes from
the automotive industry and
means that the surfaces are
tangent and have the same
acceleration of curvature.

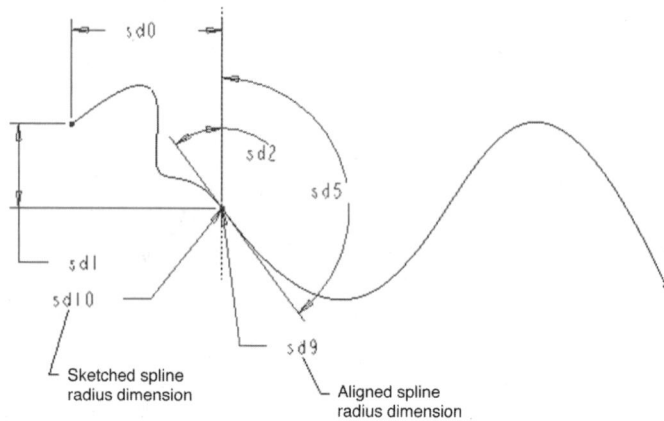

Curvature relation sketch.

Exercise 5-8: Using Control Poly

Using the Control Poly option during the initial sketching of a spline gives you another
method of modifying and controlling the shape of spline curves. Once you have finished
sketching a spline using Control Poly, a series of bounding line segments are displayed.
These line segments and their end points are what you will use to dimension and control
the shape of the spline. You will be able to select the spline's end and internal points, but
will be unable to add tangency conditions under the MOD SPLINE menu or add radius
of curvature dimensions.

The dimensioning does, however, become more straightforward, in that you are now
working with straight line segments. With the use of control polygons, you can adjust the
overall shape by selecting and moving the points along the line segments. The spline will
remain tangent to the line segments at all times, and in most cases the curve shaping is
easier to control. In this exercise you will be using control polygons to dimension and
shape a spline curve.

1. Draw a four-point spline using the **Control Poly** option from the TANGENCY menu. Once you have finished the spline, click the middle mouse button. The regenerated sketch should look similar to that shown in the illustration at right.

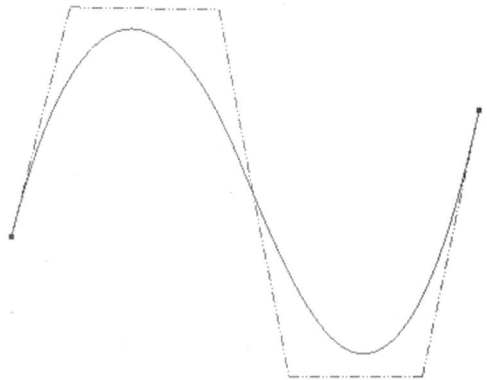

Sketched spline using Control Poly.

2. Modify the spline's shape by selecting **Modify**, picking the spline, and moving points with **Move Pnts** (as before) or by selecting the **Control Poly** option.

➡ **NOTE:** *Prior to dimensioning, you should get the rough shape of the spline close to the desired shape and use dimensions to make fine adjustments.*

3. Sketch a vertical centerline through each end point of the spline and dimension the control polylines similar to those shown in the following illustration.

Dimensioned control polylines.

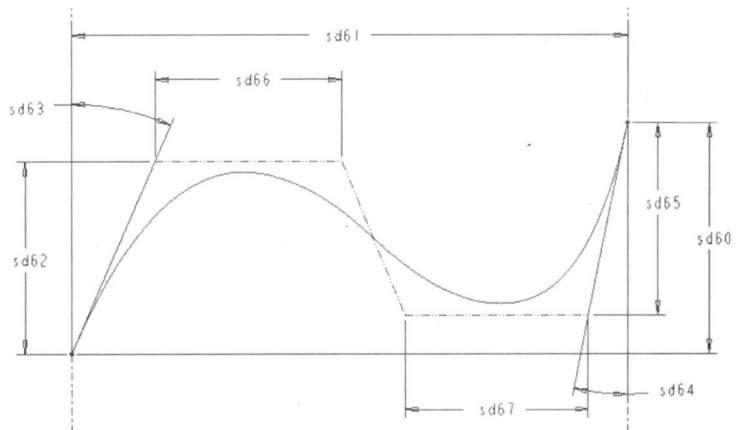

As you can see, you still have the ability to add an angle dimension to the ends of the spline. Therefore, you can still control tangency if necessary. One bonus here is that the Control Poly angular dimensions are created using half the picks necessary in the previous exercises.

In this exercise, you have completely constrained the spline curve, which means that no internal spline points may be selected or dragged. You could have left the interior legs of the control polylines undimensioned so that you could have continued free-form curve shaping. In addition, in this exercise, if you had needed to drive this spline with relations, you could have added these as well.

Conics

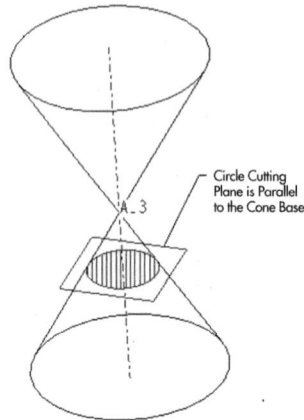

Circle.

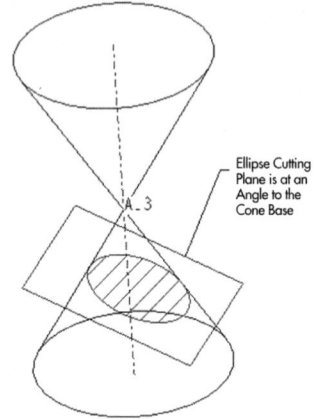

Ellipse.

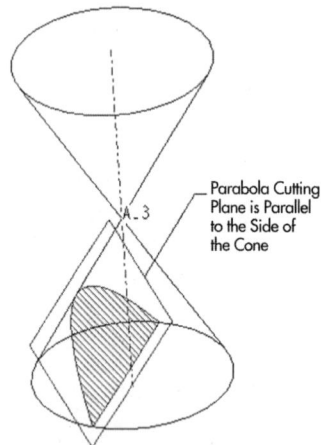

Parabola.

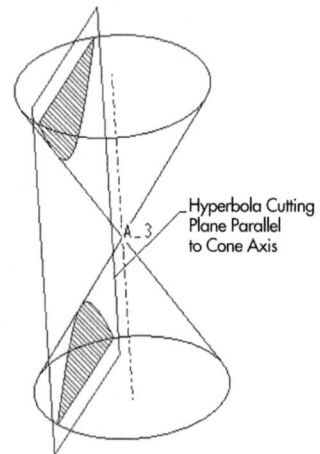

Hyperbola.

Conic entities consist of circles, ellipses, parabolas, and hyperbolas. The series of illustrations above shows how each conic section is derived.

The last three types—ellipse, parabola, and hyperbola—are the sections you are working with when you create a conic entity in Sketcher mode. Conic entities are very useful when blending curve and surface sections because they are very similar to spline curves. However, the following are two major differences between spline and conic entities.

- Conic entities must have both ends controlled with a tangent angle dimension.

- Conic entities have no more than three points (start, end, and shoulder) controlling their shape.

To sketch a conic, you pick a start point, end point, and shoulder point on the screen. Similar to sketching a three-point arc, the last point sets up the curvature. After you sketch a conic, you have to dimension it the way you do any other section, with a couple of exceptions. First, you must add tangency angle dimensions to the end points, and second you must either add a rho dimension or a shoulder point and dimensions to control the conic's curvature. In the following exercise, you will sketch and dimension an open-ended conic, substituting the rho dimension with a shoulder point.

➜ **NOTE:** *For an explanation of the rho dimension, see the Pro/ENGINEER Part Modeling User's Guide, Release 20/2000i.0, pages 2 through 33.*

Exercise 5-9: Creating an Open-ended Conic with "Rho" and the Shoulder Point

1. To sketch a conic curve entity, select **Sketch ➜ Adv Geometry ➜ Conic** and select start, end, and shoulder points on the screen, similar to those shown in the illustration at right.

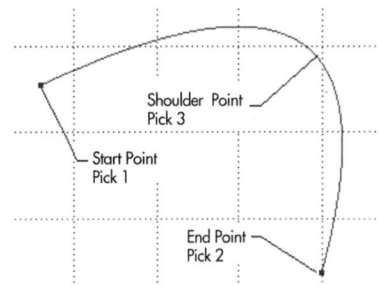

Conic sketch points.

2. Draw two centerlines—the first vertical through the start point and the other horizontal through the end point. Dimension the conic. First locate the end points with vertical and horizontal dimensions, then add the tangency angle dimensions to the end points. To dimension tangency, pick the conic, pick the centerline, pick the end point, and place the dimension. The following illustration shows the pick sequence.

To complete the conic dimensioning, you need to add a rho dimension.

3. Select **Dimension**, pick the conic, and use the middle mouse button to place the dimension. The completed dimensional scheme should look like that shown in the second illustration at right. By default, the rho value is 0.5.

As the rho value of a conic varies from 0.5 it becomes more elliptical or hyperbolic. The following are the rho values for three of the conic types.

- Ellipse: 0.05 to 0.5

- Parabola: 0.5

- Hyperbola: 0.5 to 0.95

4. Replace the rho dimension with a shoulder point. To have the shoulder point drive the conic's shape, you first have to delete the rho dimension and then sketch a shoulder point. The shoulder point could be an existing datum point, but in this case you will be sketching it on the conic and then dimensioning it.

5. Delete the rho dimension and sketch a point close to the conic. Dimension the point to your centerlines. Your regenerated sketch should resemble that shown in the third illustration at right.

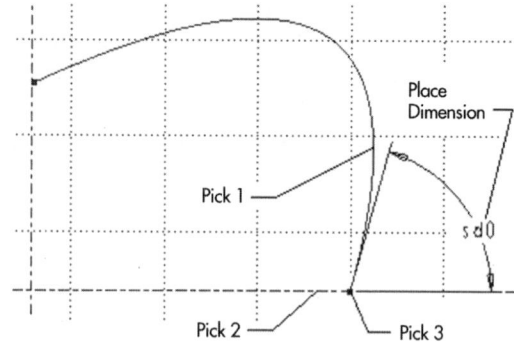

Conic tangency dimensioning.

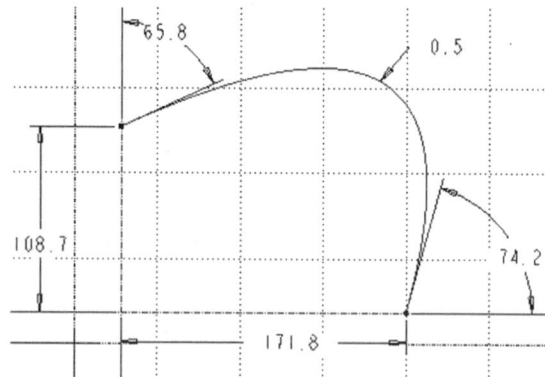

Conic dimensional scheme.

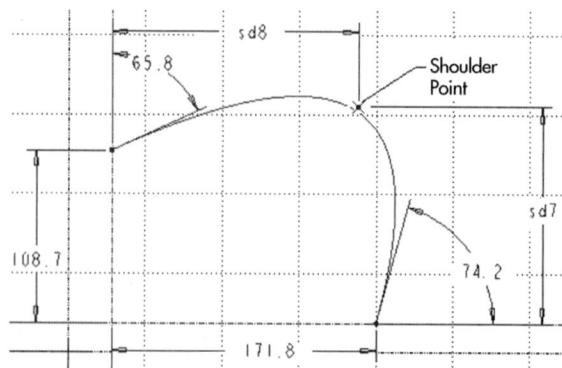

Dimensioned shoulder point.

Because the rho dimension has been deleted, the conic will snap to the dimensioned shoulder point after regeneration, and those dimensions will drive the conic's curvature. If you were to again place a rho dimension, it would take precedence over the existing shoulder point dimension and drive the conic's shape.

Summary

Through the past several exercises and introductory text, you have gone through the creation of numerous Adv Geometry curves and control techniques. It is important to note that in most blending situations you will not need to build curves with as many undulations as in these examples. As a rule, you are better off building splines with fewer internal points because they are easier to control. Determining how many internal points are enough is often a matter of experience. Unfortunately, there is no absolute rule for number of internal points.

You have also now seen how to blend curves and assign tangency control and radius of curvature constraints. These techniques will come up later in the book, and you will have the opportunity to use them in more depth. The key thing is to become familiar with how spline and conic curves are shaped and how to control them. Through the use of tangency, curvature, and dimensional schemes, you can add design intent to splines and really snap them into shape.

Part III

Flexible Part Design

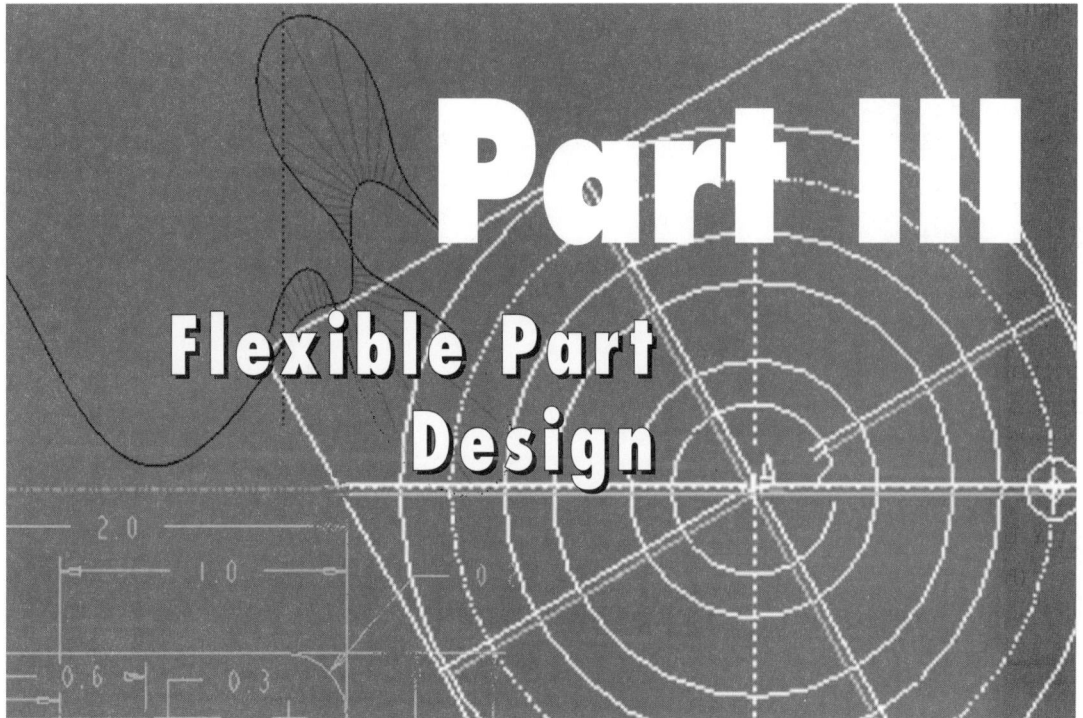

At one point or another we have all had to work on somebody else's model—or even one of our own models returning to haunt us—that just did not seem to make sense. Poorly thought-out models raise questions and comments such as "What were they thinking?," "What did he do that for?," "Where is she going with this?," and everybody's favorite, "I wouldn't have done it that way!" These are the questions that need to run through your mind while you are creating a model. You need to create a model with a specific plan in mind, one that is obvious to anyone that will be making changes or using the model in downstream applications.

In Part III you will explore several rules and techniques that will not only help the next guy but help you build more flexible models in the first place. The example chapter contains numerous how-to examples of sweeps and blends, copying features with patterns, and draft and round features. There is a section devoted to Pro/PROGRAM and how it can be used to create several different models from a single model. A step-by-step method of building a Pro/ENGINEER model from imported geometry is also included, as well as some tips for exporting models out for rapid prototyping.

Rules and Techniques for Flexible Part Design

In this chapter you will learn about some of the general rules to follow to create flexible part designs, such as creating simple features instead of complex ones wherever possible, naming features, creating features in the proper order, and understanding parent/child relationships. You will also walk through examples of the techniques employed when using the Redefine, Reroute, and Reorder commands. The chapter also discusses avoiding the undesirable cut-and-paste method of modeling.

Simple Versus Complex Features

Some of the same principles used in Sketcher mode are also used in flexible part design. First, and probably most important, is to follow design intent. The second is to keep the features as simple as possible.

Obviously, most of the models needed to create today's complex designs will require complex features. You need to understand how complex features work and when to use them. If you are struggling with the creation of a more complex feature, take a step back and look at it from another angle. See if it can be broken down into several simpler features. A series of simple features is often better than a single complex one. Simple features tend to be more robust and are more easily modified. It is also easier for the next person that works on the design to follow your thought progression as you built the part.

The part in the first illustration at right embodies some common things to look for when trying to simplify the creation of a feature. If a feature has a common material thickness, as evident in the illustration, create the feature with a thin section instead of solid. Thin sections are easier to create, and the thickness is controlled by a single thickness dimension, rather than by the method of trying to create a section with the thickness sketched into it. The solid section often requires that several dimensions of thickness be placed in the section. By having several dimensions controlling the thickness of the material, it is easy to miss one when modifying the thickness.

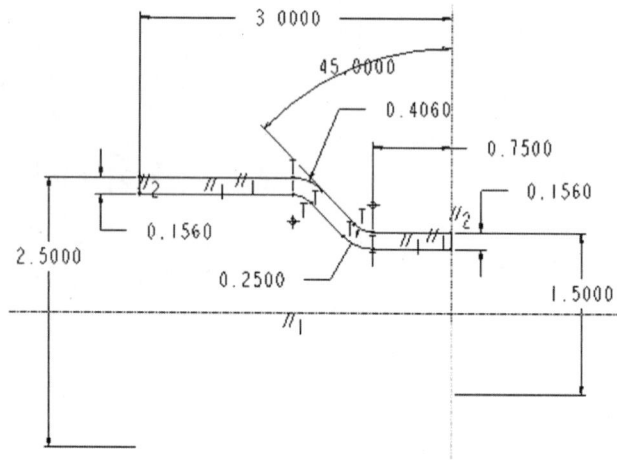

Formed tube feature section using the Solid option.

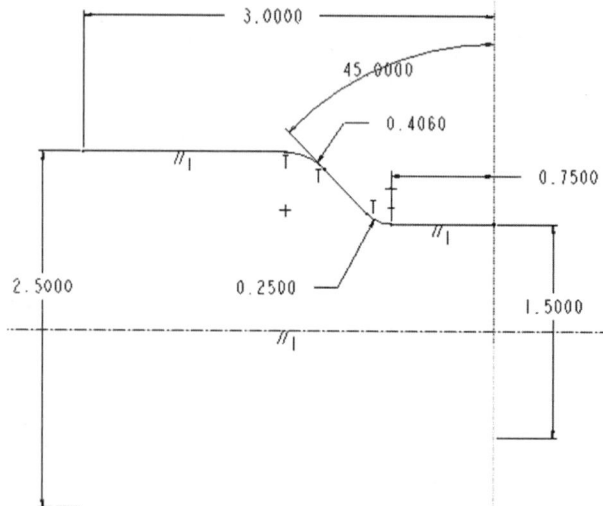

Same formed tube design intent but using the Thin option.

There are two main things to notice about the section shown in the first illustration on the previous page. The first is that the section has the undesirable condition of two dimensions controlling the material thickness. The second is that the design intent calls for all of the dimensions to define the outside of the part. This is a perfect situation for using the Thin option to create the feature. The second illustration on the previous page shows the same formed tube using the Thin option.

Always keep in mind that you must first satisfy design intent. In the illustration at right, a thin section cannot be used due to the design intent. Design intent requires that both the inside diameter of the inlet and the outside diameter of the outlet be controlled. In addition, the radii are called out as a common size fillet rather than a .25 fillet and a .406 round.

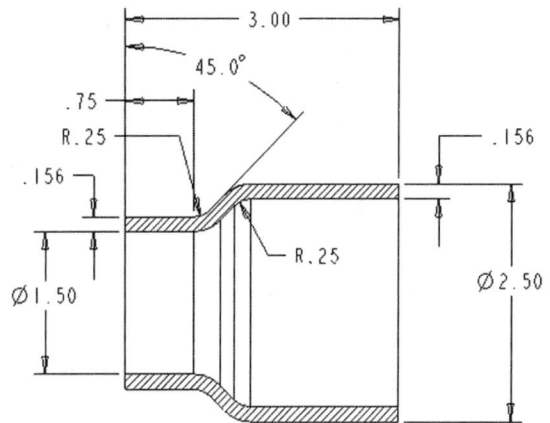

Section A-A

New design intent of the formed tube.

◆ **NOTE:** *Sketching solid feature sections of thin wall parts has gotten much easier with the use of Intent Manager. By adding Equal Length constraints to the section, thickness can often be controlled with a single thickness dimension, as indicated in the following illustration.*

Section of new tube design intent using Intent Manager. Material thickness can be controlled with a single dimension.

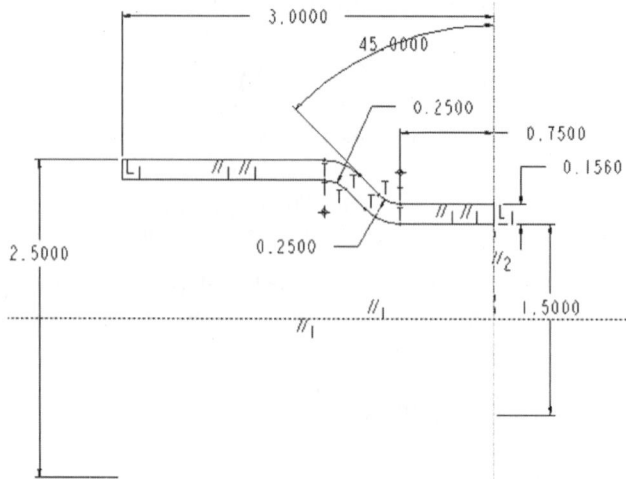

Remember that the use of simple features does not mean that the design or the model will be simple. Employing simple features is a means of achieving a model that meets the design intent.

Avoiding the Cut-and-Paste Method of Modeling

Now that you are ready to create a lot of simple features instead of one large, complex one, here is the second rule: Avoid cut-and-paste modeling. Cutting and pasting has nothing to do with taking three features to create a shape that some users would create in one. There is nothing wrong with doing it that way as long as the features and model meet the requirements of design intent. The cut-and-paste method of creating part geometry refers to cutting away part of a model only to fill the cut back in with a feature later on. An extreme example is a hole cut into a part that is subsequently filled in with a protrusion feature. (See the following illustration.)

Sequence showing cut-and-paste method of modeling and how it can lead to confusion and a model that does not follow design intent.

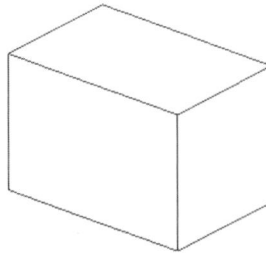

Block created as first feature

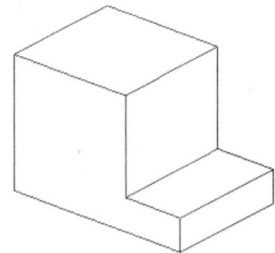

Through all cut is added

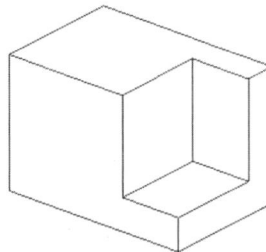

There is no diemnsion to control the depth of the cut, only the thickness of the back wall.

Rear wall protrusion is pasted on over cut

There is little chance that the next person using the model will know these types of features exist in the model until the features fail. This style of modeling does nothing to improve the model or help meet design intent; it only drives up the feature count needlessly. It can also lead to problems for downstream users such as for use in making STL files (files used to create parts from a rapid prototyping machine such as an SLA) for rapid prototyping, or for use in programming numerically controlled (NC) tool paths from the model for machining.

→ **NOTE:** *One way of avoiding cut-and-paste modeling is to use Redefine and Reroute, which are discussed in Chapter 7.*

Follow a Logical Plan

As you conceptualize the design of a component, it tends to start out as a rough shape and then becomes more and more refined. The same holds true for feature and model creation in Pro/ENGINEER. You create the basic shapes and build in the desired design intent first, then move on to add more and more detail, such as draft and rounds, to complete the design. Analyze the design intent of the part. Picture in your mind the order in which you are going to create the features. Look at how each feature will be dimensioned. Which edges do you want to dimension from, or, more important, which edges do you *not* want to dimension to? The first of the following illustrations shows the design intent for a hole location. The second of the following illustrations shows a dimensioning scheme that does not match the design intent of the first illustration.

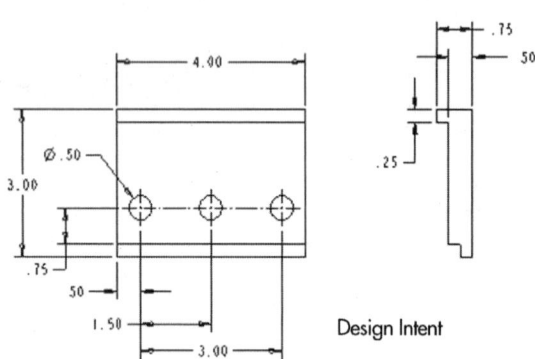

Design Intent

Design intent of hole location in guide block.

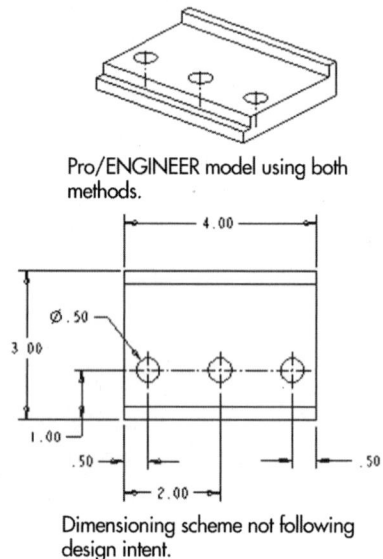

Pro/ENGINEER model using both methods.

Dimensioning scheme not following design intent.

Choice of dimensioning scheme does not match design intent.

If you work in an orderly fashion, you will help the downstream user understand the thought process that went into building the part. Work on one area of the part (or related features) at a time until complete (except for draft and rounds) as possible, then move on to the next area. Skipping around the part is confusing and can often lead to unwanted parent/child relationships that make modifying tricky, if not impossible.

Naming Features

Naming features is an excellent way of providing downstream users of the model important information about how the model was constructed. Each feature created in the model needs to have a purpose for being created. By naming the feature, you can inform others as to what that purpose is. Naming features also clarifies what features are what and speeds up the identification process of what features need to be modified to make the required changes. The following illustration at left shows a catalyst shell half model whose model tree is shown in the illustration at right, before and after the features have been named.

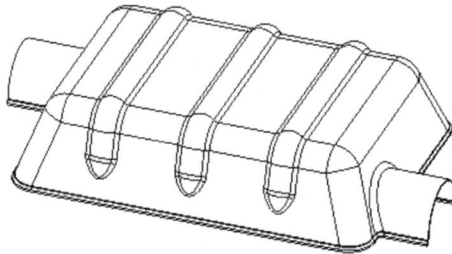

Catalyst shell half model.

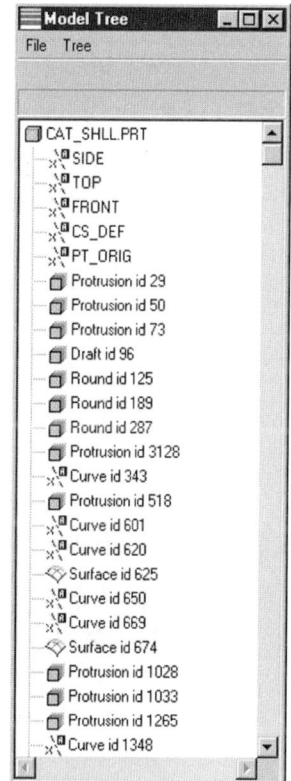

To change the name of a feature, select the Set Up option from the PART menu. There are two options to choose from at this point: Feature and Other. The Other option is to be used for items such as axes, curves, coordinate systems, points, edges, surfaces, and user volumes. However, you can rename these items with the Feature option as well if they were created as features. For example, a datum axis or curve created as a standalone feature can be named by selecting Set Up ➡ Feature, selecting the axis or curve, and entering the new name. If the axis or curve is created during the creation of a different feature, you must use the Other option to name only the axis or curve. Otherwise, the entire feature will be named. The following illustration shows the model tree of the catalyst shell after feature names have been changed.

Model tree of catalyst shell half with default feature names.

In most cases you will have a good idea of how the design intent is likely to be changed. Manufacturing may ask for bosses to be moved farther apart, wall thicknesses may need to be varied, or features modified for better molding. Work through some likely "what if" scenarios on an area of the model when it is done to make sure the model is flexible enough to handle the revisions. The model does not have to regenerate without failures (the best case), but you need to be aware of and understand any failures that do occur. If fixing failures at this time makes sense, go ahead and do so.

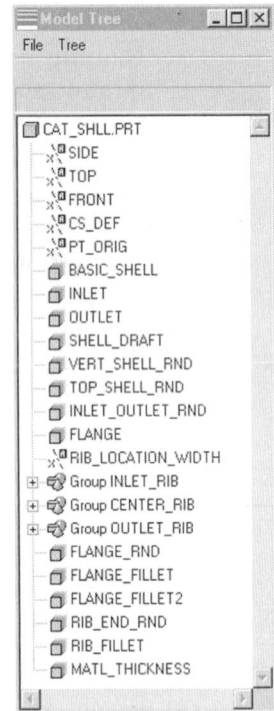

Model tree of catalyst shell half after feature names changed.

Understanding Parent/Child Relationships

Paying close attention to parent/child relationships is always important, but you need to be particularly careful when working on the areas of a design you know to be in a state of flux. It is in these areas that most of the changes will take place, and it is the parent/child relationships that are built into the model, more than anything else, that determine the modifiability of the model. Every item picked as a reference in Pro/ENGINEER creates a parent/child relationship. Therefore, it is very important that you know what item is being selected for that reference. Putting the model in an isometric view to make the selection makes it easier to visually determine what feature is being selected.

Selecting the Query Bin option in the ENVIRONMENT menu also helps. The query bin displays additional information about the feature highlighted, and allows an item to be selected right from the list without having to Next pick through all of the available features. The block shown in the first of the following illustrations demonstrates the difficulty in determining what feature is being selected. The second of the following illustrations shows the actual Query Bin list from the selection of the top edge of the block as a sketcher reference. The list contains a great many features from which to choose.

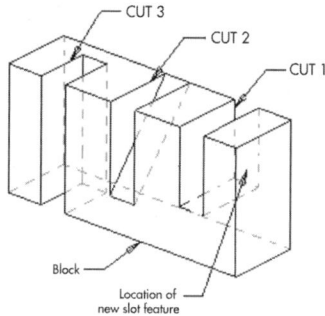

Selecting entities using
an ISO view

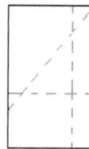

Selecting entities using
standard sketch view

The difficulty in determining which feature is being selected from a sketch view (right) rather than from an isometric view (left).

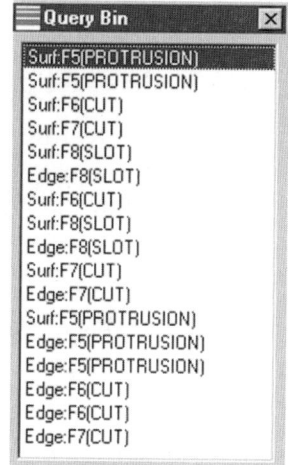

Query Bin list.

Use of Feature Info to Help with Parent/Child Relationships

One of the tools Pro/ENGINEER provides in helping you determine the relationship between features is the Parent/Child option under the INFO menu. This menu pick allows you to analyze not only the parents and children of a feature but the references used by the feature.

Parents and Children

When you select INFO ➡ Parent/Child, you are presented with the following options in the PARENT/CHILD menu. These options are discussed in the sections that follow.

- Parents

- Children

- References

- Child Ref

Parents and Children

The PARENT/CHILD menu can be broken down into two groups. The first is Parents and Children, and the second is References. The Parent and Children options work exactly the same way but in opposite directions. To find the parents of a particular feature, select Info ➡ Parent/Child ➡ Parents and then how the results are to be displayed on the screen. The File option creates an .inf file in the working directory, and displays the file on the screen. The Highlight option highlights all parent features on the model. Now select the feature you want to see the parents of and the results are displayed, as previously described. As stated previously, the Children option works the same way but in the opposite direction. To find the children of a feature, select Info ➡ Parent/Child ➡ Children ➡ File or Highlight and select the feature you wish to see the children of.

References

The other half of the Info Parent/Child menu is centered on finding the references of features. The references of a feature are any features or edges used to create the feature in question. For example, the sketch plane and the entity used to orient the sketch plane are references, as are any edges or planes the section is dimensioned from. Select Info ➡ Parent/Child ➡ References and then select the feature you want to see the references of. The first reference will highlight. This will be the sketch plane for sketched features. The menu options now available are Next, Previous, Info, and Done/Return. The Next option steps to the next reference. If the Next option is selected, the second reference to highlight will be the entity used to orient the sketch plane. Select Next again and the first reference used for dimensioning will highlight, and so on. The Next option will be available until the last of the references is highlighted.

After Next has been selected, the Previous option becomes available. This option simply steps through the references backward. The other menu option available is the Info option. This option displays the feature info for the highlighted reference in an info window. Any of these three options can be chosen at any time to step forward or backward to highlight any reference of a feature and then to get information about that reference feature.

> ✓ **TIP:** *When stepping through the references of a feature, look at the Message window to help determine what is using the highlighted entity as a reference.*

Child Ref

The Child Ref option high-lights a child of the selected feature and then steps through the selected feature's entities that are used as references by the child of the selected fea-ture. The slot feature shown in the first illustration at right makes the concept clearer.

Select Info ➡ Parent/Child ➡ Child Ref and then select the slot. The child feature will highlight. In this case, it is the edge round. The first refer-ence entity of that edge round also highlights. More informa-tion is again placed in the Message Window as to what entity is highlighted, how many entities exist in total, and in what manner the entity is highlighted. The same menu choices as for the Refer-ences option are available

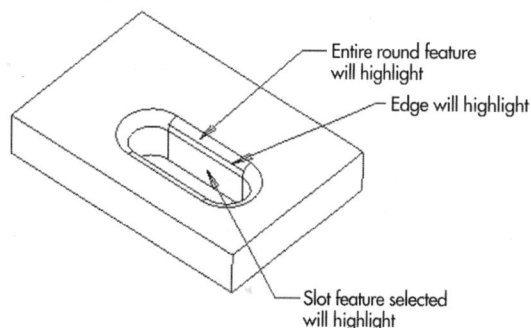

Entire round feature will highlight

Slot feature selected

Slot feature with an edge round as a child.

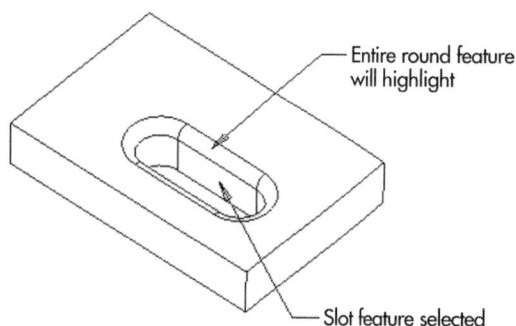

Entire round feature will highlight

Edge will highlight

Slot feature selected will highlight

Reference entities of child edge round.

here: Next, Previous, Info, and Done/Return. These options also work in the same way as for the References option. Continue to select Next and notice how each of the edges needed to create the edge round are highlighted. The second illustration at right shows the edges that will highlight as reference entities of the child edge round.

Datum Planes Versus Model Surfaces

There are no real hard and fast rules about when to use a datum plane or a model surface when choosing a sketching plane. As long as the datum plane or model surface meets the design intent, it is okay to use. However, there are some guidelines that can be followed. Regardless of what has been selected as the sketching plane, select one of the default datum planes as the orientation plane as much as possible. The default datum planes are rarely modified and are therefore very robust. In other words, a feature that has been created using one of the default datum planes as an orientation plane can never fail due to a loss of sketch plane orientation.

Make Datums on the Fly

One of the options available for sketch and orientation planes is the Make Datum option. Make Datum allows you to create a datum feature to use within the feature you are creating. All of the methods used to create a datum plane are available with the Make Datum option. A datum plane made on the fly will not be displayed as the other datum planes, but any dimensions used to create it, such as offset or angular dimensions, will appear with the finished feature.

For the boss shown in the illustration at right, the design intent calls for a dimension of the boss height to originate from the base of the part. Note that the dimension originates from the base of the block to the top of the boss. It is wrong to select the top of the block and create the boss as a blind feature up to the correct height. The better technique is to create the sketching plane on the fly (first of the following illustrations) by offsetting it from the base to the desired height, then create the boss feature down to the block using the Thru Next or Thru Until options. The second of the following illustrations shows the finished boss feature when selected to be modified. Note no datum plane is displayed at the top of the boss. Only the dimension of the offset is displayed.

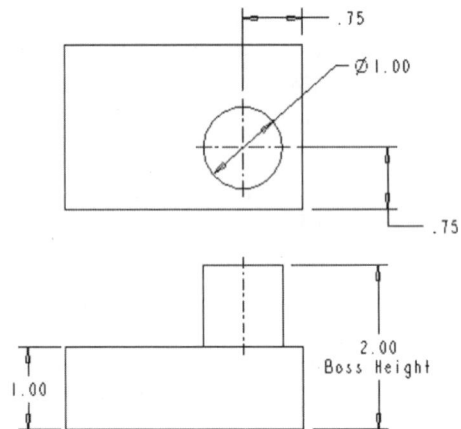

Design intent for a boss.

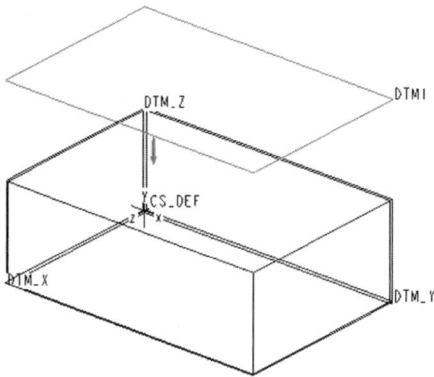

Sketching plane created on the fly and the direction the boss protrusion feature will be created.

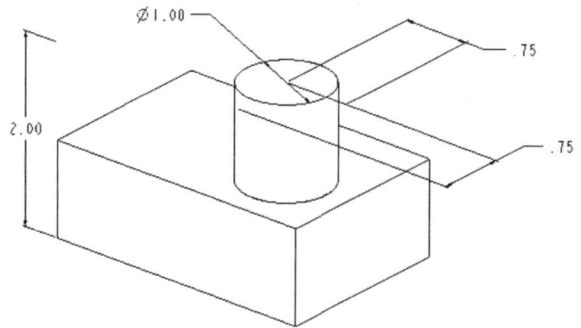

Finished boss feature.

The other situation that calls for a datum to be made on the fly is for sketch plane orientation. In fact, you must create the orientation plane on the fly if you plan to use the feature in a rotational pattern. It is the best way to create the angular dimension required for the pattern, and it is the only way each sketch plane can be set up to recreate the sections of the pattern instances without the sketch plane failing.

Redefine, Reroute, and Reorder to Control Design Intent

Redefine and Reroute are the most powerful tools you have available to you for modifying the parent/child relationships of existing features. Once you are comfortable working with them, they can both help you out of situations when things start to fail, and can help you maintain your design intent as changes occur.

Redefine

The redefine menu will be different depending on the type of feature selected to be redefined. The following items may be available to be selected for redefinition.

- **Attributes:** Redefine feature attributes.

- **Direction:** Change the direction a feature was created in relation to the sketching plane.

- **Section:** Modify the actual geometry of the section.

- **Flip:** On thin features and cuts you can change which side the feature is created on.

- **References:** Change the way a feature is placed (adding or removing edges for rounds creation) or surfaces selected to determine feature depth.

- **Boundaries:** Modify surface boundaries.

- **Scheme:** Modify the dimensioning scheme of a section.

- **Placement:** Redefine how a component has been placed in an assembly.

- **Pattern:** Redefine the type of pattern.

The most common type of redefine is the redefining of a section. There are three options to choose from when redefining a section: Sketch Plane, Sketch, and Scheme.

When you select the Section option, you are placed back into Sketcher mode, where you can make any modification to the section that is required, including adding or deleting geometry, adding or deleting dimensions, and adding or modifying alignments and sketcher constraints. After all modifications are made, the section must then be regenerated successfully before continuing with the redefine of the feature.

When a sketcher entity is being removed that is a parent of another feature, Pro/ENGINEER will issue a warning that other features may fail. You must respond to this warning with a Yes or No before continuing. A No (carriage return) will not delete the entity, and will return you to Sketcher. Answering Yes will delete the entity and suspend child features until finished with the redefine, then return you to Sketcher. If child features fail regeneration, Pro/ENGINEER will bring up the RESOLVE menu.

An alternative to deleting the sketcher entities and dealing with the failed features later on is to replace one sketched entity with another. The Replace menu choice is in the Geom Tools menu. The procedure is very simple; you simply sketch the new entity, select Geom Tools and Replace, and then select the new entity, then the old one. The old entity will be deleted and all references of the old entity are transferred to the new one automatically. This method can save a lot of time resolving failed features later on.

Scheme

The Scheme option differs from the Section option in that it does not enter the full-blown Sketcher mode. Sketched geometry and dimensions cannot be modified using the Scheme option. Dimensions can only be added and deleted. Because the section will not be modified, it will not need to be regenerated. Therefore, this is the method to use when the feature geometry is correct, but the dimensioning scheme needs to be modified to match design intent.

> ↝ **NOTE:** *As dimensions are deleted and new ones replace them, the dimension symbols will change. Anywhere these symbols are used, such as relations or Pro/PROGRAM, will need to be updated so they will work properly.*

Sketch Plane

The Sketch Plane option allows you to pick a new sketch plane and orientation reference. The new sketch plane must be co-planar to the original plane. A datum can be made on the fly for either of these two references, and will automatically be reordered to the proper location in the model. The Sketch Plane option is a specialized reroute. It is not as powerful as the actual Reroute option, but it will allow you to use a new sketch plane quickly and easily.

Reroute

Reroute allows you to modify parent/child relationships of features by enabling you to select new references for sketch plane, section placement, and dimensioning. The best use of reroute is for changing the sketch or placement plane for a given feature. Use one of the Redefine options, Redefine Section, or Scheme to modify dimensioning references. An advantage of Reroute is the option to reroute all of the children of the rerouted feature to the new references.

Redefine, Reroute, and Reorder Example

This example is one you may run into on any given day in most engineering departments. It is basically an Engineering Change Notice. It starts with a completed part that needs to be changed per a new set of design requirements, shown in the following illustration. By using the Redefine, Reroute, and Reorder commands, you will modify the original part to the new design intent.

Views show the original design intent.

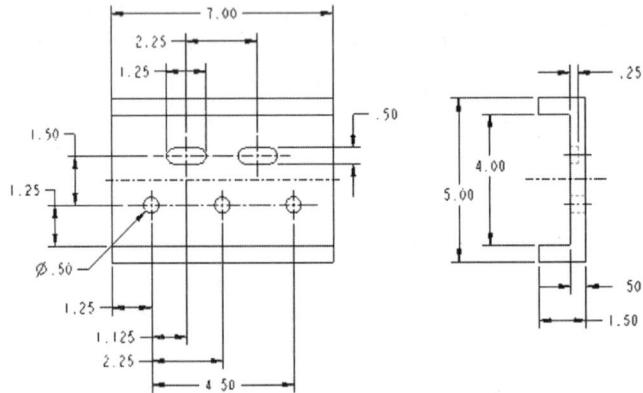

There are several changes necessary to update the model correctly, as outlined in the following list and shown in the illustration that follows.

- The shape of the center track cut changed to include a step. The cut feature section will be redefined to add the step.

- Adding the step will remove the sketching plane for the slots. Reroute the slot sketch plane to the new step surface.

- The slots are now dimensioned from the side of the track instead of from the holes. Redefine the dimensioning scheme of the slots.

- The holes are now dimensioned from the slots. Therefore, the slots must be reordered so that they are created before the holes. Reorder the slots.

- Change the dimensioning scheme of the holes to be located from the slots. Redefine the dimensioning scheme of the holes.

- Revise the depth of the slot to .375.

Changes required for the new design intent.

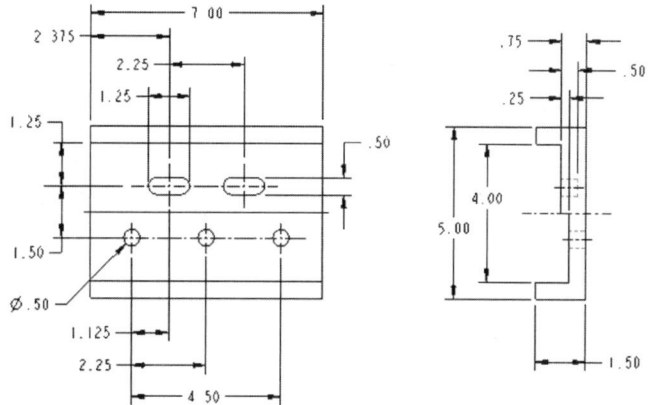

Redefine a Section

The first step is to redefine the section of the cut feature to include the step. Select Feature ➡ Redefine and select the center track cut feature, then select Section ➡ Define. This places you in Sketcher mode. Make the needed changes to the section to add the step. Regenerate if required (regeneration will be automatic if Intent Manager is used). Select Done to exit Sketcher and OK to finish the feature, shown in the following illustration.

Cut after the section was redefined to add the required step.

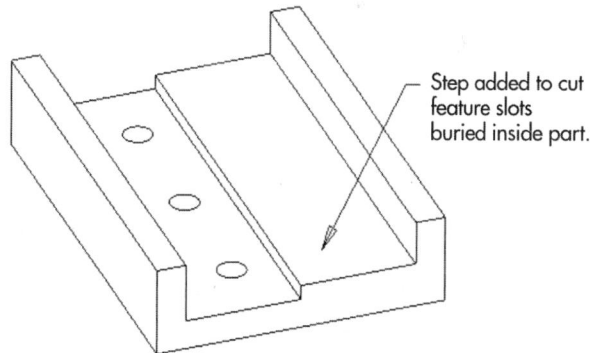

Step added to cut feature slots buried inside part.

Reroute the Sketching Plane

The slots are now buried inside the part because the original sketch plane for the slots still exists. However, now it is visible on only half of the part. To reroute the slots to the new surface, select Feature ➡ Reroute Feature and select the slots. (You will need to use Query Select or select the slots from the

Model Tree.) Pro/ENGINEER will ask if you want to roll back the part. This means that Pro/ENGINEER will temporarily suspend all features created after the selected feature. Answer Yes to this request.

Rolling back the part to the feature being rerouted will prevent you from creating an illegal parent/child relationship with a feature that was created after the feature being rerouted. Pro/ENGINEER now steps through each reference in the order it was created and gives you the option to keep it as is (Same Ref) or modify it (Alternate). Read the message window to get information as to which reference is highlighted and can be changed. The first reference is the sketching plane. The line in the message window is "Select an alternate sketching plane?" The Alternate option will already be highlighted. Therefore, you can now select the new surface created by the step in the track cut.

The second reference that can be changed is the sketch orientation plane. The line in the message window will be "Select an alternate horizontal reference plane for sketcher." Select the Same Ref option. Continue selecting the Same Ref option for the remaining references. The slot feature now looks correct. It starts on the correct surface and

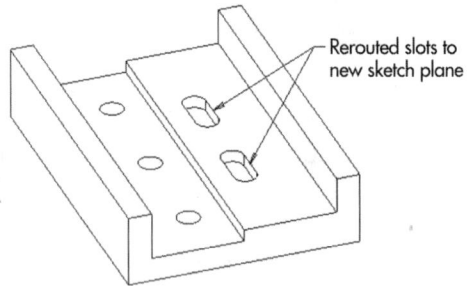

Visible slot features.

is in the correct location. However, the dimensioning scheme is still coming from the hole rather than from the side of the track. The illustration at right shows that after the slots are rerouted to the surface of the step, the slots' features again become visible.

Redefine the Scheme of the Slots

To redefine the dimensioning scheme of the slots, select Feature ➡ Redefine. Select the slot, and then select Section ➡ Define ➡ Scheme. You are now in what looks like Sketcher mode, but some of the menu options are not available. Geometry cannot be created or deleted when redefining the scheme of a feature. Delete the dimensions that locate the slots from the hole and create new dimensions from the edges of the track as required by the new design intent. Regenerate the section if needed and select Done to exit Sketcher. Click on the OK button to complete the feature. The slot is dimensioned per the new design intent, as shown in the following illustration. You

can now move on to the next step, which is to reorder the slots to come
before the holes so that the holes can be dimensioned from the slot feature.

*New dimensioning
scheme required for the
slot feature.*

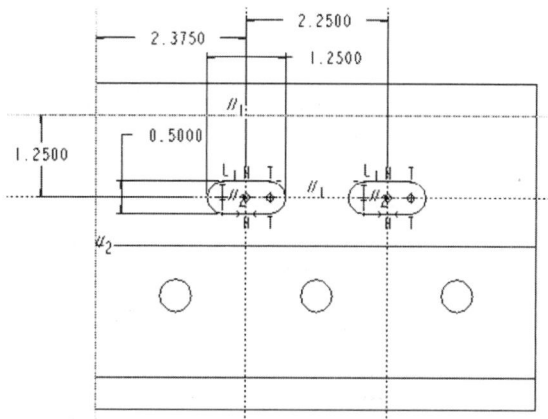

Reorder the Slot

Select Feature ➡ Reorder and select the slots. Because the slots are a child of
all the other features except the holes, the only possible position the slots can
be reordered to is before the holes. Pro/ENGINEER understands this and
will display a message in the message window telling you that the selected feature can only be moved to one location and will ask you to confirm the reorder. Select Confirm and the slots will be reordered to come before the holes.

Redefine the Scheme of the Holes

Now that the slots are in the
right order, you can redefine
the scheme of the holes to
match the new design intent.
As previously, select Feature ➡
Redefine. Select the holes, and
then select Section ➡ Define
➡ Scheme. Delete the dimensions locating the holes to the
part and create new dimensions from the slot centerline.
Regenerate if required and
select Done to exit Sketcher.

*New dimensioning scheme required for the holes to
meet design intent.*

Select OK to complete the feature, shown in the illustration at right.

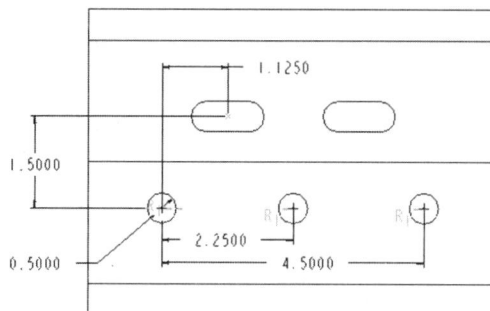

Modify the Slot Depth

The model has been completely corrected for the new design intent, with the exception of the depth of the slot change. Select Modify from the PART menu and pick the slot feature. The dimensions for the feature will be displayed. Select the .25 dimension that represents the depth of the slot and enter .375 as the new depth. Regenerate the model and the change is complete. The model now reflects the new design intent and updating the drawing will require only that the new dimensions be shown and moved into proper position. The finished model is shown in the illustration at right.

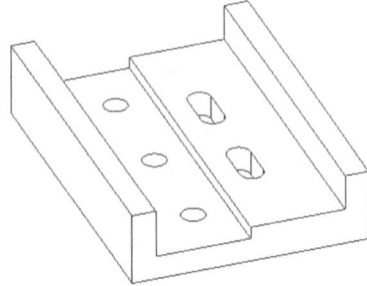

Finished model after all modifications.

Proper Order of Feature Creation

Creating features to satisfy design intent is probably the most important aspect of modeling in Pro/ENGINEER. However, stringing features together in an order that satisfies design intent and keeps the model as flexible as possible is also important. Determining the order of features early in the model is usually pretty straightforward but tends to get more difficult as the modeling progresses.

A Game of Chess

In *Thinking Pro/ENGINEER* (OnWord Press, 1995), David Bigelow compares building the Pro/ENGINEER model to playing a game of chess. The analogy is a good one. At any given time there are several moves the chess player can make. The player must decide what the move is meant to accomplish. Is it to create a block or to attack? This is the "design intent" of the move. The move must first accomplish its design intent. Beyond this goal, the chess player must also look at the move and determine how it will affect the rest of the game. Will the move prevent critical moves later in the match? Will it create an opportunity for the opponent to attack? The same holds true in Pro/ENGINEER. Placing some features before others can at best make creating additional features difficult, if not impossible.

Model in Stages

Creating a model in Pro/ENGINEER can be broken down into four basic stages.

↔ NOTE: *Keep in mind the question "How will this be assembled?"*

- The first stage is to determine the basic or foundational features of the model. These features define the outline shape of the model, as represented by the first illustration at right.

Stage 1: Start with the very basic shape of the model. In this case, half of a catalyst shell.

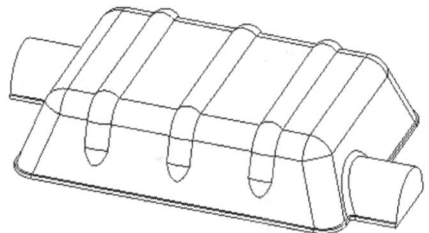

- The second stage refines the shape and adds the secondary features to complete the design of the model. At this point, all of the design elements exist in the model but it is far from being a complete engineering model. Representative of this stage, in the second illustration at right, the flange has been added along with the strengthening ribs. Only the draft and rounds required to be able to create these additional features have been added to the body.

Stage 2: Add detail.

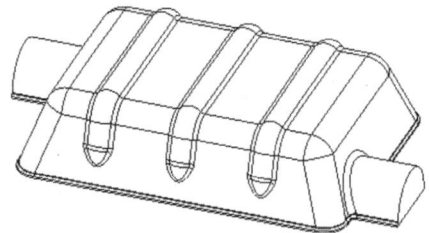

- The third stage is the creation of the detail features that complete the design to satisfy the requirements of manufacture, assembly, and inspection. These are the draft features, rounds and chamfers, and shells, as represented in the third illustration at right.

Stage 3: The last of the draft and round features are added in preparation for the final shell feature.

When this stage is complete, the model should be complete.

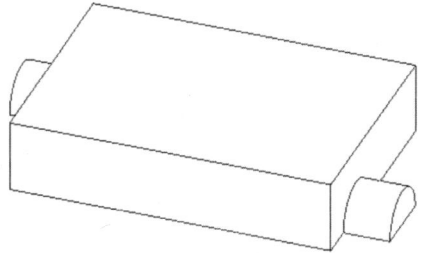

- The fourth stage uses the Redefine, Reroute, and Reorder commands to make sure the model reacts to modifications as expected. Of course, most models make use of these three commands through-out the building process and you would not expect to have to make a lot of changes in this area if no modifications are being made to the design. In the illustration at right, with the shell feature added, the model is complete and is ready to be modified as required.

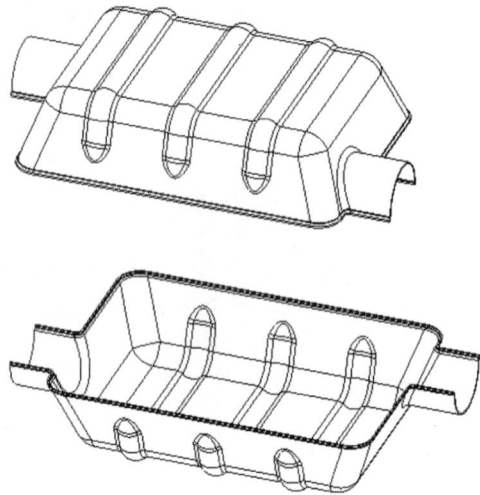

Stage 4: Modifiable model.

Create detail features such as draft and rounds as late in the modeling process as practical. Draft and round features tend to fail when their parents are modified. In addition, any children of these features will also fail. Therefore, adding draft and round features toward the end of the modeling process will keep the number of children of these features to a minimum. This keeps the model as flexible as possible; and should the rounds fail, a little easier to redefine or replace them.

Summary

The focus of this chapter was on ideas and guidelines for building more flexi-ble or modifiable Pro/ENGINEER models. Create simple features and sec-tions as much as possible. This can be very difficult at times with today's models becoming more and more complex, but look for ways to keep things simple where you can. Simple features tend to be more robust than complex ones. Try to make sure that every feature supports the design intent—that features are not added just to cover another feature. This is called the "cut-

and-paste" method of modeling, which makes it difficult for other users to understand and manipulate a model. They simply do not know about all the features that need to be modified to make what should be a straightforward change.

What makes a model difficult to modify is the web of parent/child relationships created as the model was built. The key here is to understand how and when each one is created, and to make the ones you need and avoid the ones that are unnecessary. The last point stressed was the order in which features are created. In summary, the following are the four stages of model building. Following this model building scenario will help keep the model flexible and, if and when features do fail, help keep damage to a minimum.

- Stage 1: Start with the most basic shapes of the design.

- Stage 2: Add features to refine the design.

- Stage 3: Add detail features such as draft and rounds.

- Stage 4: Test the model for likely "what if" scenarios, and use Redefine, Reroute, and Reorder to fix any problems.

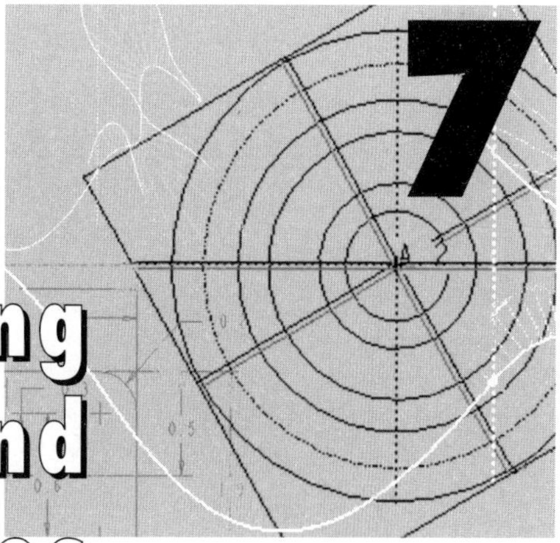

Naming Dimensions and Features

7

Family tables in Pro/ENGINEER can be great time savers when creating a series of similar parts. The family table functionality allows you to add dimensions, features, components, and parameters to a table, and then create instances by varying the table values. The standard example is a family of screws or bolts. The parts have similar features, but the sizes vary. Therefore, instead of building five to fifty independent bolts, you could use a family table and create only one model and drive the others with it.

Typically, when creating a family table, you add features and dimensions by selecting them from the screen. Those items are added to the table in the order you select them, and the names assigned to them are the default names from the system. Dimensions typically show up as their symbolic values (e.g., d26), and features are listed as the feature type and number (e.g., cut F264). These nomenclatures are not very descriptive or intuitive, and without being very familiar with the part it would be difficult at best to figure out which dimension or feature does what from looking at a table. This is why naming features and dimensions will prove highly beneficial. The following illustration shows a basic table file containing a generic instance.

```
▓ Pro/TABLE  TM  Release 20.0  (c) 1988-95 by Parametric Technology Corporation  All Rights Reserved.        _ □ ✕
 File  Edit  View  Format  Help
```

	C1 C2	C3	C4	C5	C6	C7	C8	C9
R1	▓							
R2	! FAMILY TABLE VALUES							
R3	!							
R4	! 1) This family table is for inspection ONLY; any changes made will							
R5	! NOT be saved.							
R6	! 2) Rows beginning with '@' are family table comments.							
R7	! 3) Rows beginning with '!' and empty rows are ignored.							
R8	! 4) Rows beginning with '$' contain locked instances.							
R9	! 5) '*' is used for the default value.							
R10	! 6) Generic names of features if appear are enclosed in [].							
R11	! 7) Feature identifications are their internal ids.							
R12.	!							
R13	! Generic part name: FAMILY_PART							
R14	! Name	d26	d60	d65	d63	F538	F264	
R15	!					[PRTRSN]	[CUT]	
R16	! ========== ========== ========== ========== ========== ========== ========== ======							
R17	! GENERIC	3.4	0.5	0.36	45.0	Y	Y	
R18								
R19								
R20								
R21								
R22								

```
C1R1 :
```

Basic family table.

In the material that follows, you will learn how to rename some of a part's critical dimensions and features and then creating a family table for the part. By naming a part's dimensions and features, you can clarify entries in a family table and simplify the modification of geometry. In many cases, modifying dimensions on a complex part is not as difficult as locating the features to which dimensions are attached.

A Timing Cam

The following part is a simplified timing cam. Create it using default datum planes, following the dimensional scheme shown in the following illustration. For this example, build the .36-inch-wide lobe and its four radii as two separate features because you will want to turn them on and off using the family table.

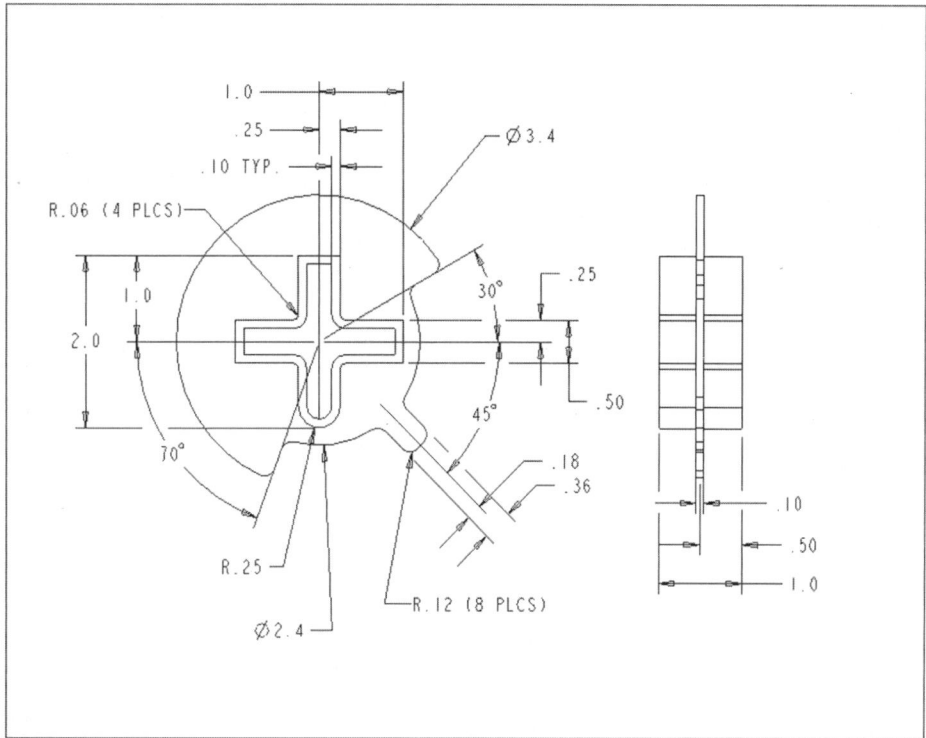

Timing cam base part.

As the basic family table shown in the first illustration of this chapter shows, if you were to start adding dimensions and features to the family table, you would see a series of unclear variables in the table. To avoid this, you want to name some of the part's features and assign new variable names to some of the symbolic dimensions. The features and dimensions are shown in the following illustration.

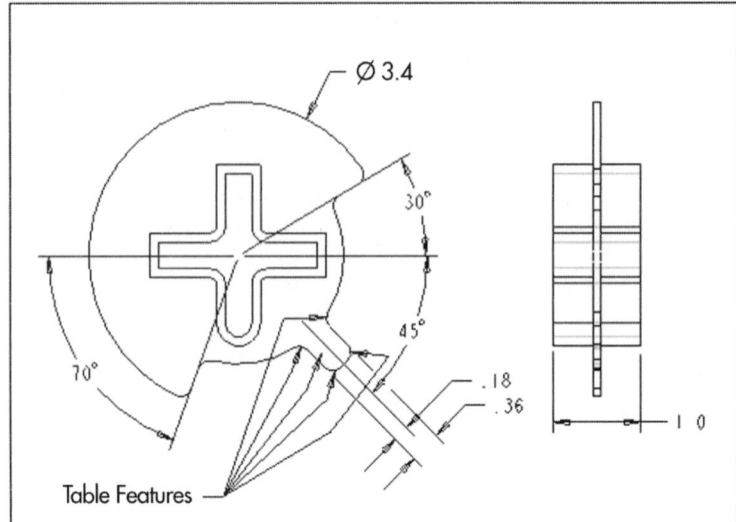

Critical features and dimension of the family table.

The .36-inch lobe is a standalone feature from the rest of the Ø3.4-inch lobe, and the four radii are a single feature as well. To name features, select Set Up from the PART menu and Name from the PART SETUP menu. You are now given the options Component (grayed out), Feature, and Other. Because you are in Part mode, there are no components available to name. However, you do have features. Under the Other option are the options Curve, Point, Edge, Surface, Csys, and User Volume.

For this example, select the protrusion feature that makes the small lobe. By using Query Sel, you can make sure you get the feature you want. After selecting the lobe, enter the name *small_lobe* for it, press the Return key and select the radii on the small lobe, and name this collective feature *lobe_radii.* You should also take the opportunity to name the large lobe and hub feature. For a part containing as few features as this, naming is not as critical, but it is a good practice. Your model tree should now look like that shown in the illustration at right, with the two named features easily identifiable. The names you assigned to those features will also show up if you use the Sel By Menu and Name options from the GET SELECT menu.

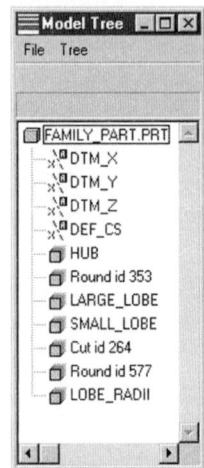

Model tree.

You now want to name the important dimensions. First, display the dimensions by selecting Modify from the PART menu. Then select the feature the dimensions belong to. For this example, you would select the large_lobe, the small_lobe, and the hub from the model tree (or by using Sel By Menu ➡ Name). Next, without repainting the screen, you would select DimCosmetics from the Modify menu and Symbol from the Dim Cosmetics menu. Select the ⌀3.4-inch dimension and enter the name *lobe_diameter*. Change the symbolic dimension names per the following table.

Symbolic Dimension Name List

Dimension	Symbolic Dimension Name
⌀3.4	Lobe_Diameter
308	Lead_Angle
708	End_Angle
458	SmLobe_Angle
.18	SmLobe_Lead
.36	SmLobe_Width
1.0	Hub_Width

Once you have renamed the symbolic dimensions you can display those names by selecting Modify, picking the features the dimensions are attached to, and then selecting Relations from the PART menu. The dimensions will all switch to their symbolic names. Your display should look similar to that shown in the illustration at right, which shows only the renamed dimensions.

The first part of this exercise is the organization and planning session for the family table you will build for the timing cam.

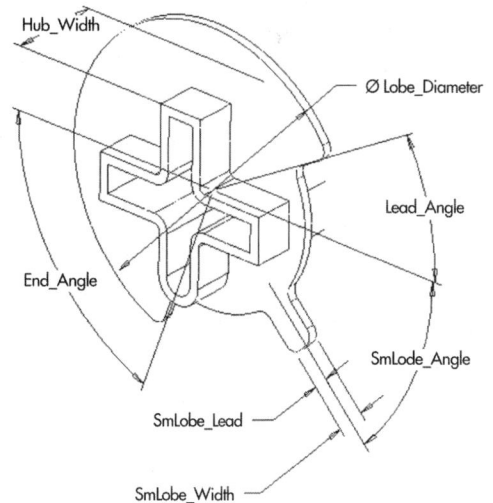

Symbolic dimensions displayed.

The dimensions and features you have already named are the ones you will use in the family table. The dimensional values for the large lobe's lead and end angles will be modified in a few instances, and the small lobe with its radii will be eliminated or shifted in other instances.

Family Table Creation

First, select Modify from the PART menu and display all dimensions to be added to the family table. Next, select Family Tab from the PART menu. Make sure you do not repaint the screen because you will be adding some of these dimensions to the table in a subsequent section. When you select Family Tab, the FAMILY TABLE menu, shown in the following illustration, pops up. The following is a brief description of the options under the FAMILY TABLE menu.

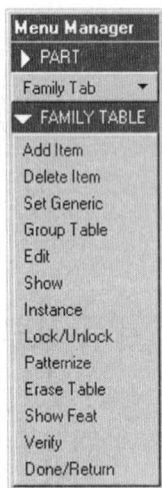

- **Add Item:** Add dimensions, features, or parameters to a family table.
- **Delete Item:** Delete dimensions, features, or parameters from a family table.
- **Set Generic:** Make an instance within the family table the driving configuration.
- **Group Table:** Add a UDF or grouped set of features to the family table.
- **Edit:** Edit the current family table, adding instance names and variable values.
- **Show:** Display the current family table on the screen.
- **Instance:** Retrieve an existing instance from the family table.
- **Lock/Unlock:** Lock or unlock an instance to deny or allow modification to its control variables.
- **Patternize:** Automatically create instances in a family table by selecting dimensions or parameters values to vary incrementally.
- **Erase Table:** Erase all entries in a famliy table.
- **Show Feat:** Identify features you have added to the family table if you have not named them.

Family Table menu.

- **Verify:** Check all instances in the family table to make sure they have valid geometry.

  ➥ **NOTE:** *For a more in-depth explanation of the FAMILY TABLE menu options, see Chapter 13 of* Introduction to Pro/ENGINEER Release 20/2000i.

Select Add Item from the menu and then Dimension from the ITEM TYPE menu. Pick the dimensions you renamed earlier in this example. Be aware that the order in which you select the dimensions and features is the order in which they will appear in the family table, so if need items to appear in a specific order select them that way. Once you have selected all the dimensions for your table, select Feature and then pick the small_lobe and lobe_radii features to add. Next, pick Show or Edit to display the Family Table you have created. It should look similar to that shown in the following illustration. Notice that the items are listed in the order you selected them and that this first instance is by default the generic, which means that its variables will be used to create all table instances.

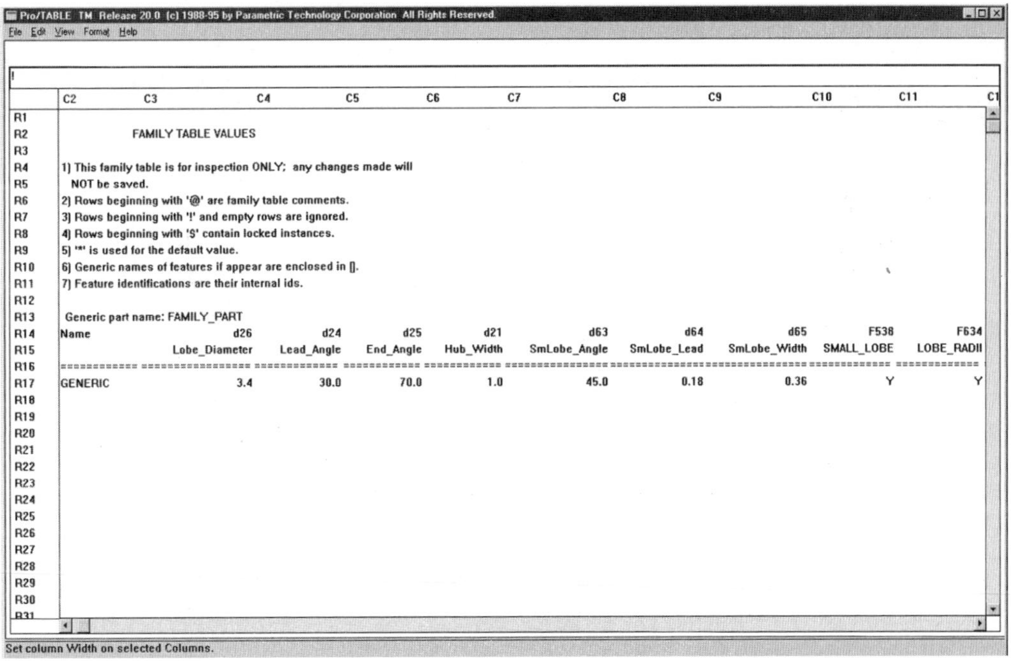

Timing cam family table.

To add instance names and values for the variables within the table, you must use the Edit option from the FAMILY TABLE menu. The table editor works like a typical window-based spreadsheet program. You will need to place the cursor in the cell you want to modify. The current data in that cell will be displayed along the top of the table (it will be blank for new cells). Then type in the data you want in that cell. Use the arrow keys on your keyboard or select with the mouse to move to a new cell.

> ↦ **NOTE:** *When entering instance names and table values, do not line them up directly below the variable names. To enter data, start with the instance name in the cell furthest left and enter data in each consecutive cell. After you have saved and exited the edit table window, Pro/ENGINEER will align the information.*

The following illustration shows an example of the data entry screen. The dimensional values are the same as the generic instance so that you can easily see where you would expect them to appear. Also notice that numeric values are right justified in each cell, whereas alphabetical characters are left justified. If you try to line up the cell information, you will get an error message when you exit the edit screen.

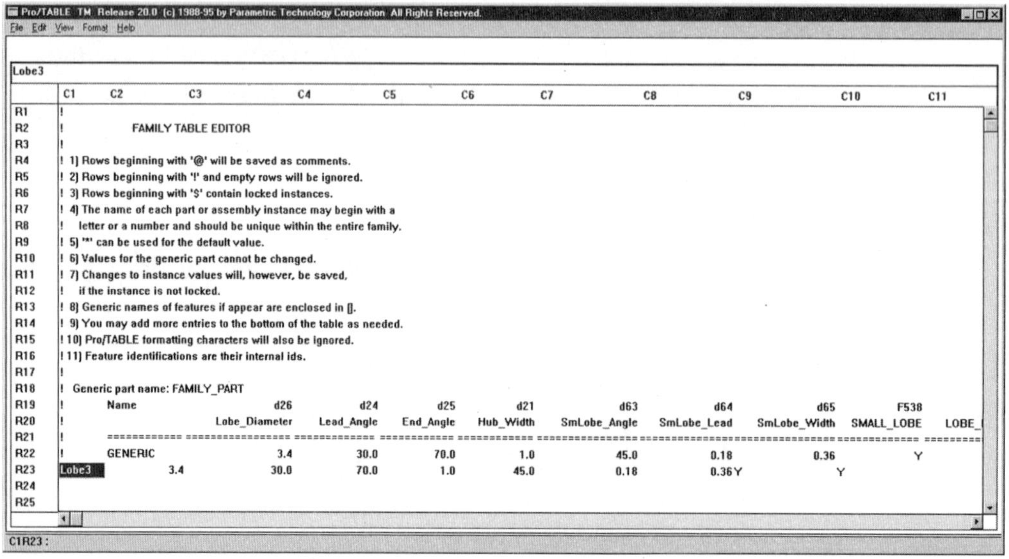

Data entry misalignment.

Timing Cam Continued

When creating instances you must enter dimensional values into to each dimension cell. However, in cells controlling features, you can leave them blank. To turn on or off features ,you need to enter either a Y for displayed or an N for suppressed. One reason you added the small lobes radii to this table is to prevent the system from giving you a message that because the small lobe is suppressed the radii cannot be built. The following table contains the instance names and variable values to enter in your family table.

Name	Lobe_ Diameter	Lead_ Angle	End_Angle	Hub_ Width	SmLobe_ Angle	SmLobe_ Lead	SmLobe_ Width	SMALL_ LOBE	LOBE_ RADII
Generic	3.4	30.0	70.0	1.0	45.0	0.18	0.36	Y	Y
Cam1	3.4	50.0	60.0	1.0	45.0	0.18	0.36	N	N
Cam2	3.4	90.0	90.0	1.5	30.0	0.2	0.5	Y	Y
Cam3	3.6	45.0	10.0	2.0	20.0	0.1	0.3	Y	N

Once you have entered the information from the previous table into your family table, exit the edit screen and select Verify from the FAMILY TABLE menu. Pro/ENGINEER will now go through and regenerate each table instance to make sure you did not enter values that would make the part fail. If any instance had failed regeneration, you would see a message in the message window. You could then bring up the failed instance using the Instance option and redefine the failed section, or reedit the family table to correct the problem.

In this example, if you have entered all of the values from the previous table, the Verify option should have had no errors and you should be able to use the Instance option to bring up each model. The following illustration shows what your family-table-driven parts should look like.

Timing cam instances.

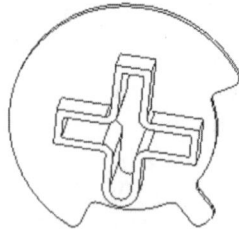

Instance: Generic   Instance: Cam1

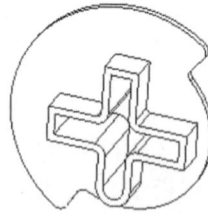

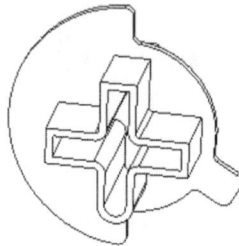

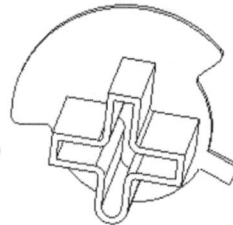

Instance: Cam2   Instance: Cam3

Summary

In this chapter you have learned how to name features, rename symbolic dimensions, and create a family table. Assigning names to features and dimensions is not required in Pro/ENGINEER. However, you can see how handy names can be in locating important geometry, not just from the model tree but inside tables as well. Capturing design intent does not happen accidentally. You need to plan for it in your models.

Using family tables, you have the ability to generate entire families based on a common part by building a generic part and varying data in a table. Each instance of the part is driven by the data in the table, and by the features that make up the generic part. However, you have the ability in Pro/ENGINEER to add features to each instance that are completely independent of the generic or other instances. Later in this book you will be using family tables again, but this time in Assembly mode to control assembly configurations. The model in the example you worked through in this chapter was used in the creation of timing cams for a water softener. The lobes are used to open flappers for varying lengths of time as the cam shaft rotates.

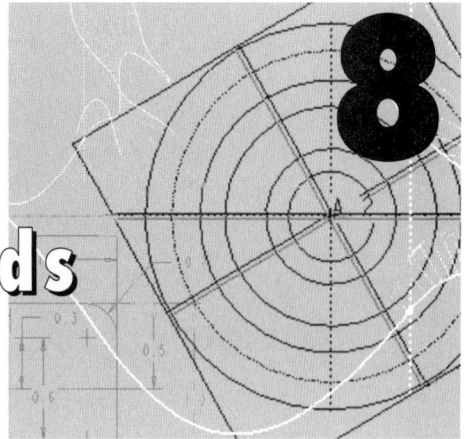

Blends

8

This chapter explores the Parallel, Rotational, and General Pro/ENGI-NEER blend types, as well as the tangency control of nonparallel blends. These blends and their geometry creation methods are a little more involved than the basic extrusion or protrusion, and therefore have a few more rules to be aware of. With all of these geometry creation methods, it is important to keep in mind that the geometry you create will be complex from a dimensioning standpoint. Therefore, standard dimensioning may not be possible for these features. Standard dimensioning refers to the fact that at the beginning and end sections of these features you have a defined dimensioned section, but along the feature the cross sections vary and true radii as well as X and Y locations are not easily defined.

Using Blends

Blends serve the purpose of creating solid or surface geometry between two or more sections of varying size, shape, and location. The flared boss in the illustration at right is an example of such geometry. This same result can be obtained by adding draft to the boss sides. However, if the angle is greater than 30 degrees, you need another way of creating it.

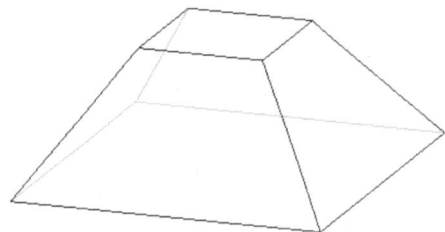

Flared square boss.

The BLEND OPTS menu contains three blend options: Parallel, Rotational, and General. The example in the previous illustration was created using a parallel blend. All sections were sketched and extruded parallel to one another, and each section is drawn in the same sketcher window, toggling between each section. One of the key rules with blending is that each section needs to have the same number of entities unless you are blending vertices or to a point.

Exercises

In the next two exercises you will be creating several parallel blends. The first exercise helps you become familiar with how blends work and what the blend options will do for you. The second exercise takes you through the process of blending between circular and square features, as well as blending vertices and to a point.

Exercise 8-1: Creating Parallel Blends

1. Start a part using default datum planes and select **Feature** ➥ **Create** ➥ **Protrusion** ➥ **Blend** ➥ **Smooth** ➥ **Done**. This will bring you to the BLEND OPTS menu. Select **Parallel** for the blend type.

You will notice two other options in this menu: Regular Sec and Project Sec. These options define section constraints. Regular Sec means that the sections and feature will be sketched and created from the sketching plane. Project Sec sets up a sketching plane for creating sections but allows you to select existing part surfaces for each section from which to blend in creating the feature.

2. Project Sec is very useful for building blends on rounded surfaces. For this exercise, select **Regular Sec**. Then select **Done** to get to the next set of options, Straight or Smooth.

These options control the blending between sections. Straight will blend between each pair of sections as if they were separate entities. In a three-section parallel blend, the center section will have a sharp, nontangent union, as shown in the illustration at right.

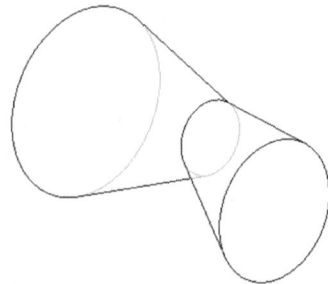

Straight blend; formerly known as a "sharp blend."

The Smooth option allows the system to create a smooth transition between the sections and gives you a nice tangent blend, as shown in the illustration at right.

3. Select **Smooth** ➥ **Done**, and select a sketching plane and a horizontal reference. This will place you in Sketcher mode.

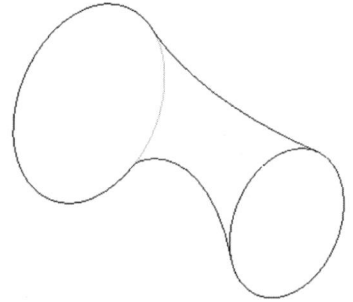

Smooth blend.

The main thing to remember when building a parallel blend is that you will be sketching all feature sections in this sketch window. Each section will need to be dimensioned and regenerated, and you will need to toggle from section to section as each is regenerated. The Toggle command is found under the Sec Tools menu option, which contains another important menu pick: Start Point. Toggle allows you to switch between sections, and Start Point allows you to set each section's start point.

4. Create the following sketch, using the **Toggle** command after you sketch, dimension, and regenerate each section. The current section will turn gray and you will be ready to draw the new section. When you have completed the last section, select **Toggle** twice to return to the first sketched section. The completed blend sections are shown in the following illustration.

Completed blend sections.

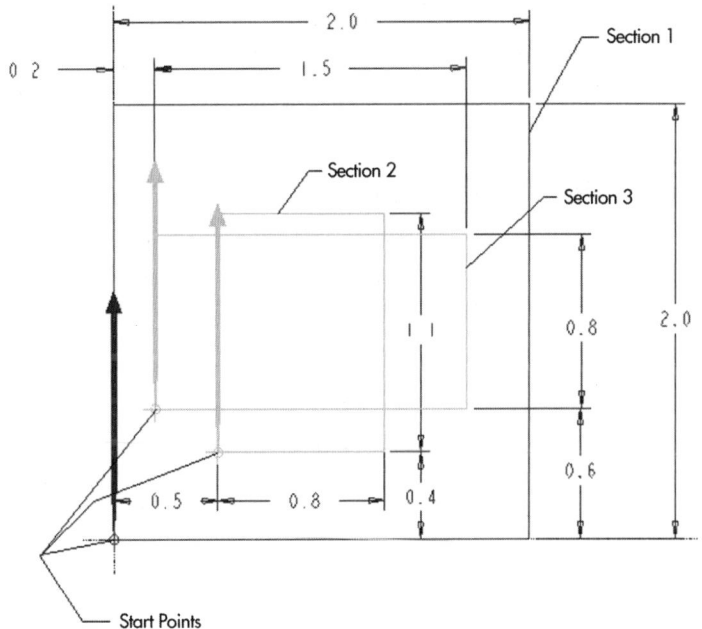

➥ **NOTE:** *Make sure to Toggle after regenerating each section.*

After regenerating each section, you will notice an arrow at the end point of one entity. This is the start point of a section. Each section has a start point. When the geometry is created, each section will blend from start point to start point. Using the Start Point option, you can change where each section start position is. This affects the resulting blend feature.

5. Select **Done**. You will be prompted for the Depth to Section 2 and 3. Enter the values *2* and *3* at these prompts. Define the feature's attributes and select **OK**. The feature should look like that shown in the illustration at right.

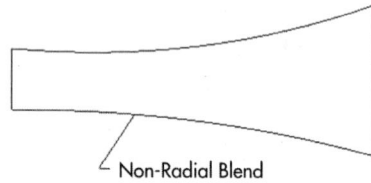

Blending curvature.

Notice that because the feature blends smoothly from section 1 through section 3 you do not see an outline for section 2. Therefore, in Drawing mode, unless you build a cross section at the depth of section 2, your dimensions will float in space. In addition, blending is controlled by the size of the sections and the tangency of the blend. Therefore, the arc of the blend is not a selectable dimension (see the previous illustration).

6. To see what control the start point has over your blended feature, **Redefine** the current feature and clockwise, starting with section 2, move each section's start point to the next vertex. To move a section's start point, select **Sec Tools** ➥ **Start Point**, then pick the vertex for the new start position. Remember that you will need to **Toggle** to each section to move its start point. When you have finished moving start points, the sketch should look like that shown in the following illustration.

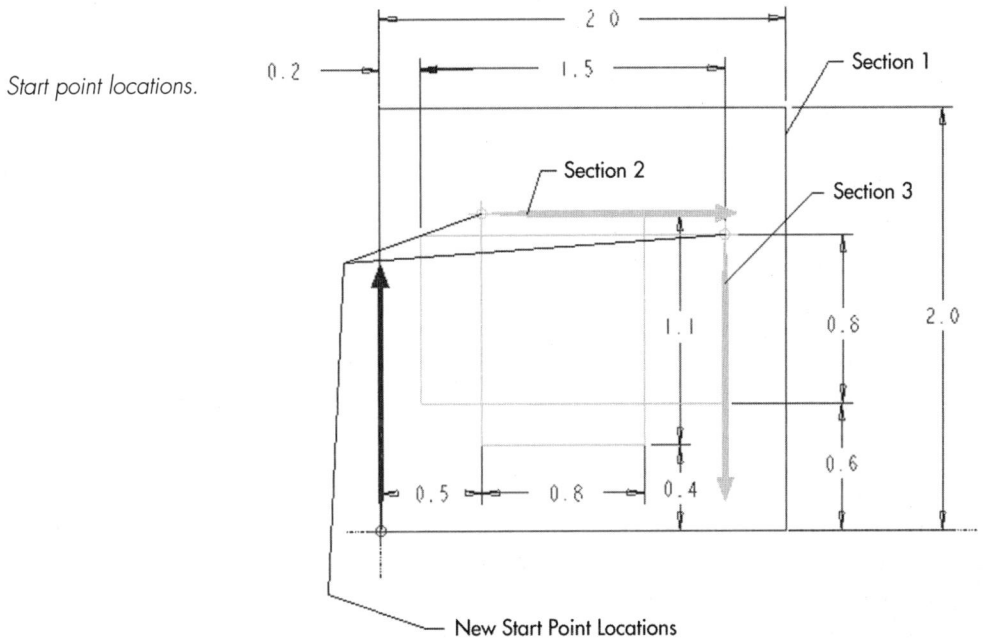

Start point locations.

New Start Point Locations

After moving the start locations of sections 2 and 3, the redefined feature should look like that shown in the illustration at right. By varying the distance between sections or the start positions, you can vary the amount of twisting for this feature.

> ↦ **NOTE:** *This can be an easy way of representing a visual effect, but if specific geometry is needed, make sure you define the sections accurately.*

In most cases when you end up with a feature that looks like that shown in the illustration at right, you will need to redefine the start point locations to get each section to blend correctly.

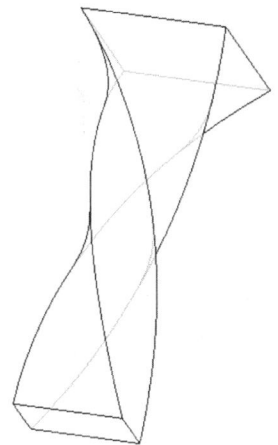

Twisting the feature.

Exercise 8-2: Blending Circles to Squares, and Blending Vertices

So far you have created some pretty straightforward parallel blends. All sections have had the same number of entities. In this exercise you will be blending a circle to a square, which is like putting a square peg in a round hole. However, it is not as difficult as it sounds if you know how to do it (or have a big enough hammer).

1. Set up a blend feature, as you did in exercise 8-1, by selecting **Feature** ➡ **Create** ➡ **Protrusion** ➡ **Blend** ➡ **Smooth** ➡ **Done**. For your first section, sketch a 1 x 1 inch square, regenerate the section, and then **Toggle** to the next section and sketch a 2-inch diameter circle. The sections should look like those shown in the first illustration at right.

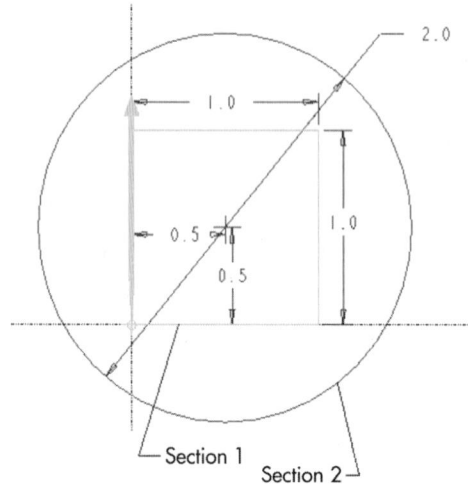

Square-to-circle sections.

Upon regeneration of the circle section you will be told that both sections must have an equal number of entities. This means that the circle in section 2 needs to be broken into four pieces. To break the circle into four entities, you will draw two centerlines through the circle. The centerlines will be used with the Intersect option under Geom Tools to split the circle and control the split locations. Controlling the location of the new arc entities' start and end positions is necessary in controlling the shape of the blend. This is because each line segment from the square will be blending to each arc segment of the circle.

2. Draw two centerlines through the center of the circle, and dimension them as shown in the illustration at right.

3. Use the Intersect option to split the circle where it is intersected by the centerlines. You may need to move the start point of your circle section, because the start point by default will appear where you first intersected the circle. The following illustration shows what the regenerated section might look like. Notice that the start point's arrow is pointing counter-clockwise.

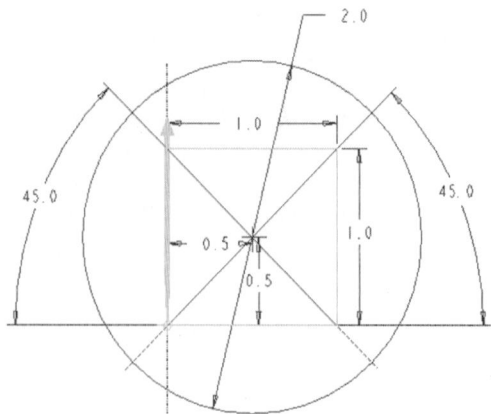

Centerlines for splitting the circle.

In this case, the arrow direction does not affect the resultant feature geometry. If it does, you can change the arrow's direction by selecting Start Point using Query Sel. The query option allows you to select either Accept or Next, which flips the arrow direction. The arrow direction becomes important during the creation of surfaces and most often when creating surfaces using boundaries.

4. Once you have successfully regenerated the sections, select **Done**, enter a depth of 2, and complete the feature. It should look like the feature shown in the second illustration at right.

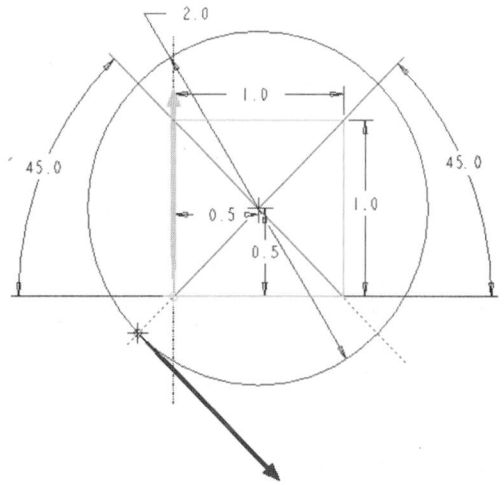

Circle section start point.

Having the ability to define and control the vertex locations of a blend section lets you control how the blend feature will look. In this exercise, you created a blend between a square and a circle. The technique you used to split the circle into four pieces could have been used to split any shaped section into the necessary number of segments needed to create a blend to it. In the third illustration at right, an eight-segment section is blended to a square. This was accomplished by breaking the square's line segments in half.

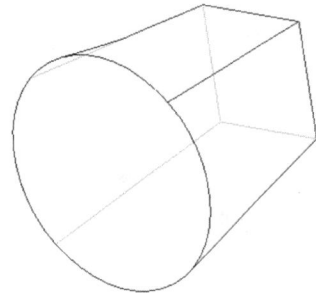

Square-to-circle blend feature.

In the third illustration at right, the four corner radii are blending to straight line segments. This shows you that you can blend between pretty much any shaped entity. However, what you would most likely want in this example is the four radii blending to the corner vertices of the square.

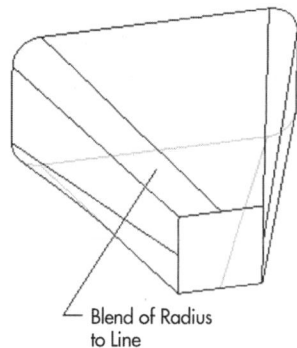

Blend of Radius to Line

Eight entities blended to a square.

5. Create a parallel, smooth blend using the sketch segments shown in the illustration at right. The default datum planes go through the center of the sections.

Again you have more entities in one section than the other. However, this time, instead of splitting all the line segments of the square, you will use the Blend Vertex option under the ADV GEOMETRY menu.

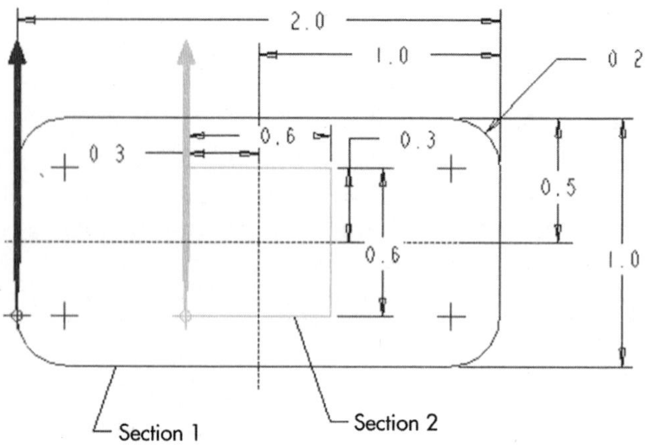

Section 1 contains eight entities, whereas section 2 contains four.

6. Make sure to **Toggle** to section 2, then select **Sketch ➥ Adv Geometry** and pick the **Blend Vertex** option. Select each corner of the square section. The Blend Vertex option will place a circle around each vertex. Each of these blend points adds to the entity count of the square, so you will now have eight.

After you select Regenerate, however, you still have a problem, indicated by the message "Cannot have blend vertices at the start point of a closed loop." This means you have to move the start point of the square section off the corner.

7. To move the start point, use **Intersect** to split one side of the square into two entities and then use the **Start Point** command under **Sec Tools** to move the start point to that position. Now you have nine entities in the square section and eight in section 1. Use **Toggle** to return and activate section 1, then use **Intersect** to split the side corresponding to the start point of section 2, and move its start point to that new entity. The sketch should look similar to that shown in the following illustration.

Blended
vertices and
split bottom
segments.

2.0

1.0

0.2

0.3

0.6

0.3

0.5

0.6

1.0

Blend Vertex Circle

Start Points

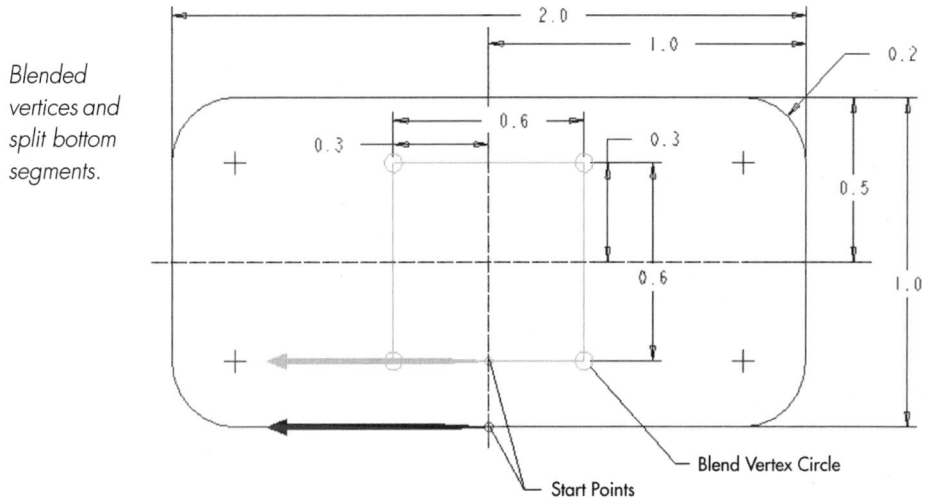

The completed feature, with its blended vertices, should look like that shown in the following illustration.

Completed blended
vertices model.

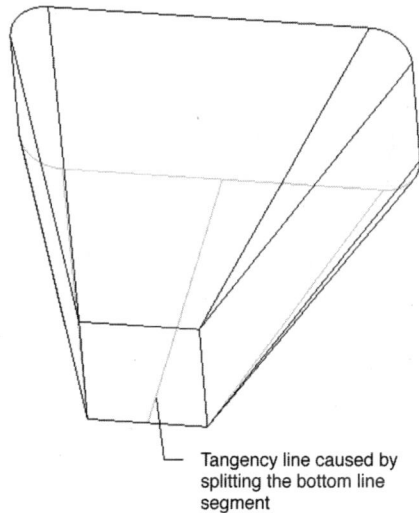

Tangency line caused by
splitting the bottom line
segment

One last thing to consider when creating a parallel blend is "capping" the blend off if necessary. Another way of terming this is "blending to a point." A blend point is drawn like any other section, but the point section can be at either or both ends. You cannot, however, have a point section as section 2 of a three-section blend.

8. Create a parallel, straight blend, with the first section being a 1 x 1 inch square, then **Toggle** to the second section and sketch a point at the center of the square section. The sketch should look like that shown in the first illustration at right.

With the point blend section, you do not specify a start point location because the software will be blending all line segments of the square to the point.

9. After regenerating the sections, complete the feature by entering a depth of 2. The capped blend will look like that shown in the second illustration at right.

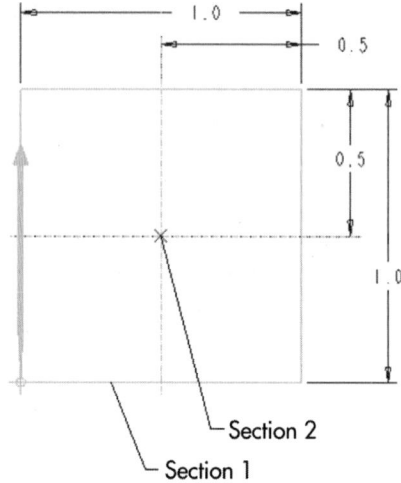

Point blend sections.

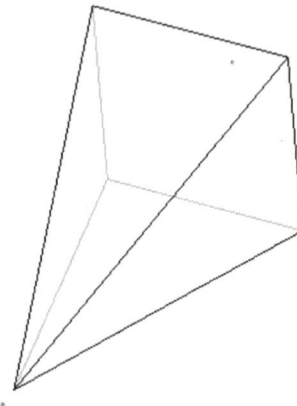

Blending to a point (blend capping).

Rotational and General Blends

The previous exercises give you a pretty good understanding of how blending works. You have seen how the start point location affects which entity blends to which, and you have created several blends with different types of segments (i.e., with circles, lines, and points). The sections that follow, and their exercises, take you through processes of working with rotational and general blends.

When working with nonparallel blends, remember that you will still be sketching multiple sections and blending between them, and that the orientation of each section will now be on a different sketching plane. In addition, each section will be sketched in a new window because the initial sketching plane will not be valid for consecutive sections. Each section you sketch in these nonparallel blends will need to have a coordinate system sketched as well; the coordinate systems are used to orient the sections to one another to create the blend.

Nonparallel blends also open up a few more menu options under the initial setup menus. You now have the option of creating a Closed or Open blend. A Closed blend uses the first section as its last section; for example, as with a donut. The Open option will not join the blend's first and last sections. You also have the options Select Sec and Sketch Sec. The Select Sec option lets you pick existing geometry as your blending sections instead of sketching each one. With Sketch Sec you will draw each blend section. The last of the options, Tangency, is located in the Protrusion dialog box. The Tangency option lets you pick surfaces for either the first or last section to be tangent to. The tangent surfaces must exist prior to the blend creation.

When building a rotational blend, you have to enter an angle of rotation between each section. The new sketch window then pops up to allow you to draw the new section. The angle of rotation is relative to only one plane, but you can still vary each section location in X and Y directions by the dimensions you set relative to the local coordinate system.

> ⇝ **NOTE:** *The angle of rotation you enter is relative to the previous section, not the initial sketching plane.*

The General blend option allows you to vary the orientation angle of all three planes. Both nonparallel blend types are limited to a 120-degree maximum angle for sketching plane orientation.

In the next three exercises you will be creating rotational and general blends to become familiar with how these blend types work. You will also be creating a General blend with tangency to existing geometry.

Exercise 8-3: Using Rotational Blends

The part shown in the illustration at right is a good example of where you would use a rotational blend feature.

In this part, you have a ⌀1.0-inch opening tapering down to ⌀0.70-inch through a 90-degree elbow. This elbow, minus the threaded end, is what you will be building in this exercise. In Chapter 9 (Sweeps), you will be completing this part by adding the helical sweep and capped thread end.

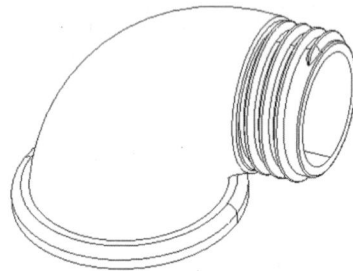

Rotational blend part example.

1. Create a new part consisting of the default datum planes. Then select **Feature** ➡ **Solid** ➡ **Create** ➡ **Protrusion** ➡ **Blend** ➡ **Thin** ➡ **Rotational** ➡ **Smooth** ➡ **Open** and set up your sketching plane and horizontal reference.

In this exercise, you are using the Thin option for your sketch because you are building a part in which the internal diameter is important and the external geometry is controlled by a wall thickness. The Thin option simplifies your sketch dimensioning and captures the design intent of a wall-thickness-driven part. The Open option has also been selected for this case. The Open option is used when you do not want the blend's first and last sections to join.

2. Create a sketch similar to that shown in the illustration at right.

➡ **NOTE:** *For nonparallel blends, remember to sketch a coordinate system for locating each section to each other.*

3. After you have regenerated the section, select **Done**. Because this is a Thin feature you will first be prompted to "Indicate which side of entity to create feature."

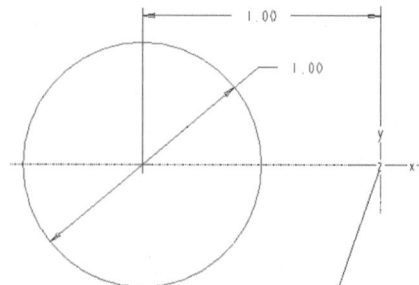

Sketched Coordinate System
Aligned to Default Datum Planes

Rotational blend section 1 sketch.

Your options are Flip, Okay, and Both. Because you want to add material to the outside, select **Flip** to get the arrow to point toward the outside. Then select **Okay** to finish this section. You will then be prompted to "Enter y_axis rotation angle for section 2 [Range 0–120]." Enter *90* for the rotation angle.

4. Once you have entered a value, the system will orient the next section. Sketch the section shown in the following illustration. After you successfully regenerate the section, you will be prompted to indicate the material side. Flip the arrow to the outside of the section.

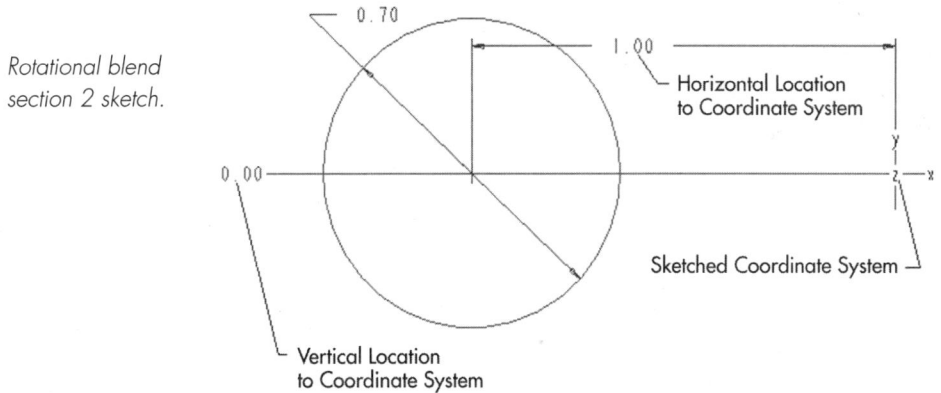

Rotational blend section 2 sketch.

For this section, you will have to dimension both the vertical and horizontal locations to the coordinate system because you no longer have any reference geometry to align to. For nonparallel blends, the system will align all the sketched coordinate systems and then create the feature by blending the sections. Therefore, if you wanted an offset between sections, you could vary the horizontal or vertical dimension.

5. After you have selected the material side and clicked on Okay, you will receive the prompt "Continue to next section [Y/N]." Enter **N**. The prompt "Enter width of thin feature" will appear. Enter *0.12* and complete the feature by selecting **OK** from the dialog box. Your rotational blend should look like that shown in the illustration at right.

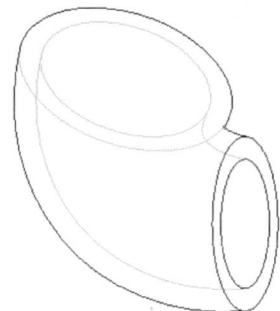

Rotational blend feature.

To complete this exercise, you will add a flange to the large end of the part. There are a number of ways to do this. The easiest is to select the end surface of the part and add the protrusion. However, keep in mind that this part is tapering down to a smaller diameter. Therefore, if you intend to use the large outer edge as part of your section, the resulting feature may not be attached to the thinner area, as indicated in the first illustration at right.

This type of gap can occur in any geometry that tapers to a smaller size, including draft entities. There are a couple of ways you can avoid such a gap, one of which is the following. You will need to bury the protrusion by offsetting the outer edge of the existing geometry inward, as shown in the second illustration at right.

6. To complete this exercise, add a Ø1.45-inch flange that is .125 inch thick and then add a .06-inch radius to its top edge. The model should look like that shown in the third illustration at right.

7. Save this part for later use, in the helical sweep exercise.

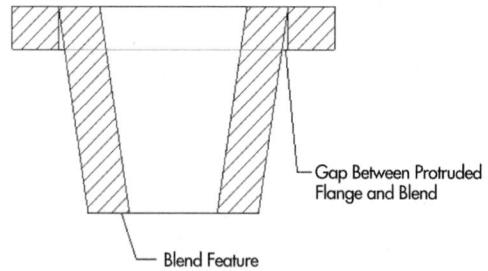

Gap between blend and flange.

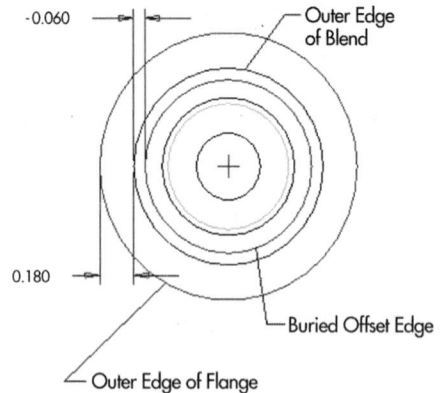

Flange section buried in blend wall.

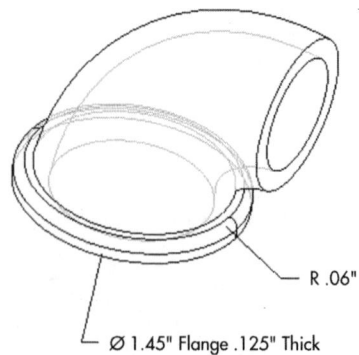

Flange on rotational blend.

Exercise 8-4: Creating a General Blend

As previously stated, the main difference between a general blend and a rotational blend is that a general blend allows you to control the X, Y, and Z rotation, as well as the depth between each sketched section, whereas a rotational blend does not When you first try to create one of these features, the sketch orientation is difficult to visualize. This exercise is aimed at clarifying how each section is oriented. The illustration that follows shows views of the feature you will be creating. Yes, you will be creating a piece of elbow macaroni.

Three views of the general blend.

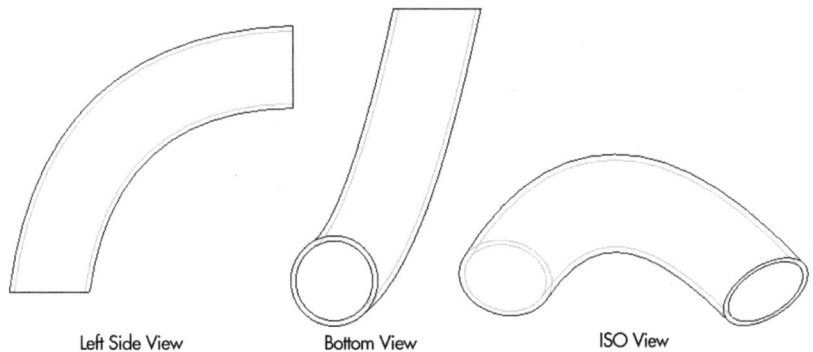

Left Side View   Bottom View   ISO View

1. Start this general blend with your default datum planes and a coordinate system. Select **Feature ➥ Create ➥ Solid ➥ Protrusion ➥ Blend ➥ Thin ➥ General ➥ Done ➥ Smooth**; then set up your initial sketching plane. Keep in mind that you will need to sketch a coordinate system for each section. The coordinate systems are what are used to locate one sketch to another for the blend creation. Create the sketch shown in the illustration at right as section 1, and align your local coordinate system to the default datum planes.

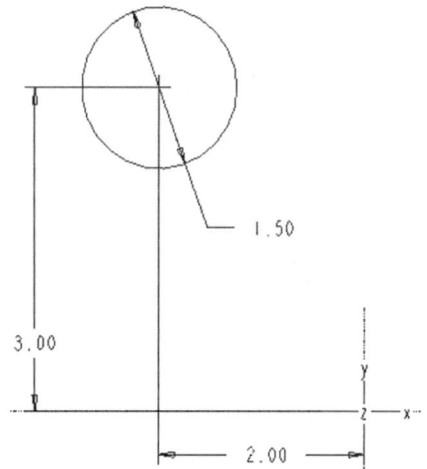

1.50

3.00

2.00

Section 1 sketch.

2. After successful regeneration of this section, select **Done**. Again, because this is a Thin feature you will first be prompted to "Indicate which side of entity to create feature." Your options are Flip, Okay, and Both. Because you want to add material to the outside you will need to select **Flip** to get the arrow to point toward the outside. Then you must select **Okay** to finish this section.

3. The next series of prompts are to set up the X, Y, and Z orientation of section 2. For the X axis rotation angle enter *45*, and for Y and Z enter *0*. This will open a new sketching window for you to work in. Sketch the section shown in the first illustration at right.

Section 2 sketch.

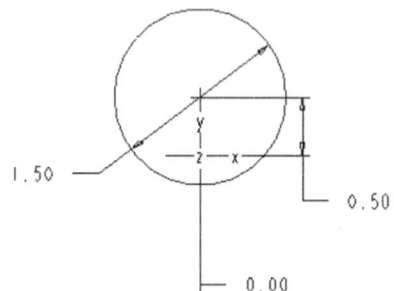

4. After completing section 2 and selecting **Done**, you will again be prompted for a material side. Flip the arrow to the outside and select **Okay**. Next, you will be prompted "Continue to next section? [Y/N]." Enter or select **Yes**. Because all blends consist of at least two sections, you never see this prompt after the first section but it always pops up after the second, and then after each consecutive section until you have finished the feature.

5. When prompted for an X axis rotation, again enter *45*, then enter *0* for both Y and Z. As with section 2, a new window will pop up for you to work in. Create the section shown in the second illustration at right.

Section 3 sketch.

6. Again, after you complete the sketch and select **Done** you will be prompted for a material side. Set that to the outside. When prompted to continue to the next section, enter **No**.

7. Now you will be prompted "Enter width of thin feature." Enter a value of *0.10*.

8. Next, you are prompted "Enter DEPTH for section 2." Enter a value of *2.0*. Then for the depth to section 3, enter a value of *3.0*. Once you select **OK** from the Protrusion dialog box, your completed general blend should look like that shown in the following illustration.

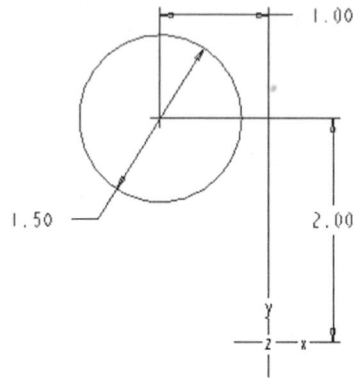

9. To see what effect these rotational values had, select **Modify** from the PART menu. Then select the feature and all sections. Once you select **Done**, you should have an image that looks like that shown in the illustration below, which has been clarified through the use of layered section dimensions.

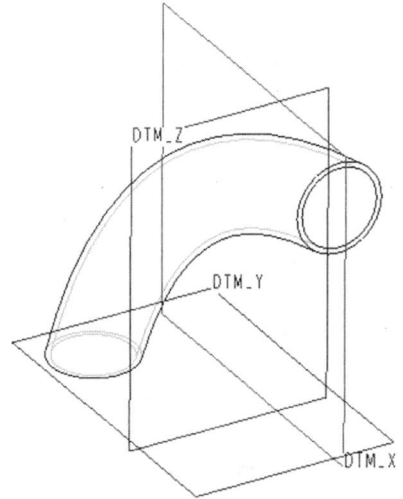

Completed general blend.

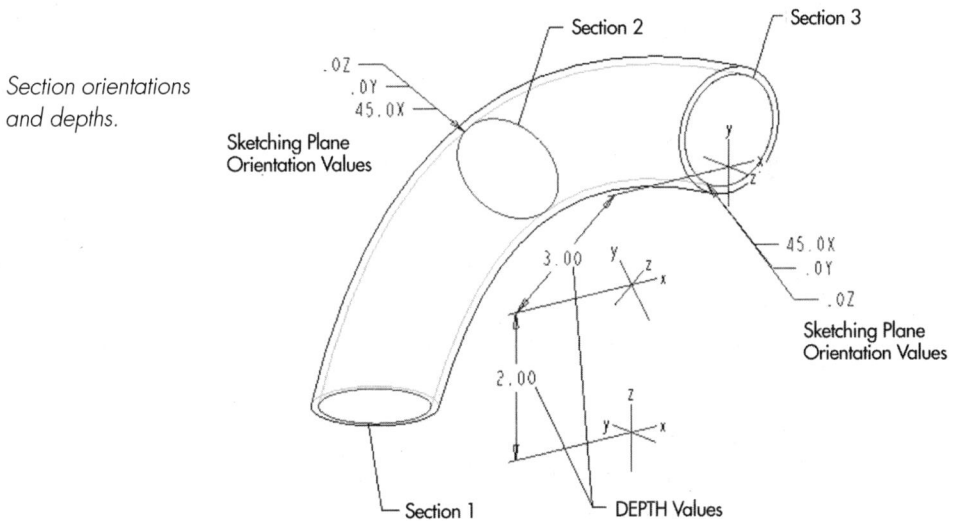

Section orientations and depths.

10. As is evident in the previous illustration, each section is oriented relative to the previous section, and the DEPTH values are also relative to the previous section. To become more familiar with the orientation control between each section, modify the orientation values of X, Y, or Z for the second or third section and look at the results. The rotation you add to each section's sketching plane also changes the orientation of your section's X and Y placement dimensions. This can add greater flexibility to your blend, but results in less control over the blend's shape.

Exercise 8-5: Using the Tangency Option

In this exercise you will be creating another nonparallel blend, with tangency to existing geometry.

1. Create the part shown in the illustration at right.

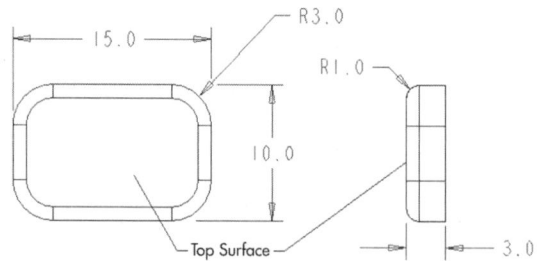

Base part drawing.

2. Now that you have created the base part, you will build a general blend to the top surface and make it tangent to the 1.0-inch radius. The reason you will be building a general blend instead of a parallel blend is that you cannot control tangency on a parallel blend. Create a smooth, general blend using the top surface as your sketching plane.

> *NOTE: Nonparallel blends are not drawn on the same initial sketching plane. Therefore, when blending to existing geometry you should make the first section on the existing geometry's surface. Failure to do this will make section location very difficult.*

3. For section 1, select **Use Edge** from the GEOM TOOLS menu and pick the inner tangent edges of the radius along the top surface. The following illustration shows the completed section.

4. Select **Done** and enter *0* for all rotation orientation prompts. This will bring up a new sketching window. Sketch the section shown in the second illustration at right in this window.

5. Once section 2 has regenerated, select **Done** and enter **No** when prompted for the next section. You will then be prompted for the depth to section 2. Enter *2.0*.

6. Now you are ready to add the Tangency control to your blend. From the Protrusion dialog box, select **Tangency** and **Define**. You will then be prompted "Should the blend be tangent to any surface at the first end?" Select or enter **Yes**. Then select the surfaces that correspond to the highlighted entities to make them tangent. In this exercise, you will be selecting all top radii surfaces. If for some reason you wanted to skip an entity, simply select Done Sel and the next section entity will be highlighted.

7. Once you have selected all tangent surfaces, click on the **OK** button in the Protrusion dialog box. The completed blend should look similar to that shown in the third illustration at right.

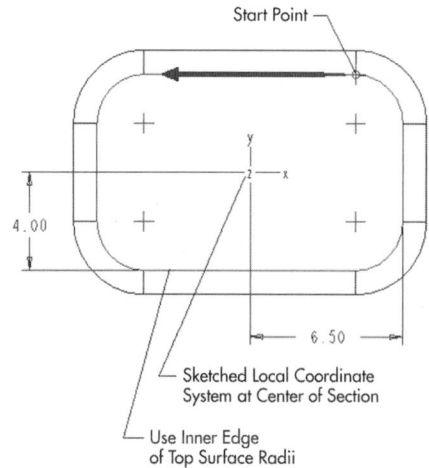

Tangent Edge section 1.

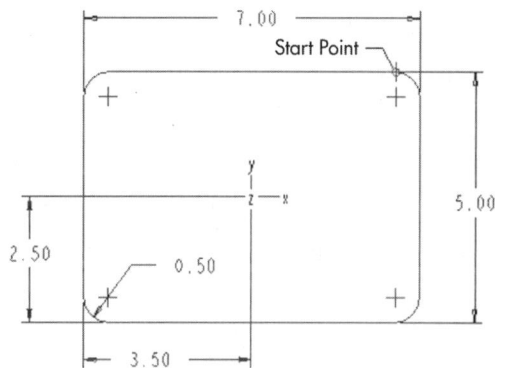

Section 2 sketch.

Completed blend, with tangency.

Summary

In this chapter you have explored parallel, rotational, and general blends, as well as how to assign tangency to nonparallel blends. These types of features give you an added degree of flexibility in your designs, but also an added level of complexity. This makes it even more critical that your design intent is clear and easy to follow.

In the first two exercises, you saw how blended sections create "standard dimensioning" challenges and learned what effects the start point position has on a blend feature. Complexity in your blend features should come as a result of several simple sections blending together, not from the blending of several complex sections. Exercise 8-2 served as an example of this. In that exercise, the resulting geometry is complex but the sections are relatively simple. Modification to these features is easy to accomplish and understand.

As you progressed to rotational and general blends, you added another level of difficulty to your feature geometry. In both nonparallel blend techniques you are able to control the sketching plane between each section. For the rotational blend, this is pretty straightforward. However, as evident with the general blend, you need to be very careful with how you define section orientations. In the rotational and general blend exercises in this chapter, you have only limited control over how your feature blends between each section. This is also apparent with the general blend.

In the last exercise, you controlled the tangency between a blend and existing geometry. This control is only possible with nonparallel types of blends, and in the last exercise you could have used either a rotational or general blend to achieve the same result. The difference between nonparallel blends in the tangency exercise is the number of entries to orient each section's sketching plane.

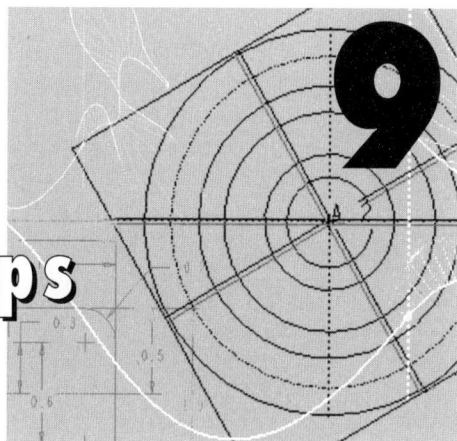

Sweeps

9

In this chapter, you will be creating several sweeps, the first of which will be fairly basic to familiarize you with how sweep features behave. Then you will be creating several more complicated sweeps from the ADVANCED feature menu. The first will be a helical sweep used to create the thread for the adapter elbow you built in exercise 8-3 of the previous chapter. In the helical sweep exercise you will also be capping the thread using a rotational blend. The last two exercises explore using swept blends and multi-trajectory sweeps.

Sweeps and Sweep Exercises

One absolute rule on sweeps is that they all have two components: a trajectory and a section. The trajectory for a sweep can be either 2D or 3D, both of which can be created on the fly or selected from existing geometry. As with the trajectory itself, the section driven along the trajectory can also be

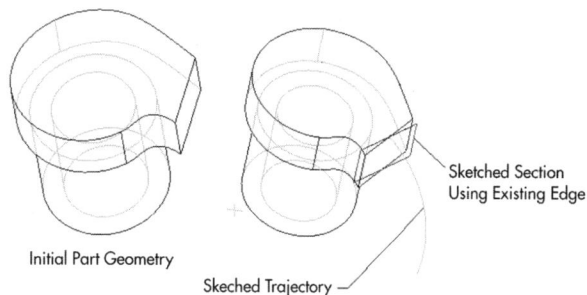

Initial Part Geometry

Sketched Section Using Existing Edge

Sketched Trajectory

Sketched section misalignment.

sketched or selected. Typically when problems occur during sweep creation, it is due to a section's sketching plane not being oriented exactly as expected. The trajectory you define is also used to define the section's sketching plane, and that plane is normal to the trajectory. The previous illustration exaggerates what may happen when a trajectory is not normal to existing geometry.

In the previous illustration you can see that the sweep trajectory is drawn on the top surface plane. The sweep trajectory is one arc segment aligned to the rectangular face. In this case, there is no control over the angle the trajectory meets that face. Because the trajectory does not meet the face normally, the sketched section appears misaligned, and the resulting sweep will have a gap between it and the existing face, as shown in the example in the illustration at right.

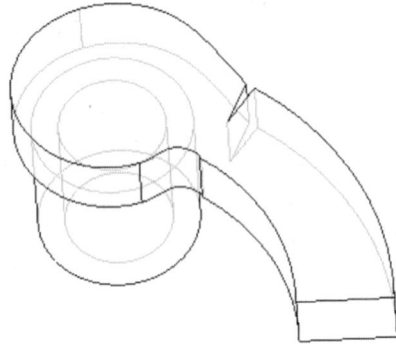

Gap between sweep and part.

The easiest way to avoid this problem when creating sweeps is to use the Merge Ends option during the initial creation of your sweeps. The Merge Ends option causes the sweep to meet up cleanly with existing geometry, eliminating this type of gapping. By default, the system is set to Free Ends, shown in the previous illustration.

Exercise 9-1: Performing Sweep Basics

Basic sweep features are used whenever you have need for a constant section to be swept across geometry. The following illustrations show two simple examples of the use of sweep features. The first is a pipe with two 90-degree bends, and the second is a glue joint on a plastic part.

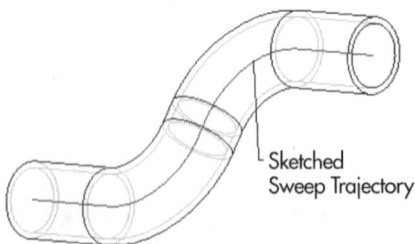

Sketched
Sweep Trajectory

Pipe.

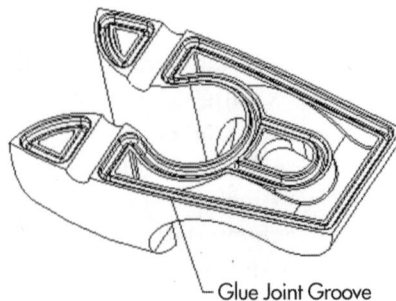

Glue Joint Groove

Glue joint.

The first thing you need to define for a sweep is the trajectory. Here you will have two choices. The first is to sketch a 2D section, as in the previous pipe illustration. The pipe example consists of three line entities and two radii, and the completed pipe has a tangency line at the end of each entity. The second choice is to select existing geometry to use as a trajectory. Existing geometry can be part and surface edges or existing datum curves. All selected trajectories can have their lengths adjusted by trimming or extending them. The ability to lengthen a selected trajectory can be very convenient when sweeping surface geometry, because you often need that extra length to close the surface with a merge.

In this exercise, you will be cutting a T-shaped groove in a block.

1. Create the geometry shown in the illustration at right.

2. Create a cut and select **Sweep** from the SOLID OPTS menu. Select **Done.** From the SWEEP TRAJ menu, select **Sketch Traj,** set up your sketching plane as the top surface, and create the section shown in the following illustration.

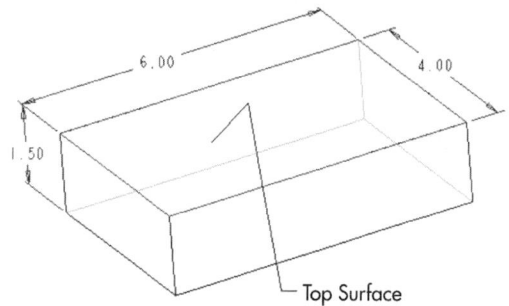

Base block.

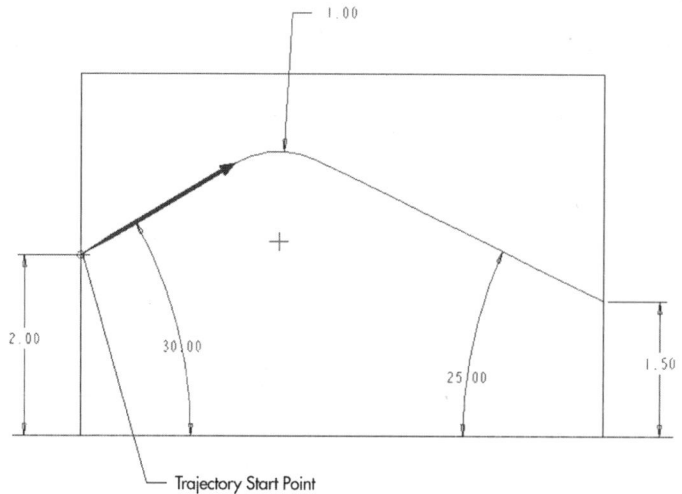

Trajectory section sketch.

As with sections in a blend feature, sweep trajectories have a start point you can move using the Start Point option under the SEC TOOLS menu. In the case of an open trajectory sketch, such as that in the previous illustration, the start point must be either end point.

3. Once the trajectory is complete and you have selected **Done**, the ATTRIBUTES menu will pop up giving you the options Merge Ends or Free Ends. Select the **Free Ends** option. The system will then create the section sketching plane normal to the trajectory. The trajectory start point will look like a large set of centerlines on the screen. The centerlines are what you will be dimensioning your section sketch to. Create the section sketch shown in the first illustration at right.

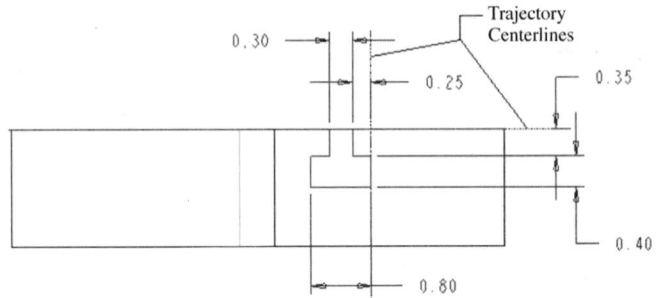

Sweep section sketch.

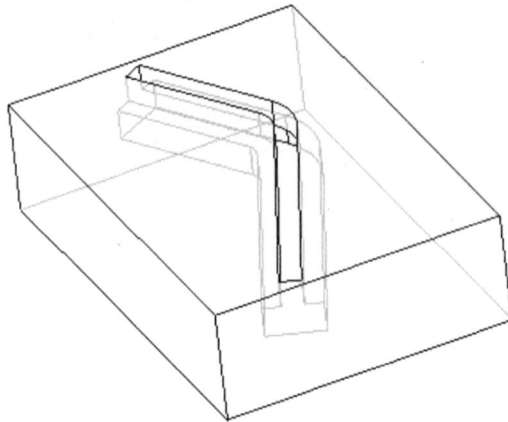

Completed sweep.

4. Once you have completed the sketch and have selected **Done**, you will be prompted "Arrow points TOWARD area to be REMOVED. Pick FLIP or OKAY." You want to remove the area inside the section; therefore, **Flip** the arrow accordingly. The completed feature should look like that shown in the second illustration at right.

5. As you can see in the previous illustration, the T-shaped cut does not extend through the side surfaces of the block. If this model meets your design intent, the model is okay, but if you had intended for the cut to go through the side walls, you will have to Redefine the sweep feature, changing the Free Ends attribute to Merge Ends. The illustration at right shows the redefinition.

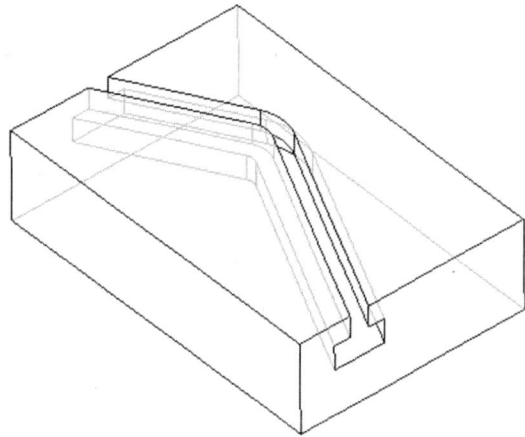

Redefined to Merge Ends.

Included in the functionality of basic sweeps is the ability to add inner faces to a swept section. This functionality is available only in sweeps that have a closed trajectory, as shown in the illustration at right.

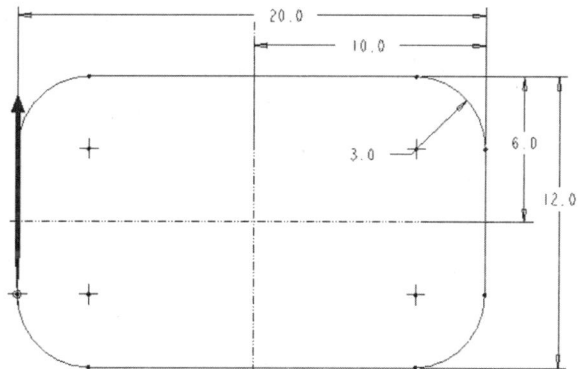

Closed trajectory.

6. Create a swept protrusion using the closed trajectory shown in the previous illustration as your sketched trajectory. Once you have completed the trajectory sketch, the ATTRIBUTES menu will pop up and you will have two options: Add Inn Fcs and No Inn Fcs. The Add Inn Fcs option allows you to sweep an open section around your closed trajectory, and the system will add geometry to the swept feature. Care needs to be taken when sketching the open section so that you get the effect you are looking for.

•• **NOTE:** *This is true anytime you let the system approximate or assume your design intentions.*

7. Select **Add Inn Fcs** from the ATTRIBUTES menu and create the section shown in the illustration at right. The section is to be swept along the closed trajectory.

The completed feature is shown in the following illustration. Notice that the underside is completely filled in.

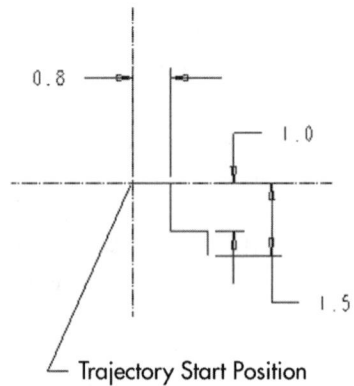

Open section to be swept.

Completed Add Inn Fcs feature.

8. Redefine the swept section sketch and add one vertical line, as shown in the following illustration.

Add vertical line segment.

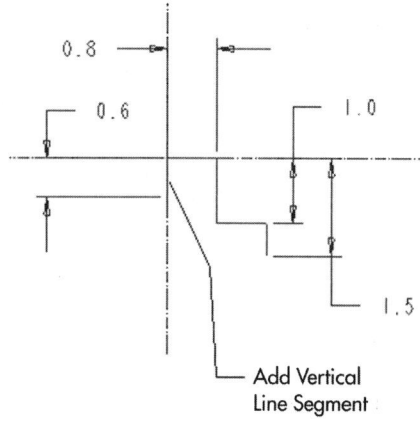

Add Vertical
Line Segment

The new feature geometry will look like that shown in the following illustration.

New sweep feature.

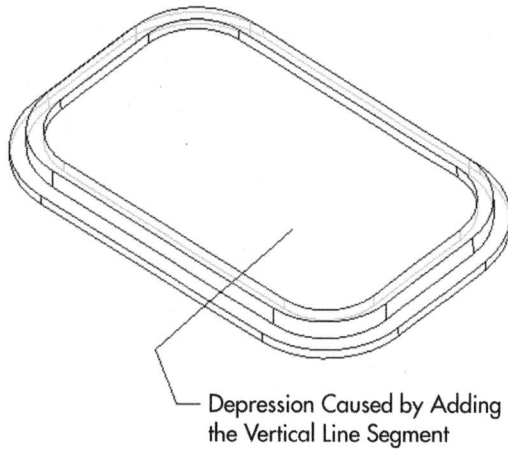

Depression Caused by Adding
the Vertical Line Segment

As you can see in the previous illustration, the Add Inn Fcs option can help simplify the section sketching process. However, you need to make sure to include all necessary geometry segments to get the desired result.

Exercise 9-2: Using a Helical Sweep

In this exercise, you will be completing the elbow adapter part from exercise 8-3. The completed part is shown in the illustration at right.

The Helical Swp feature is found under the ADV FEAT OPT menu. The most common uses for helical sweeps are in the creation of threads and springs. This is most evident when you look at the following helical sweep attribute options.

- **Constant:** The pitch value is constant and coils will be evenly spaced.

- **Variable:** The pitch value varies and is driven by a graph, and the resulting coils will have a variable distance between them.

Elbow adapter with a capped helical sweep thread.

- **Thru Axis:** The section is oriented relative to a plane that passes through the axis of revolution.

- **Norm To Traj:** The section's orientation is relative to a plane normal to the trajectory.

- **Right Handed:** The trajectory is generated using the right-hand rule. (Your right hand thumb points from the trajectory's start point toward the end point and your fingers wrap in the direction the sweep will travel.)

- **Left Handed:** The trajectory is generated using the left-hand rule.

  1. Either retrieve the part from exercise 8-3 or create a ∅0.8-inch diameter by 0.4-inch protrusion to the tapered end of the adapter, as shown in the first of the following illustrations. (If you are starting in this chapter, create the revolved protrusion shown in the second of the following illustrations.)

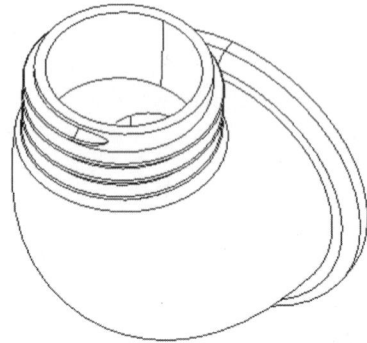

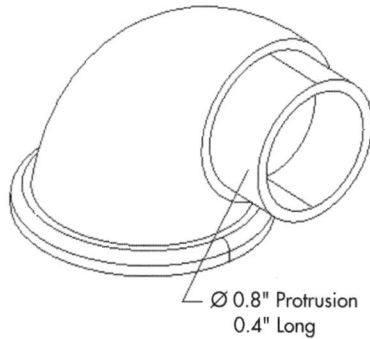

Protrusion on adapter part.

Ø 0.8" Protrusion
0.4" Long

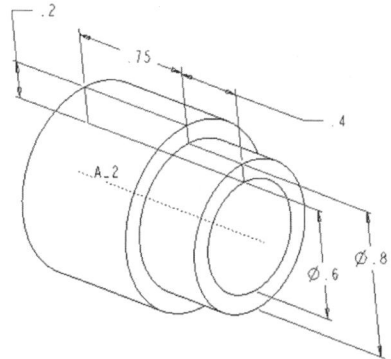

Helical sweep base revolve.

2. To create the thread, you will need to select **Advanced** from the SOLID OPTS menu and **Helical Swp** from the ADV GEOMETRY menu. Then from the ADV FEAT OPT menu, select **Constant ➥ Thru Axis ➥ Right Handed** and set up your trajectory profile sketching plane. Next, create the profile section shown in the illustration at right. The part for this illustration is the adapter part from exercise 8-3. The profile section is created as a **Geom Tools ➥ Use Edge** function. This will work the same on the "base revolve" as well.

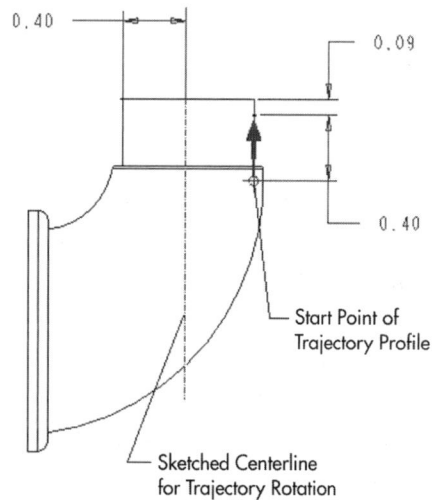

Start Point of
Trajectory Profile

Sketched Centerline
for Trajectory Rotation

Sketched trajectory profile.

In the section shown in the previous illustration, notice that the start point is buried into the adapter and the end is 0.09 inch below the adapter end. This will cause the start point end of the sweep to be buried and leave the other end to be capped later, very much as this thread would be cut onto a pipe in the real world.

3. Once your profile is complete, you will be prompted "Enter pitch value." Enter *0.10.*

The pitch of a thread is calculated by dividing 1 by the number of threads per inch. Now the model will be oriented for you to sketch the tooth profile. This is where the options Thru Axis and Norm To Traj come into play. The first illustration at right shows the section you will need to create. The orientation is for the Thru Axis option you selected. If you had selected Norm To Traj, the model would be skewed to match the trajectory profile. The second illustration at right shows the same section oriented to the Norm To Traj option.

Once you have completed the section sketch, the system will create the helical sweep. Your model should look like that shown in the following illustration, with the start of the thread buried into the shoulder of the adapter and the other end squared off. To cap the squared-off end of the thread you will need to use a nonparallel blend.

     ⊷ **NOTE:** *For more information on non-parallel blends, see Chapter 8.*

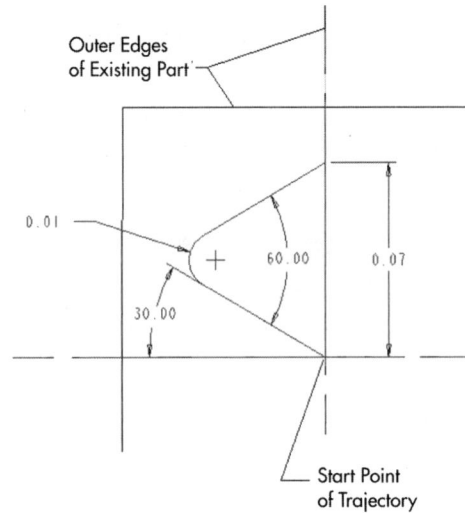

Thru Axis section orientation.

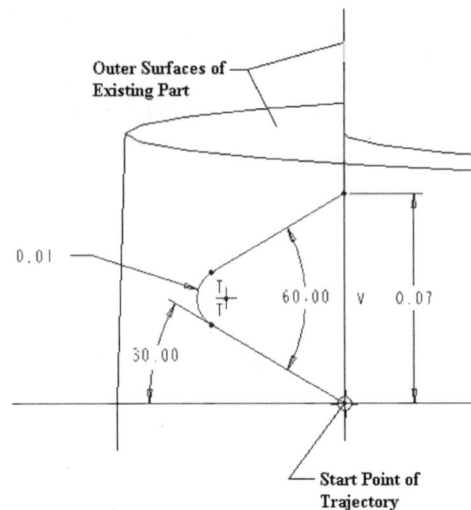

Norm To Traj section orientation.

4. Select **Create** ➥ **Protrusion** ➥ **Blend** ➥ **Rotational** ➥ **Regular Sec** ➥ **Sketch Sec** ➥ **Smooth** ➥ **Open**, and select the sketching plane to be the squared-off end of the thread. For section 1 of your blend, select **Use Edge** and select the squared-off thread end. Do not forget to create a local coordinate system. Because you are blending to a point, the start point of your section is not critical. The sketch of section 1 should look similar to that shown in the second illustration at right.

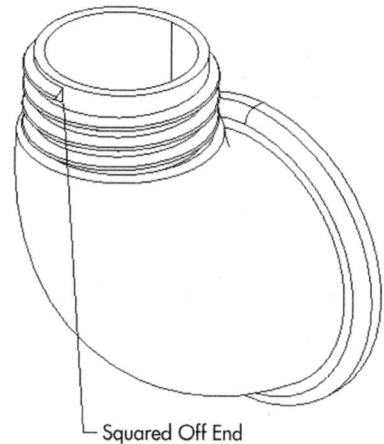

Completed helical sweep feature.

5. Complete section 1 and proceed to section 2, "the point." Here you need to decide an angle of rotation for section 2. The angle you choose is important for determining the location of the point section, as well as the length of the cap. Enter *20* degrees for the rotation. This angle is important to a calculation of the point position relative to the pitch of the thread. Here your pitch is .1, which means that the thread coils will be 0.1 inch apart in one 360-degree rotation. In this exercise, you have entered *20* degrees. Therefore, your

Thread capping section 1.

percentage of pitch is approximately 0.055. The pitch percentage multiplied by the pitch of .1 gives your point offset from the sketched coordinate system of 0.0055 inch. The following illustration shows the sketch for section 2.

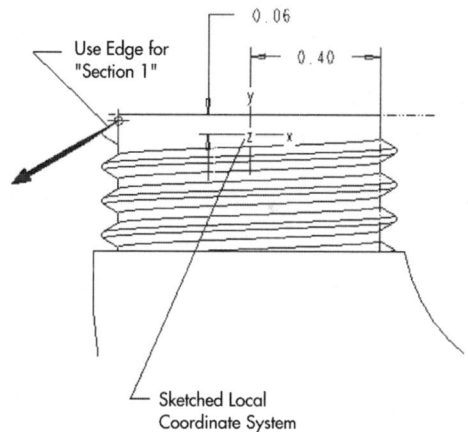

6. Once you have completed section 2, you can complete the feature and select **Smooth** for the CAP TYPE. Now you have the opportunity to assign tangency between this blend and the squared-off end of the thread.

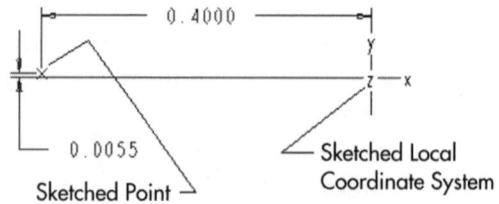

Section 2 sketch showing the blend point.

⤏ **NOTE:** *See exercise 8-5 for more details on tangency.*

The completed part with tangency assigned should look like that shown in the second illustration at right.

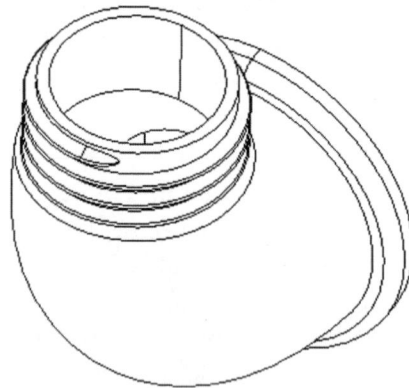

Completed thread with capping.

Exercise 9-3: Learning Swept Blend Basics

Through the last two exercises you have become more familiar with some basic functionality related to the sweep feature and how it is used. Now you will be looking at combining two types of features, the sweep and the blend. The Swept Blend feature creates a trajectory-driven blend, allowing you control over the trajectory sketch, each blend section sketch, and the location of each blend section along the trajectory. This makes for a fairly complex feature but gives you more areas of control and definition. In the first part of this exercise you will become more familiar with how a Swept Blend feature is defined. When completed, the first feature will look similar to that shown in the following illustration.

In the second part of the exercise you will build a more complex feature. It will represent the methods used to create the lip feature shown in the second illustration at right. The lip tapers inward to begin with and transitions to an outward taper through the upper radii.

The part you will build will be a simplified representation of this part. However, the sweep you will build will demonstrate the type of control the swept blend gives you. The two biggest differences between the "basic" sweep or blend and the "advanced" swept blend are the ability to control how the swept section is related to the trajectory and the ability to specify blending section sketch locations along the sweep trajectory. The following three new options for controlling a sweep's relation to the trajectory are found under the BLEND OPTS menu.

First swept blend.

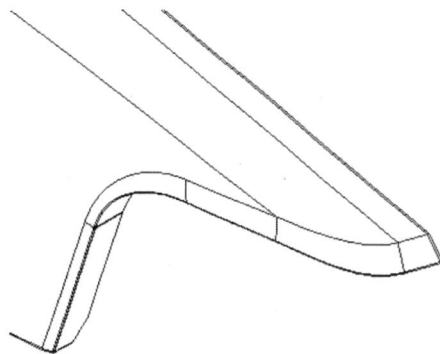

Example of complex swept blend.

- **NrmToOriginTraj:** The default option, NrmToOriginTraj, will keep the section sketch plane normal to the origin trajectory throughout the length of the sweep. This is the same as the basic sweep feature.

- **Pivot Dir:** This option will cause each section sketch plane to be normal to the origin trajectory. It also lets you select a plane that each sketch will remain parallel to throughout the length of the sweep.

- **Norm To Traj:** With this option you must select or sketch two trajectories. The first is the Origin Trajectory, which is used as the origin of each section sketch. The second trajectory is the Normal Trajectory. Each section sketch plane will be normal to this trajectory throughout the length of the sweep.

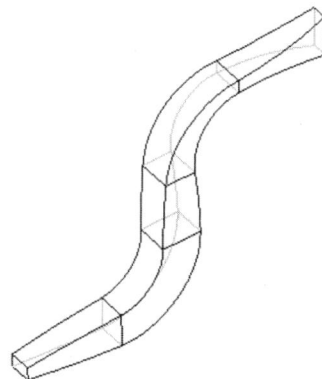

Now you are ready to build that first Swept Blend part. This feature will consist of an origin trajectory, which will be sketched first. Then you will be able to select blend section sketch locations. After the section locations are selected, you will then sketch each blend section. Each sketch origin will have a coordinate system. The active sketch coordinate system will be cyan colored, and you may need to zoom out to locate it.

> ✓ **TIP:** *A good indicator that you are not using the active coordinate system is when you cannot dimension to the coordinate system currently on the screen.*

1. Select **Feature ➥ Create ➥ Protrusion ➥ Advanced ➥ Done ➥ Swept Blend**, accept the default options, select **Done**, and set up the sketch plane for the origin trajectory. Create the section shown in the first illustration at right. The units for this part are set to inches, and both radii measure 2.00 inches.

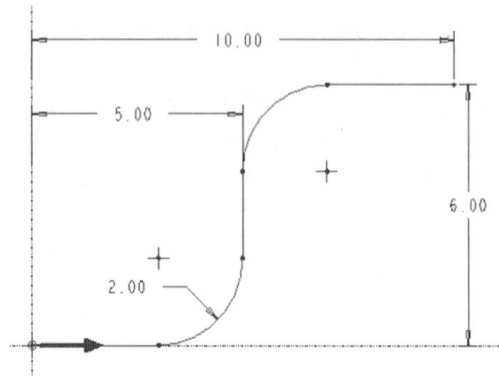

Origin trajectory sketch.

2. In the section shown in the previous illustration, the section vertices were left on to show the end points of each sketch entity. These entity end points are what you will be selecting as the sketch plane location of each section you want your feature to blend through. Once the Origin Trajectory is complete, the system will highlight each entity end point and allow you to Accept the position for a section or select Next to skip that location. You automatically have the two end points selected for a section. Therefore, here you are selecting interior section locations. For this exercise, Accept the positions shown in the second illustration at right as section locations.

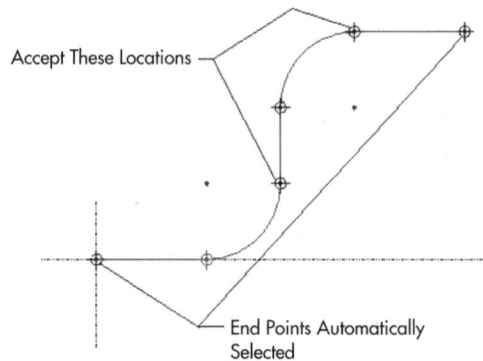

Locations for section sketching.

3. Once the last section location is Accepted, you will be prompted "Enter z_axis rotation angle for section 1. [Range 0–120]." Enter *0*. The view will orient for you to sketch the first section.

4. Create section 1, shown in the first of the following illustrations. As with regular sweeps and blends, the start point location is important. By using Sec Tools, you can relocate the start point on each section.

5. When section 1 is regenerated, select **Done**. You will be prompted for the z_axis rotation for section 2. For this exercise, enter *0* for all rotation prompts. The following illustrations show sketches for sections 2 through 4. Create each in order, and remember that you may need to Zoom Out to see the new sketch origin location.

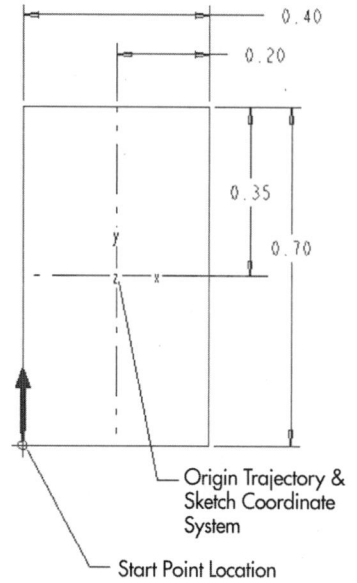

Sketch for section 1.

Sketch for section 2.

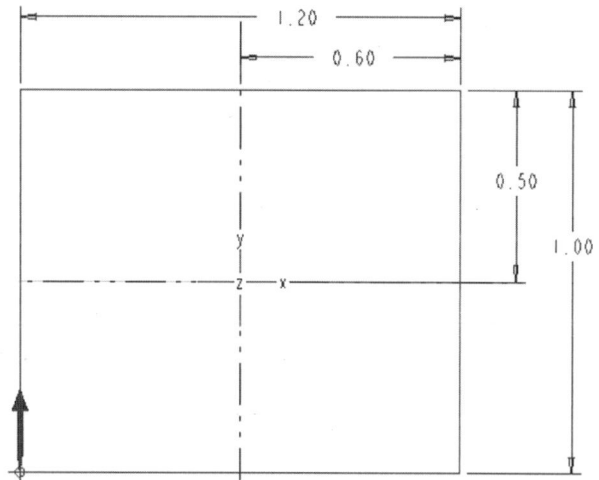

Sketch for section 3.

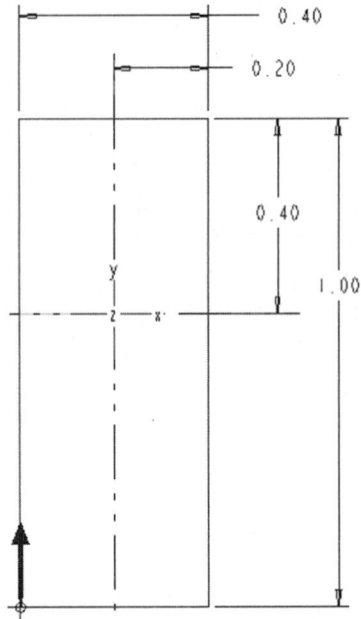

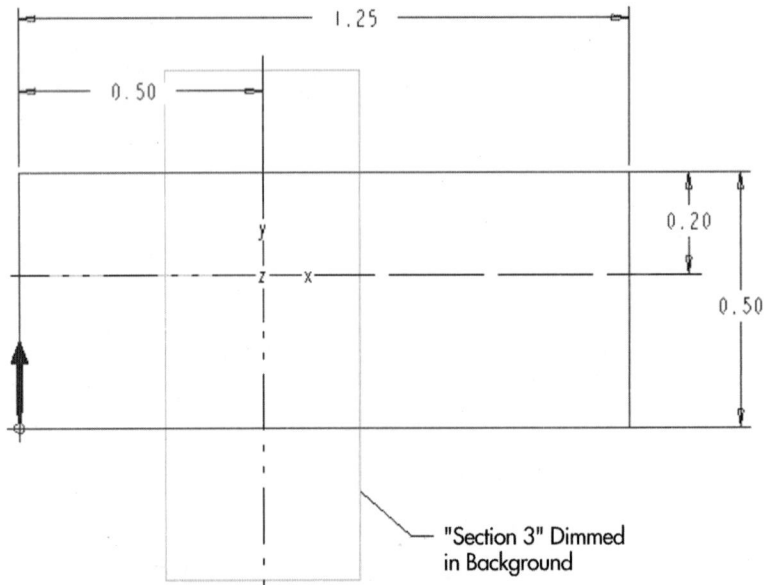

0.40

0.20

0.40

1.00

y

z — x

1.25

0.50

y

z — x

0.20

0.50

"Section 3" Dimmed
in Background

Sketch for section 4.

6. Once you have completed section 4, select **OK** from the Protrusion dialog box and the system will generate a feature similar to that shown in the illustration at right.

7. To see the trajectory and all sections in a 3D orientation, select **Modify** and pick the feature. Then select the trajectory and all sections from the SPECIFY menu. The image should look like that shown in the following illustration. Now you can select each section or trajectory dimensional value in order to change it.

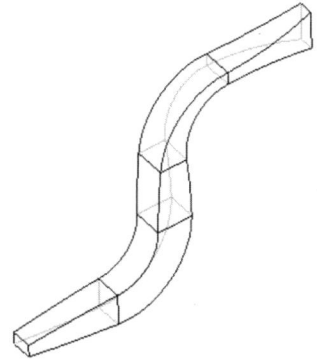

Completed swept blend.

Modify sections and trajectory.

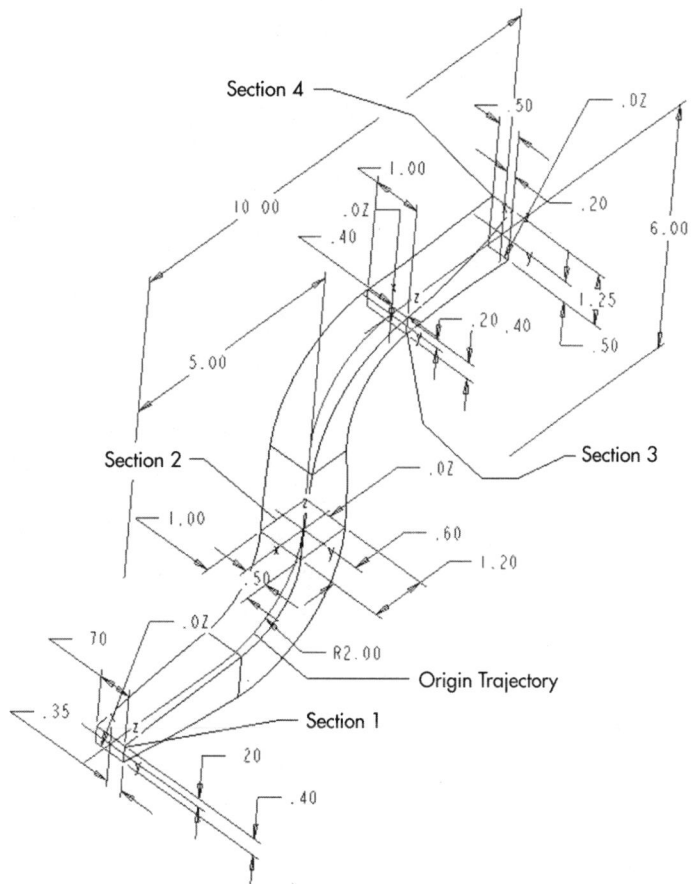

The following are swept blend rules to keep in mind.

- You will have the opportunity to create a section at all entity end points. Therefore, if you need a section in the middle of a radius you will need to split the radii in the trajectory sketch.

- You cannot redefine Swept Blend features to add more sections.

- If you are going to use the Select Traj option to select an existing datum curve or feature edge as your trajectory, you could also use datum points on that curve or edge as section locations.

Exercise 9-4: Performing a 3D Swept Blend

Now that you have some of the basics of the swept blend down and have created a sweep along a 2D trajectory, it is time to work with a 3D trajectory. In this exercise you will be creating a lip feature similar to the one in the "complex swept blend" illustration earlier in this section. The final part will look similar to that shown in the illustration at right.

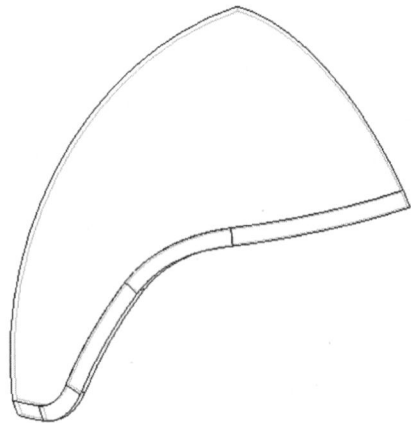

3D trajectory swept blend.

1. For the base of this feature, you will need to create a revolved protrusion and then extrude a cut through it. The extruded cut feature will form the bottom of the swept lip feature. The first two of the following illustrations show sections for the base feature. Revolve the first sketch (sketch 1) 90 degrees; then cut sketch 2 through the revolved dome. Your new feature should look similar to that shown in the third of the following illustrations. Add 2-inch radii where shown.

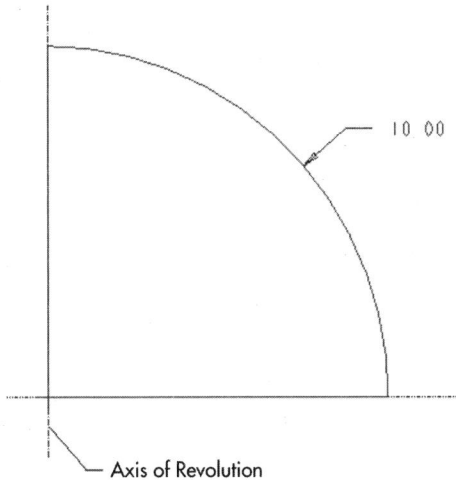

Revolve the first sketch (sketch 1) 90 degrees.

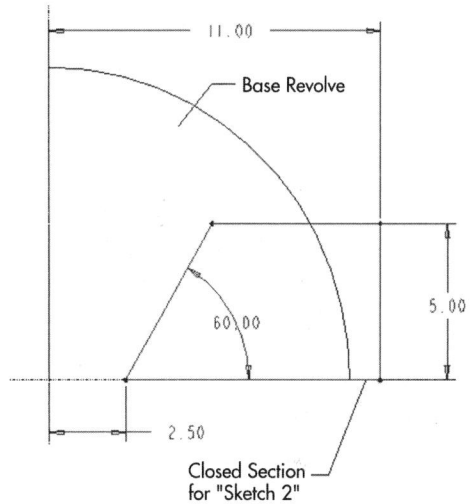

Cut sketch 2 through 90-degree revolve.

If your base part looks like the one shown here, bottom right, you are now ready to start the swept blend portion of this exercise. You next need to define the trajectory for this feature. The lip is cut into the revolved dome and is 0.5 inch tall. You have two options for the trajectory. The first option is to use the Select Traj option and pick the edge of the cut as a trajectory. This would be very easy. However, the problem would result in trying to align the top edge of each section to the outer surface of the dome. The second option is to project a datum curve onto the outer surface of the dome, which is offset from the cut edge by 0.5 inch. This projected curve will lie on the surface of the dome and give you a very consistent cut.

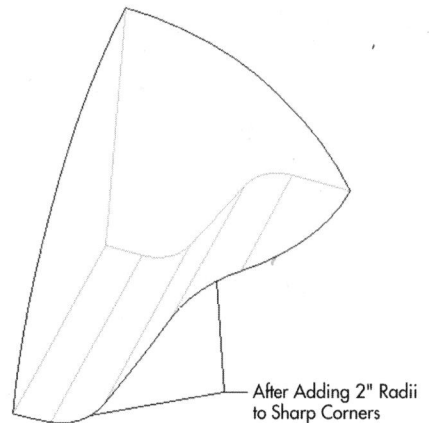

Add rounds to base part.

↝ **NOTE:** *Chapter 11 contains more information on creating datum curves.*

2. To create the projected curve trajectory for the Swept Blend feature, select **Feature** ➥ **Create** ➥ **Datum** ➥ **Curve** ➥ **Projected** ➥ **Sketch** ➥ **Done** and set up your sketch plane as the flat surface normal to the sketch 2 cut. You will need to select a direction to view the section in, and then a direction for feature creation.

➥ **NOTE:** *Make sure you flip the feature creation direction toward the rounded dome surface, as shown in the illustration at right, or your curve will have no surface to project onto.*

3. Use **Geom Tools** and **Offset Edge** to offset the bottom edge by 0.5 inch. The section and view should look similar to those shown in the second illustration at right.

4. The only dimension necessary for this section is the offset value. Once you have completed the section and have selected **Done**, you will need to select the surfaces to include for the curve projection. It will be convenient to reorient the part so that you can select the rounded dome surface for the curve to be projected onto. After you pick the surface, select all of the **Done** menu options and select **OK** to complete the feature, which should look like that shown in the following illustration.

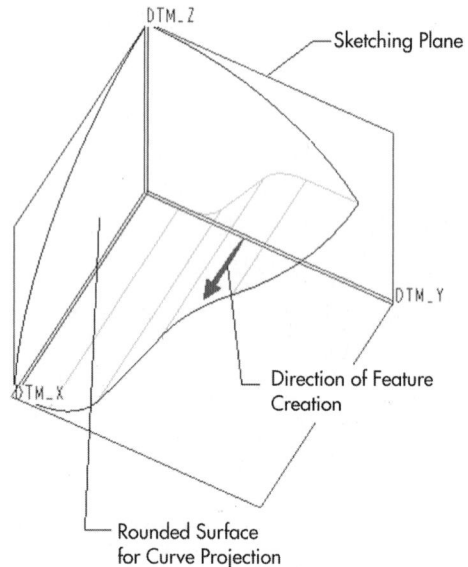

Feature creation direction.

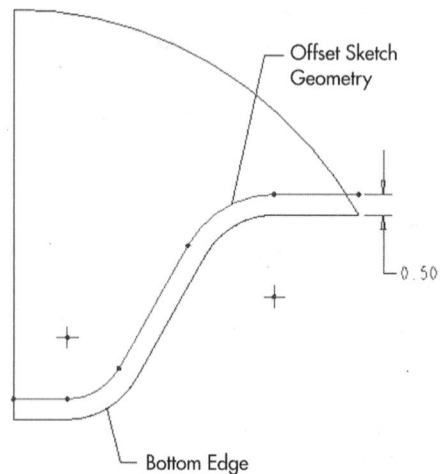

Bottom edge offset for curve projection.

Now you are ready to create the swept blend. This time you will be using the Select Traj option and picking the projected curve as your origin trajectory. Remember that as you move from sketch to sketch the part will reorient, and that because this trajectory is 3D you will have to Zoom Out to get your bearings on the new section sketch plane.

Projected curve on base part.

5. Select **Cut** ➡ **Advanced** ➡ **Swept Blend** ➡ **Done** ➡ **Sketch Sec** ➡ **NormToOriginTraj** ➡ **Done** ➡ **Select Traj** ➡ **Curve Chain**, and select the projected curve and **Select All**. Make sure the start point is at the lower left, as shown in the following illustration. If it is not, use the Start Point option from the CHAIN menu to move it there.

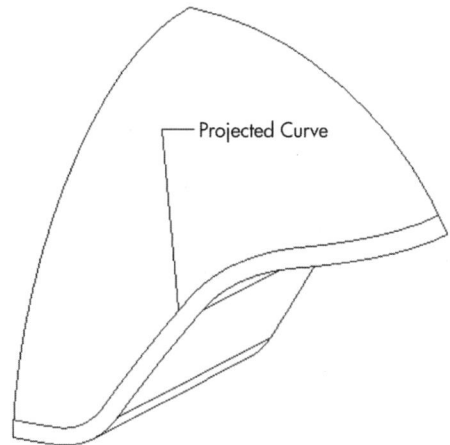

Selected trajectory and start point.

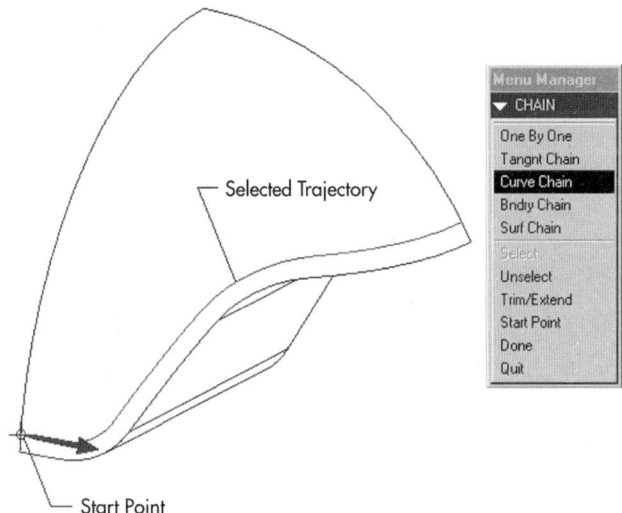

You will next be asked to set the section orientation, and presented with the following options. These options can add some complexity to the feature's section orientation but can all result in the same geometry.

- **PickXVector:** Allows you to select a datum plane, surface, axis, or curve for each section's horizontal reference.

- **Automatic:** Determines the section sketching plane relative to the origin trajectory start position.

- **Normal to Surf:** Allows you to select a surface to determine the upward direction for your section sketch plane. The surface you select here can cause the sections to look skewed relative to the part geometry.

6. For this feature, select **Automatic** and use the **Flip** option to make the arrow point toward the top of the part, as shown in the illustration at right. The same result can be achieved by selecting PickX-Vector and then picking DTM_Y as the reference. Either method works.

7. Next you will be prompted to determine additional section locations. Accept all section locations. The part will then be oriented for section 1. Create the section shown in the second illustration at right.

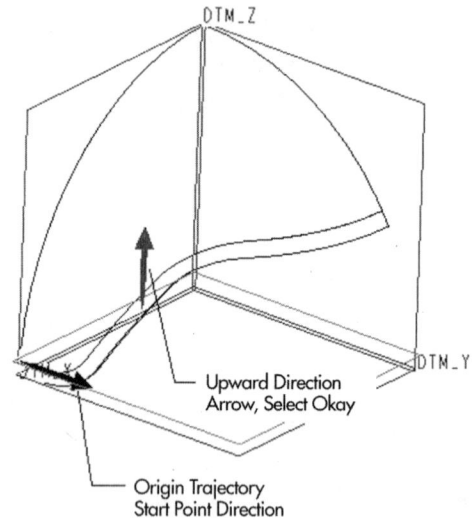

Upward direction for sections.

8. When prompted for z-axis rotation, enter *0.* Accept the **Automatic** orientation option for all sections.

9. When you have completed section 1, select **Done** and continue to section 2. Create the sections shown in the following series of illustrations.

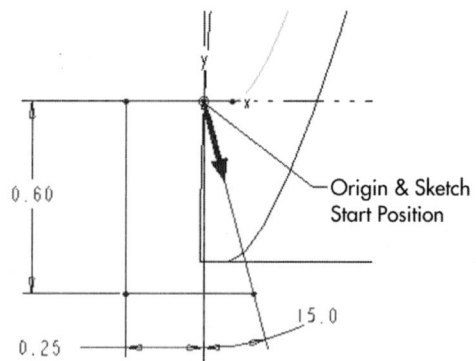

Section 1 sketch.

Section 2 Origin
& Start Point

Dimmed Section 1

0.60

0.25

15.0

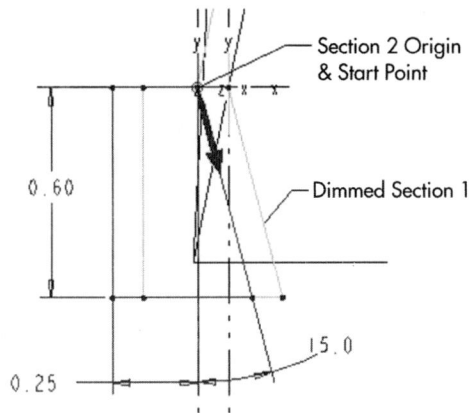

Section 2 sketch.

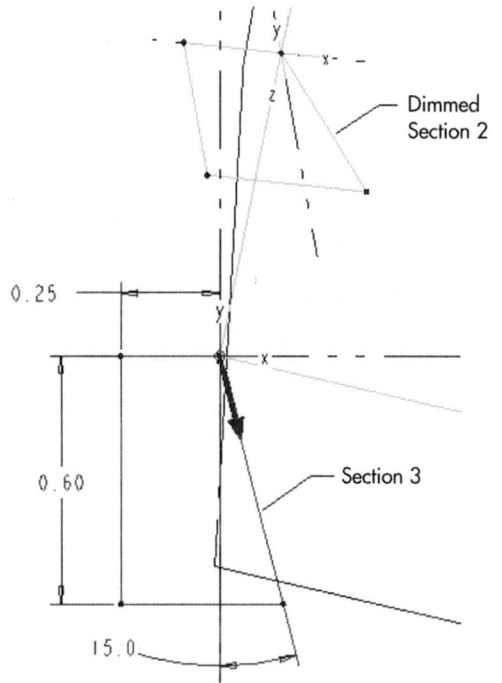

Dimmed
Section 2

0.25

0.60

Section 3

15.0

Section 3 sketch.

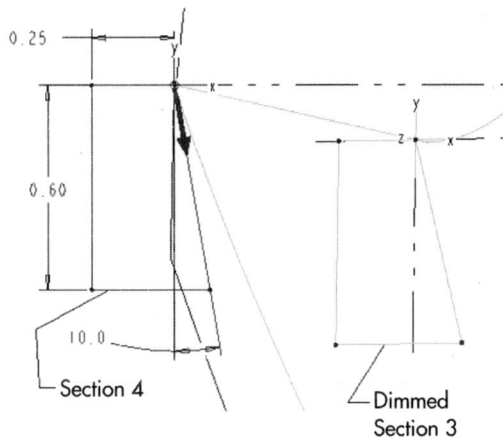

0.25

0.60

10.0

Section 4

Dimmed
Section 3

Section 4 sketch.

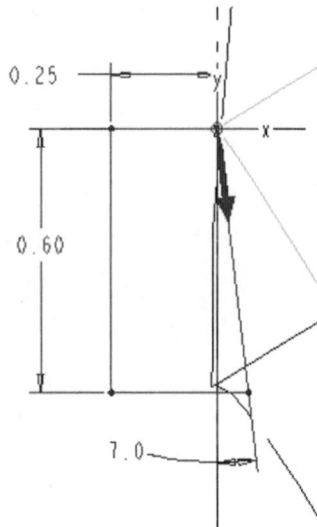

Section 5 sketch.

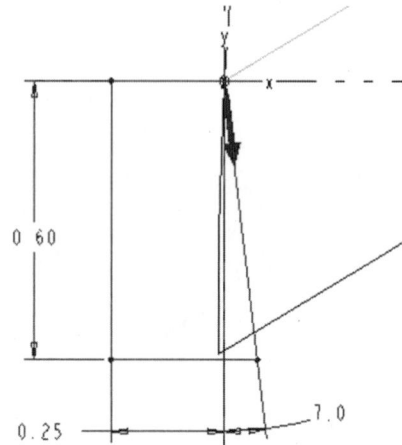

Section 6 sketch.

10. When you have finished the section 6 sketch, you will need to tell the system which side of the sketch to remove material from. Flip the arrow inside and select **OK** from the Protrusion dialog box. The completed feature should look like that shown in the illustration at right.

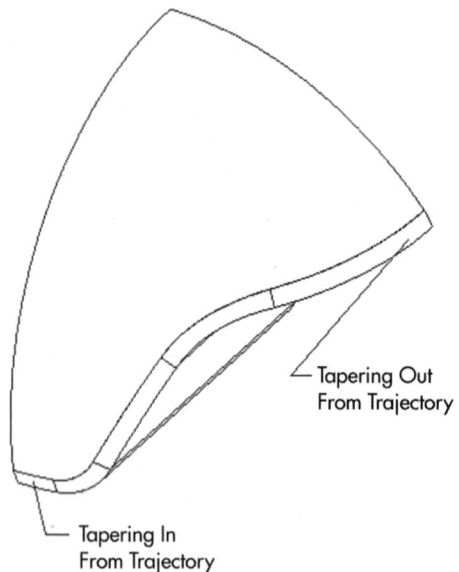

Tapering Out
From Trajectory

Tapering In
From Trajectory

Completed lip cut.

Through the creation of these two swept blend features you have gained another level of control over your blend geometry. In the previous exercise, you saw that by modifying the taper angle of the lip you can adjust a feature's appearance along the entire sweep. When creating swept blend geometry, there is no exact rule as to the number of sections needed to create the feature. It is up to you to determine where and how many sections are necessary to get the results you seek.

To see for yourself the effects the number of sections has on a swept blend, try rebuilding the lip feature using every other section and location. By selecting fewer section control positions, you allow the system to do what it wants between the control points. This can result in some pretty unaesthetic geometry.

Exercise 9-5: Learning Multi-trajectory Sweep Basics

Multi-trajectory sweeps (called Var Sec Swp under the ADV FEAT OPT menu) are very much like the basic sweep feature, with one major difference: the Var Sec Swp has more than one trajectory. With the basic sweep feature, you sketch or select a trajectory for a section to be swept along. The section will remain constant from the beginning to the end of the trajectory. With the Var Sec Swp, you again select or sketch the first trajectory, which is called the "origin trajectory." The section you will be selecting or sketching is swept along the origin trajectory.

Prior to creating the section, you have the ability to add several other trajectories that will vary the shape of your section as it is swept along the origin trajectory. All Var Sec Swp features have at least two trajectories: the Origin Trajectory and the X-Trajectory. The Origin Trajectory is the same in both Var Sec Swp and basic sweep features. The X-Trajectory is used to set the swept section's orientation, as well as vary the section's shape as it is swept along the Origin Trajectory. The following illustration shows a Var Sec Swp with only the Origin and X-Trajectories. Notice that the Origin Trajectory is shorter than the X-Trajectory; therefore, the feature stops at the origin's length. Also notice that the shape of the X-Trajectory varies the feature's shape.

Var Sec Swp with only two trajectories.

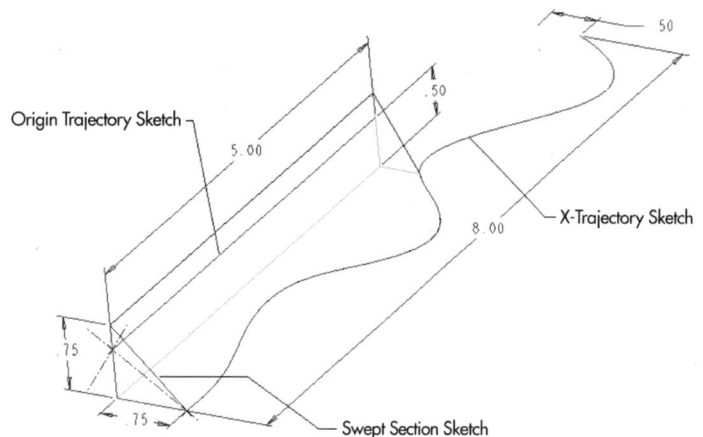

In this exercise, you will be creating a Var Sec Swp with four trajectories: an Origin Trajectory, an X-Trajectory, and two additional shaping trajectories. One important thing to keep in mind when working with multiple trajectory features is that the start points of all trajectories must be on a common plane, because the system will be using that plane as the section's sketch plane. The illustration at right shows the part you will be building.

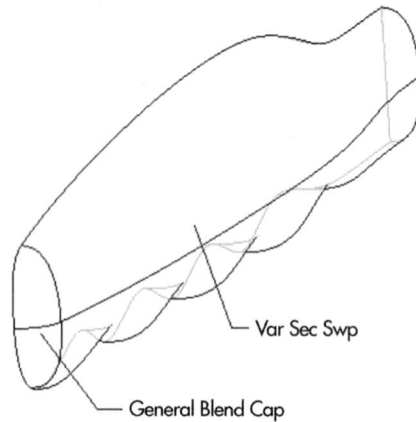

Exercise 9-5 model.

1. Start your model with the default datum planes and create a protrusion by selecting **Advanced** ➻ **Solid** ➻ **Done** ➻ **Var Sec Swp** ➻ **Done** ➻ **Nrm To Origin Traj** ➻ **Done**. You will now have the VAR SEC SWP trajectory menu open. For this exercise, you will be selecting the Sketch Traj option for all trajectories. However, if you had existing solid, surface, or curve geometry, you could use the Select Traj or the Sel Tan Traj option. The difference between Select Traj and Sel Tan Traj is that the Sel Tan Traj option allows you to tell the system you want to assign tangency conditions to the feature before it is created.

2. Select the **Sketch Traj** option and create your first trajectory, the origin trajectory. Use DTM2 as the sketch plane and DTM3 as the top for horizontal orientation. The origin trajectory section is shown in the illustration at right.

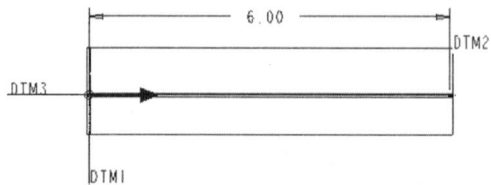

Sketch the Origin Trajectory as a Horizontal Line Aligned to DTM3 and the Start Point Aligned to DTM1.

Origin trajectory section.

3. Once the origin trajectory is complete, select **Done** and then reorient the model to the **Default View**. Select **Sketch Traj** from the VAR SEC SWP trajectory menu. Pick DTM2 as the sketch plane and DTM3 as the top for horizontal orientation. Sketch the second trajectory (X-Trajectory). In this section, you will be sketching a spline with **Tangency** set to **Both**.

The exact location of the internal points is not critical in this exercise. Simply make it look similar to the trajectory shown in the following illustration.

X-Trajectory section.

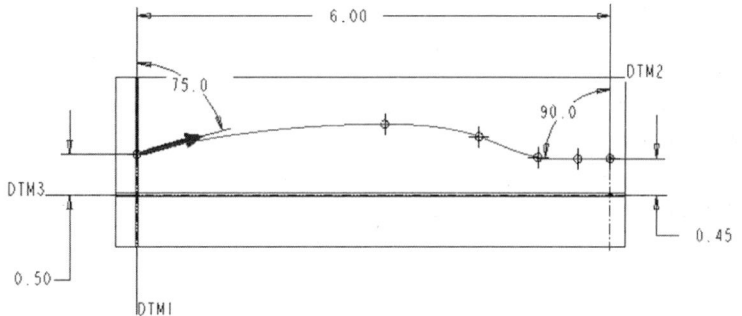

X-Trajectory Sketched on DTM2 with its Start Point Aligned to DTM1. Dimension the Spline as Shown and Move the Spline's Internal Points to Approximate the Curvature Shown.

4. Once the X-Trajectory is complete, select **Done** and then reorient the model to the **Default View**. Select **Sketch Traj** from the VAR SEC SWP trajectory menu. Pick DTM2 as the sketch plane and DTM3 as the top for horizontal orientation. Sketch the third trajectory, which will be used to shape the finger grip area along the bottom of the handle. In this section, you will be sketching a spline again, with **Tangency** set to **Both**. Again, the exact location of the internal points is not critical. Simply make it look similar to the trajectory shown in the following illustration.

Trajectory for the three finger grips.

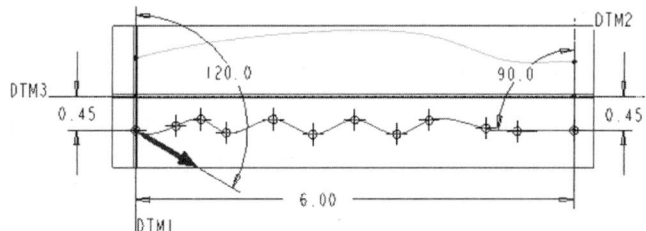

Third Trajectory Sketched on DTM2 with its Start Point Aligned to DTM1. Dimension the Spline as Shown and Move the Spline's Internal Points to Approximate the Curvature Shown.

5. Once the third trajectory is complete, select **Done** and then reorient the model to the **Default View**. Select **Sketch Traj** from the VAR SEC SWP trajectory menu. Pick DTM3 as the sketch plane and DTM2 as the bottom for horizontal orientation. Sketch the fourth trajectory, which will be used to shape the side profile of the handle. In this section, you will be sketching a spline again, with **Tangency** set to **Both**. Make the sketch look similar to the trajectory shown in the following illustration.

Trajectory four-side profile.

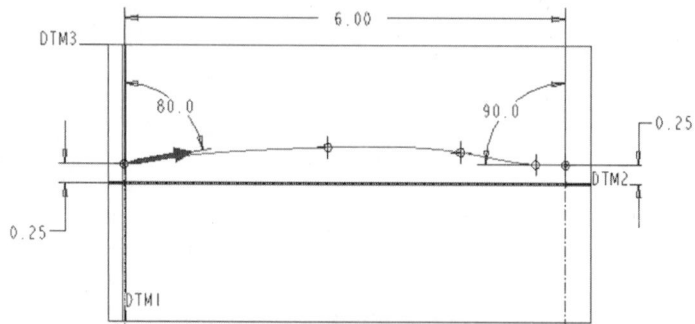

Fourth Trajectory Sketched on DTM3 with its Start Point Aligned to DTM1. Dimension the Spline as Shown and Move the Spline's Internal Points to Approximate the Curvature Shown.

6. Once the fourth trajectory is complete, select **Done**. The system will orient the model to the section sketch plane. Each trajectory will have an X at its start point location to help you locate the section you will now draw to the trajectories. The section you will be sketching will consist of two conic entities and a line. The reason for using conics in this section is to control the tangency of the feature along DTM2, as well as to control where the two conics meet. This same tangency control could be achieved using spline entities. Sketch the section shown in the following illustration.

Fourth Trajectory & Start Point

Sweep Section Sketch

Sweep section sketch.

0.50

90.0

0.50

90.0

90.0

X-Trajectory & Start Point

Origin Trajectory & Start Point

Third Trajectory & Start Point

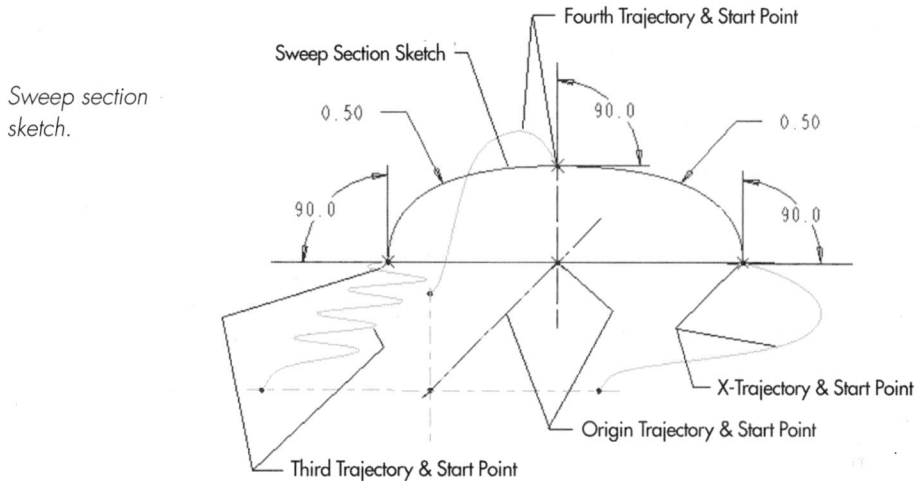

This View Has Been Rotated to Display All Trajectories and Start Point Locations.

7. When you sketch the section's entity end points close to the trajectory start point locations, the system will assume they are to be Aligned. In this exercise, that is exactly what you want. However, if necessary you could dimension your section to the trajectories. After completing the section sketch, select **Done**. Then select **OK** from the Protrusion dialog box to complete the feature. Your result should look similar to that shown in the illustration at right.

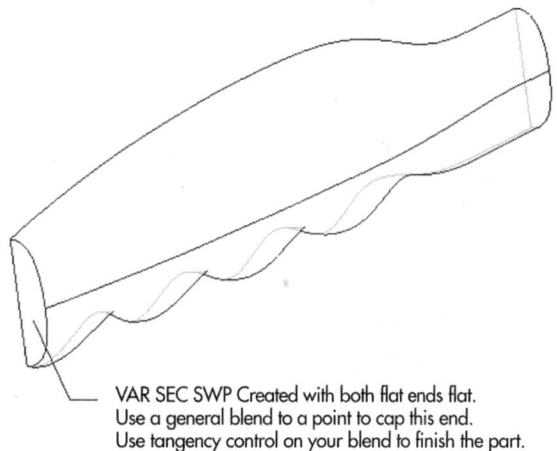

VAR SEC SWP Created with both flat ends flat.
Use a general blend to a point to cap this end.
Use tangency control on your blend to finish the part.

Completed Var Sec Swp.

➦ **NOTE:** *In this exercise, you can use a general blend to cap the end of the part. You can blend to a point in Var Sec sweeps by using Datum Graph features. For more information on using Graph features to blend to a point, see the* Pro/ENGINEER Part Modeling User's Guide, *Release 20/2000i.0, Chapter 6, pages 35–38. In addition, when using Graph features to cap more organic parts, as in this exercise, surfaces are often needed because the solid geometry may not work.*

Summary

Throughout this chapter you have been developing your skills beyond the basic functionality of Pro/ENGINEER. In general, using sweeps and swept blends allows you to create more stylized and complex geometry while maintaining some control over the feature's shape. The tools you use to control these features are not just the sections but the trajectories the sections follow. Again, while creating swept geometry in your day-to-day modeling it is important to capture the design's intent so that others who may need to modify your design have a clue as to what you were trying to accomplish.

In exercise 9-5 you created a variable section sweep in which you defined four paths and a section to create the feature. The trajectories used to shape the section as it traveled the length of the sweep are the controls and define your intent. In many cases, parts created using the Swept Blend and Var Sec Swp functionality are becoming more the norm than the exception. This is more apparent when you look at new products and redesigns to existing products. Straight lines are being eliminated, and smooth, flowing curves are taking their place. This design and styling shift, which has been going on for some time, forces everyone using CAD systems to improve their ability to blend and sweep with the times.

Copying Features

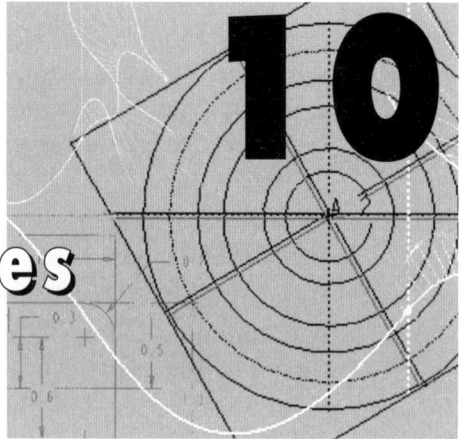

In just about every design there exists the need to copy features within a model or from one model to another. It may be as simple as a bolt pattern to a complete product that varies only slightly from model to model. Pro/ENGINEER gives you several tools to deal with all of these situations. This chapter covers several techniques of copying features, patterns, and user-defined features (UDFs), as well as mirroring models and features. The chapter also explores local groups, which are often used in conjunction with a copy or pattern.

Understanding Local Groups

Local groups are relatively simple features. They are a way of treating several features as a single feature. The local group can be used only in the current model, and must consist of sequential features. If you select features that are not sequential, Pro/ENGINEER will prompt you if you want to group all features in between. Answering No will abort the creation of the local group. You can create a pattern or copy the features of a local group within the active model, or use the group as a way of suppressing or resuming many features at once.

Patterns: The Basics

There are three types of patterns that can be created in Pro/ENGINEER. Identical is the most basic category of pattern, followed by Varying. General is the most complex. Identical patterns are used when instances are all identical in size. This means that the feature being patterned does not change size; it is simply copied to other locations on the part. Identical patterns also regenerate the fastest of the three pattern types. When instances vary in size or will be placed on different surfaces, use the Varying or General pattern types. Only the General pattern can be used if any of the instances intersect each other.

Patterns copy features by altering the values of selected dimensions by specified amounts. This can be done in a number of ways. The most common is the value-driven pattern, where the next instance is determined by the incrementation of selected dimensions by a constant amount. Just as a relation is created to determine the value of a feature dimension, it can also be used to drive the instances of a pattern.

A classic example of a relation-driven value of a pattern is the angle between the holes in a 360-degree bolt pattern. After the pattern is created, simply write a relation, using the dimensional symbol for the angle between the holes. This would be equal to 360 divided by the dimensional symbol for the number of holes in the pattern. This guarantees that the holes of the pattern will always be evenly spaced whenever the number of holes is changed. There is another way of driving the values of a pattern and that is with a table.

Table-driven Patterns

Using the table-driven pattern has several key advantages and differences over value- or relation-driven patterns. The most significant difference between a table-driven pattern and all other patterns is that the dimensional values come off a single baseline rather than from the previous instance. This means that the patterned features can be placed in a truly irregular fashion, including rectangular patterns. Dimensional values can be positive or negative. The pattern configuration can be placed in, and modified through, family tables, allowing different family table instances to be associated with different patterns. Data or dimensional values for the table-driven pattern can be imported from files outside Pro/ENGINEER.

Pattern Examples

In this section of the chapter you will go through an example of each of the four types of patterns: rotational, varying, general, and table-driven. Each of these pattern types has unique applications and advantages.

Rotational Patterns

The rotational pattern is one most engineers run into on a regular basis. They are used to create everything from bolt circle patterns and cooling slots to decorative cutouts and protrusions.

Exercise 10-1: Creating a Rotational Pattern

In this exercise, you will be creating a bolt circle pattern on a flange part. The first step is to create a flange part similar to that shown in the illustration at right.

1. Create the bolt hole feature that will be patterned later. Select **Feature** ➥ **Create** ➥ **Slot** ➥ **Done** and then select the flange surface as the sketching plane. In order to create a rotational pattern, an angle of rotation must be available to be selected. This angle will be attached to the sketch orientation plane created on the fly. To create this datum, select **Top** ➥ **Make Datum** for the sketch orientation plane, and use the **Through** option and select the center axis A_1, shown in the following illustration. Select the **Angle** option from the DATUM PLANE menu and pick the datum plane TOP. Select **Done** and then use the **Enter Value** option from the OFFSET menu. An arrow will display the direction of rotation, as shown in the following illustration.

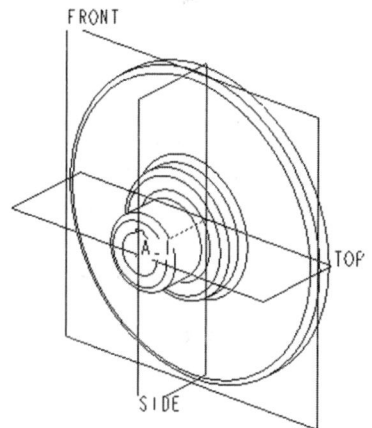

Flange part needed to start the rotational pattern exercise.

FRONT

Items that need to be selected to create a datum on the fly for the rotational pattern.

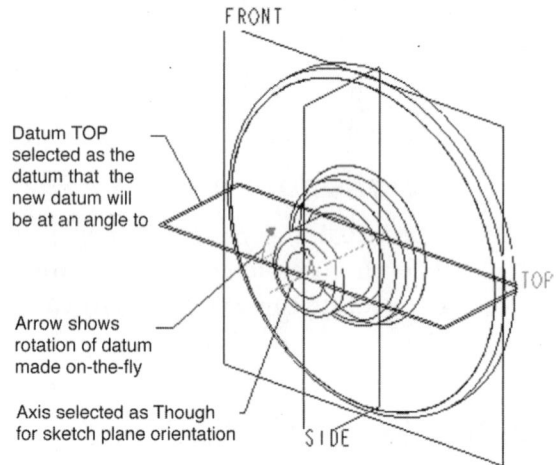

Datum TOP selected as the datum that the new datum will be at an angle to

Arrow shows rotation of datum made on-the-fly

Axis selected as Though for sketch plane orientation

TOP

SIDE

2. Enter the angle (45 degrees is the default). Enter *45* degrees so that the orientation plane will look as it does in the following illustration. (You may have to enter *–45* degrees to get the same orientation.) The advantage of the "on the fly" datum is twofold. First, the datum and its angular dimension becomes a part of the feature being patterned. Second, the datum is not visible after feature creation, which helps keep the model from becoming too cluttered.

Feature to be patterned

Bolt hole feature ready to be patterned.

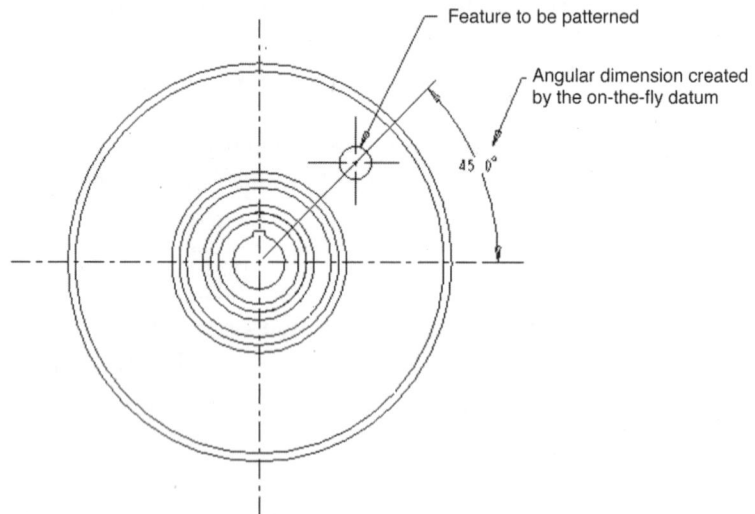

Angular dimension created by the on-the-fly datum

45 0°

3. To create the pattern, select **Feature** ➥ **Pattern** and select the hole as the feature to be patterned. Select **Identical** (this is already highlighted) ➥ **Done**. The **Value** option is by default also selected. Therefore, you can go ahead and select the angle dimension, shown in the previous illustration. Pro/ENGINEER will now prompt you for the increment value. This angle is calculated by dividing 360 degrees by the number of holes required. In this case, there are going to be 12 holes equally spaced in a complete circle. Therefore, the angle is $360/12 = 30$.

> ✓ **TIP:** *This equation can be used as a relation for the increment value of the pattern so that no matter how many instances are needed the pattern will always end up equally spaced and form a complete circle.*

4. The increment amount is 30 degrees. Enter *30* for this angle. Pro/ENGINEER prompts for a second dimension in the first direction. However, in this example there is none. Therefore, select **Done**. The next prompt in the message window will be for the **TOTAL** number of instances, including the original. Enter *12*. A new prompt in the message window asks for a dimension in the second direction. Select **Done**. The pattern will now be created and should look like that shown in the following illustration.

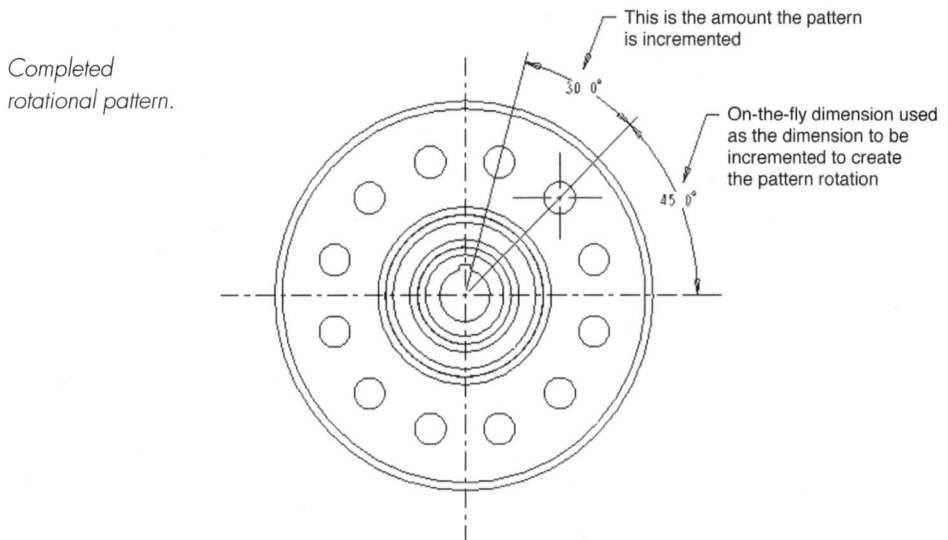

Completed rotational pattern.

This is the amount the pattern is incremented

On-the-fly dimension used as the dimension to be incremented to create the pattern rotation

30 0°

45 0°

5. Modify the on-the-fly dimension to match the increment dimension so that the pattern is oriented correctly, as shown in the following illustration. Note that the orientation of the holes in the pattern have rotated so that the first hole of the pattern lies on the horizontal centerline of the flange part.

Proper pattern orientation.

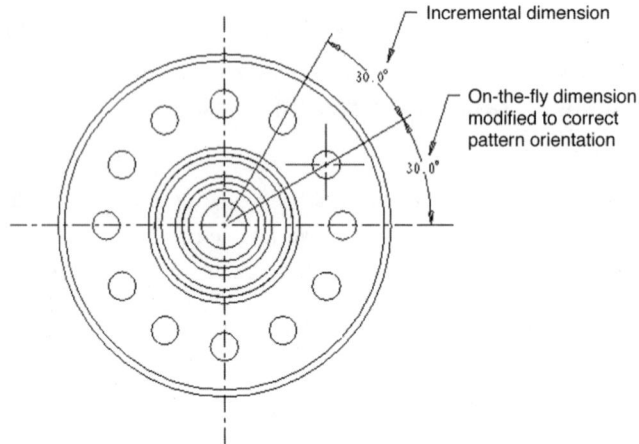

Incremental dimension

On-the-fly dimension modified to correct pattern orientation

30.0°

30.0°

✗ **WARNING:** *Do not use Feature ➥ Delete to delete a pattern. It will not only remove the pattern but will remove the original feature. Use Feature ➥ Del Pattern. This command sequence will remove the pattern instances without deleting the original feature.*

Varying Pattern

The varying pattern option is used when the size of the feature in the pattern changes. Exercises 10-2 and 10-3 deal with two different types of varying patterns. Exercise 10-2 presents an example of a simple single-direction pattern. Exercise 10-3, which builds on exercise 10-2, is an example of a two-directional pattern that demonstrates how the second direction affects the outcome of the pattern.

Exercise 10-2: Creating a Block and Hex Features

You will be using the same block and hex features for exercises 10-2 and 10-3.

1. Create a block with the following dimensions: 10 inches wide, by 6 inches high, by 2 inches deep.

2. Add a hexagonal cut feature using the **Thru All** option. The dimensions for the hexagon are shown in the first illustration at right.

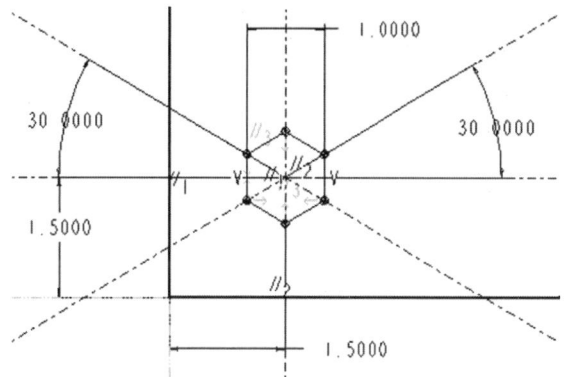

Detail of the hexagon cut and its location on the block.

3. Create the single-direction varying pattern. Now that the block has been modeled with the hex cutout, select **Feature ➡ Pattern** and pick the hexagon cut feature to pattern. Select the **Varying** option and **Done**. As in exercise 10-1, use the **Value** option (already highlighted), select the horizontal 1.50 locational dimension, and enter an increment of *2.00* at the increment prompt. Next, select the vertical 1.50 locational dimension and enter an increment of *1.00*. Select the 1.00 width of the hexagon, enter an increment of *.25* at the prompt, and select **Done**. The dimensions for the pattern are shown in the second illustration at right.

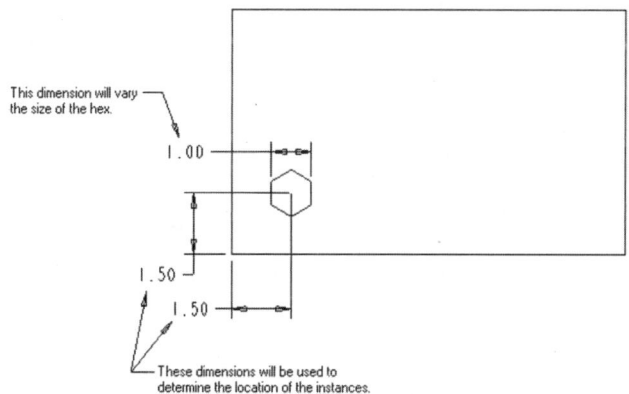

Dimensions used to create the varying pattern.

4. Pro/ENGINEER will prompt you for a second direction. For this example, select **Done**. The next prompt in the message window is for the total number of instances to be in the pattern. Enter *4*, and select **Done**. The pattern will be created, which should look like that shown in the following illustration.

Completed single-direction varying pattern.

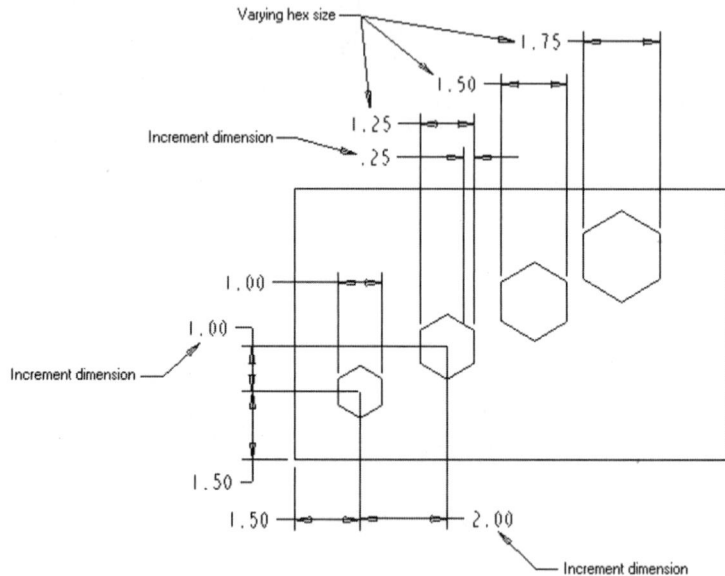

Exercise 10-3: Creating a Two-directional Pattern

This exercise demonstrates what the second direction of a pattern does and how it affects a pattern.

1. Select **Feature** ➡ **Del Pattern** and pick one of the instances of the pattern from the previous exercise to delete the pattern. You can use the block and hexagonal cut from exercise 10-2.

2. Create a two-directional varying pattern. The following illustration shows which dimensions you need to select to create the pattern, as well as whether they are for the first or second direction. Select **Feature** ➡ **Pattern** and select the hexagonal feature. Select **Varying** ➡ **Done**, and use the **Value** option as before. For the first direction, select the horizontal 1.50 locational dimension and enter an increment of *2.00*. Select the 1.00 width of the hexagon, enter *.25* for the increment value, and select **Done**.

Dimensions and
directions for the two-
directional varying
pattern.

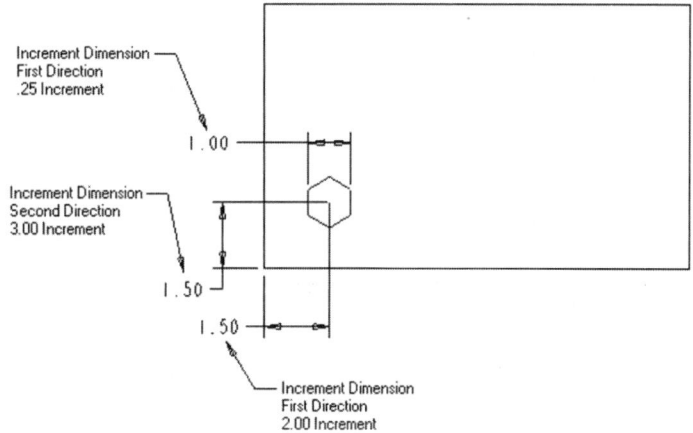

Increment Dimension
First Direction
.25 Increment

1.00

Increment Dimension
Second Direction
3.00 Increment

1.50

1.50

Increment Dimension
First Direction
2.00 Increment

3. After selecting **Done** for the first direction, Pro/ENGINEER will prompt you for the total number of instances in the first direction. Enter *4*. The next prompt is to select the dimensions to be incremented in the second direction. Select the vertical 1.50 locational dimension, enter an increment of *3.0*, and select **Done**. This is the only dimension that will be used in the second direction. When Done is selected, Pro/ENGINEER again prompts for the total number of instances. However, this time it is for the second direction. Enter *2*, and select **Done**. The pattern is created and should look like that shown in the following illustration.

4 Instances in the second direction
are created second

Completed two-directional
varying pattern.

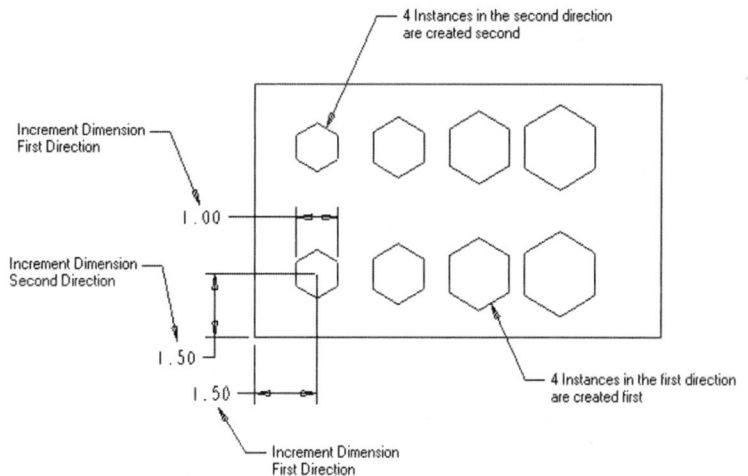

Increment Dimension
First Direction

1.00

Increment Dimension
Second Direction

1.50

1.50

4 Instances in the first direction
are created first

Increment Dimension
First Direction

You can see that the pattern is really created in two steps. The first is to create a pattern in the first direction. In this case, the hexagon is duplicated four times horizontally. Then that pattern is duplicated in the second direction, which in the previous exercise is vertically. Look for this type of pattern in the models you build every day. The pattern is quite common, and using the two-directional pattern is a good way of controlling such patterns.

General Patterns

General patterns can be used to create any type of identical or varying pattern, making it the most versatile of the three pattern types. The price of that versatility, however, is that it is also the slowest to regenerate. Always use the appropriate pattern type for the pattern you are going to create. You must use the General pattern type when instances intersect each other, as shown in the illustration at right. In the following exercise, you will create a pattern where the wedge cut will be patterned in such a way that the cuts will eventually overlap one another, requiring the use of a General pattern.

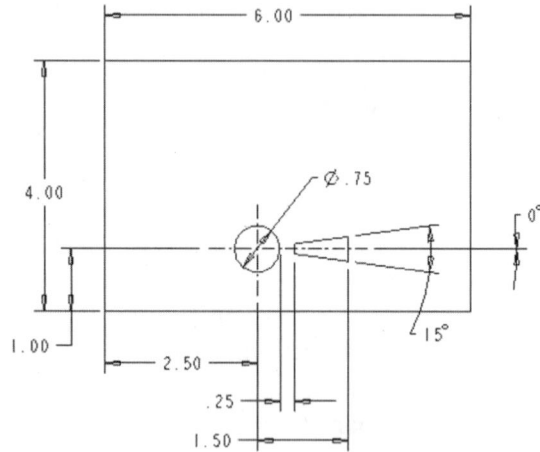

Geometry required for the general pattern exercise.

Exercise 10-4: Creating a General Pattern

1. Create the block and the wedge-shaped cut shown in the previous illustration. Be sure to use a "make" datum or an "on-the-fly" datum when making the wedge cut because this pattern will be rotated.

2. To create the general pattern, select **Feature** ➥ **Pattern** and pick the wedge cut. Select the **General** pattern option and **Done**. Select the dimensions to be incremented by the pattern. These are the angle dimensions for the sides of the wedge (15 degrees in the following illustration). Enter 5 degrees for the

amount of the increment. Select the dimension to the outside edge of the cut (1.50 dimension) and enter .25 for the increment. Select the angle dimension of the "on-the-fly" datum (0 degrees), which will set the rotation of the pattern. Enter *30* for the increment amount.

Dimensions of the wedge-shaped cut that will be incremented by the General pattern.

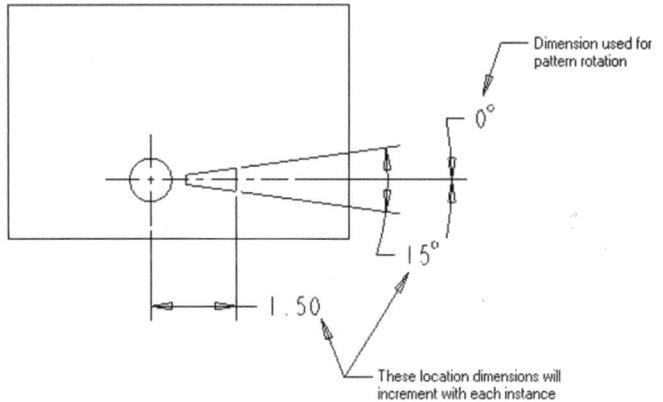

Dimension used for pattern rotation

0°

15°

1.50

These location dimensions will increment with each instance

3. Select **Done** and enter *5* for the total number of instances. At the prompt for the second direction, select **Done** and the pattern will be created. As you can see in the following illustration, the last two instances of the pattern overlap or intersect each other. This is why the General option is required for this pattern.

Completed General pattern.

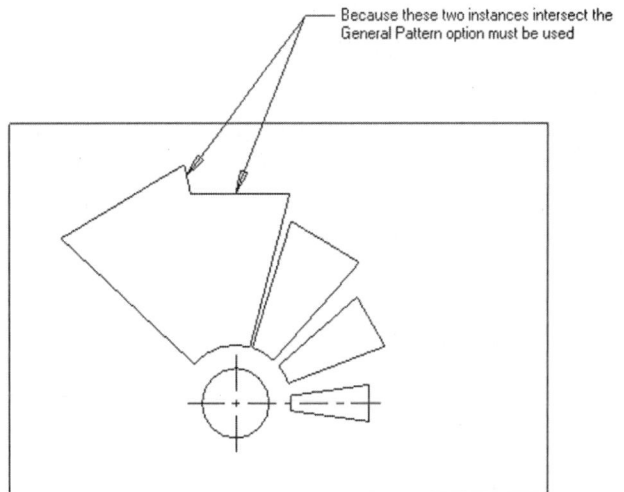

Because these two instances intersect the General Pattern option must be used

Table Patterns

Table patterns differ from all other patterns in the way instance locations are determined. In all of the other patterns, the instance location is determined by an incremental increase of a selected dimension. In a table pattern, the value of the dimension is entered in a separate table file. This difference allows for a great deal of freedom in the location of the instances. In the following exercise, you will use a pattern table to produce a bolt hole pattern around a rectangular cutout. The following illustration shows the geometry required for this exercise.

Base geometry.

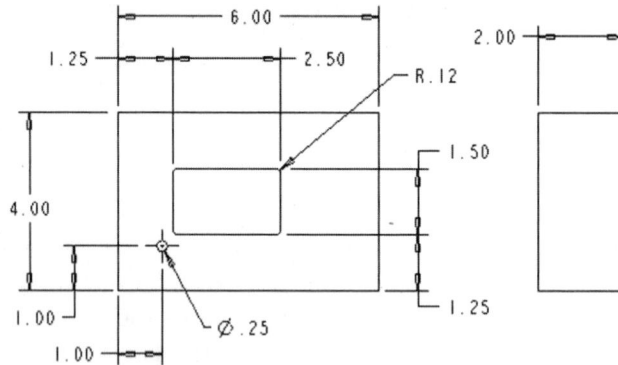

Exercise 10-5: Creating a Table Pattern

1. To create a table-driven pattern, select **Feature ➥ Pattern**, and select the feature to be patterned. In this case it is the single bolt hole. Select the **Identical** option and **Done**. Select the **Table** option from the PAT DIM INCR menu. Pro/ENGINEER now prompts you to select the Driver dimensions. Pick the two bolt hole location dimensions. Select the horizontal 1.00 location dimension first, and the vertical 1.00 dimension second. These dimensions are shown in the following illustration.

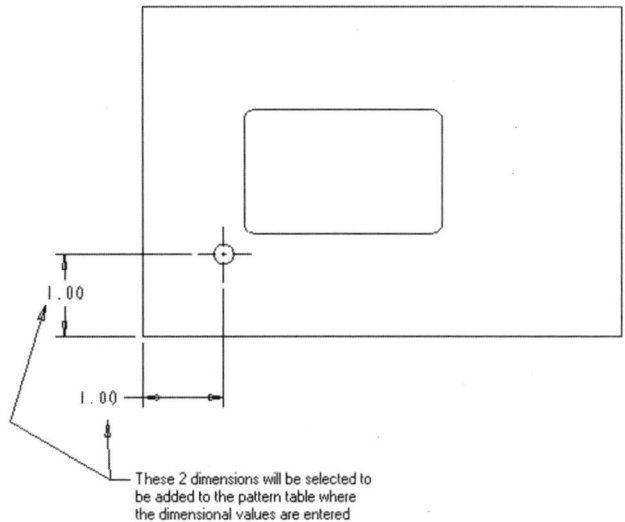

Dimensions to be selected
and added to the table
pattern.

1.00

1.00

These 2 dimensions will be selected to
be added to the pattern table where
the dimensional values are entered

2. Add values to the table. Select **Done** ➡ **Add**. The next prompt is for the name of the table of dimensional values for the pattern. Enter *Bolts*. The Pro/TAB editor will open. This is where the dimensional values for the driver dimensions for each instance will be entered. The table in this exercise requires three pieces of information to be entered. These items are listed in row 9 (R9) in the table shown in the following illustration. The first is the instance number (idx in column 1, or C1). The first instance already exists and therefore does not need to be reentered. Begin with instance number 2 (as shown in R10). After the instance has been entered, the pattern needs the dimensional values for the dimensions that were selected. The column in the table is labeled with the dimension symbol as well as the original dimensional value [d10(1.0000) in C2] for the horizontal dimension and [d11(1.0000) in C3] for the vertical dimension. This information is used by Pro/ENGINEER to place each instance in the model. Enter the information into the table as it appears in the illustration.

Table with information entered to create the pattern of holes.

	C1	C2	C3	C4	C5	C6
R1	!					
R2	! Input placement dimensions for each pattern member.					
R3	! Indices start from 1. Each index has to be unique,					
R4	! but not necessarily sequential.					
R5	! Use `*` for default value equal to the leader dimension.					
R6	!					
R7	! Table name BOLTS.					
R8	!					
R9	! idx	d10(1.0000)	d11(1.0000)			
R10	2	2.0000	1.0000			
R11	3	3.0000	1.0000			
R12	4	4.0000	1.0000			
R13	5	4.0000	2.0000			
R14	6	4.0000	3.0000			
R15	7	3.0000	3.0000			
R16	8	2.0000	3.0000			
R17	9	1.0000	3.0000			
R18	10	1.0000	2.0000			

3. Save and **Exit** the table window. Select **Done** from the Pro/ENGINEER menu and the pattern will be complete, as shown in the following illustration.

Completed table pattern with the dimension of each instance shown.

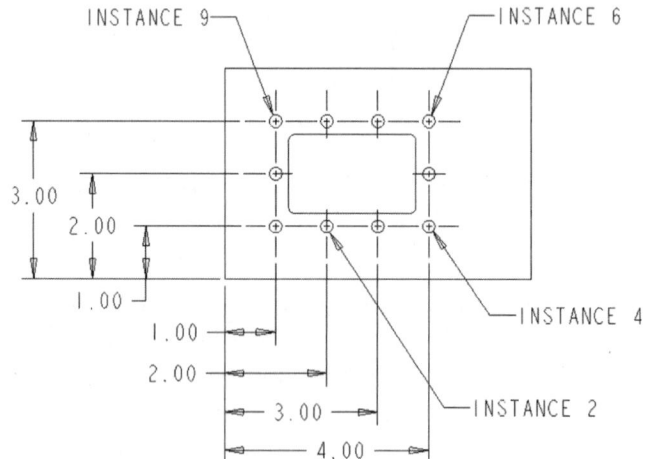

Compare the dimensional values shown for each of the pattern instances with the values in the pattern table. The values in the table are the exact values of the dimensions and not an incremental difference between the original instance and the following instances. This is what makes a pattern table different from any of the other types of patterns.

Mirroring Geometry to Copy Features

Using the Copy and Mirror commands enables you to create copies of features that are mirrored about a mirror plane (datums are used most often, but a part surface can also be used). This might involve a single feature or an entire part, the copy might be dependent on the parent feature or completely independent, and the copy can be modified or redefined without affecting the original feature.

Mirror Choices

Under the Feature ➡ Copy command sequence you can select one or more features or a group of features to copy, or you can select the All Features option. The All Features option will make an independent mirror copy of each feature created up to this point. Modifying any of the copied features will not affect the original features. Another option for creating mirrored geometry, Mirror Geom, is found under the FEATURE menu. This selection differs from Feature ➡ Copy ➡ All Features in that it creates a mirror of all features created up to this point as a single merge feature. Any changes made to the merge feature will also change the original features. The single merge feature cannot be converted into an independent copy.

Independent and Dependent Copies

If your intention is to use mirroring as a quick way of copying some features to be modified later, select Independent when creating the mirror. This will allow you to make any changes to the copied feature without affecting the original. However, if the intent is to create geometry about a line of symmetry, select the Dependent option. Using the Dependent copy option will ensure that the mirrored geometry will always be identical to the original features.

A word of caution: a dependent copy uses the section of the original feature to create its section. Therefore, section entities and dimensions that locate the section to the part cannot be added, deleted, aligned, or unaligned. Existing section entities that are not dimensioned or aligned can be deleted, and new entities can then be sketched and dimensioned.

Exercise 10-6: Creating Base Geometry and Mirroring Single Features

1. Create the base geometry shown in the first illustration at right. Similar geometry will be used in exercise 10-7.

2. Mirror single features. To mirror single features of model geometry, select **Feature ➡ Create ➡ Mirror ➡ Select ➡ Done**, and select the features to be mirrored. Select the feature shown in the second illustration at right.

3. Select **Done Sel ➡ Done** and Pro/ENGINEER will prompt you to select the mirror plane. Alternatively, you can use the Make Datum option to create the mirror plane. The types of geometry that can be selected are datum planes, a model surface, or a surface feature. For this exercise, select the datum plane DTM_X. After you select the mirror plane, the feature will be created, as shown in the following illustration.

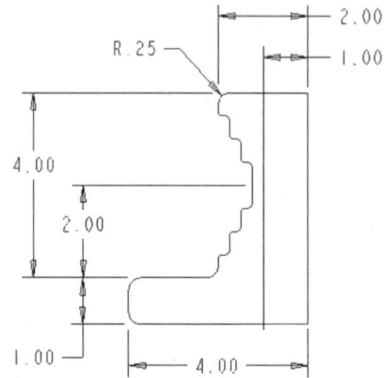

Create this geometry.

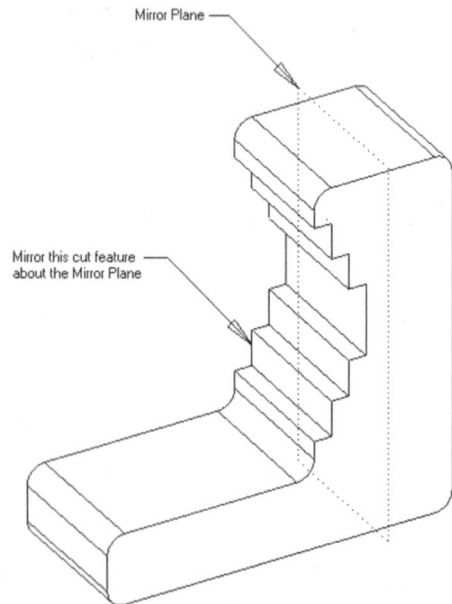

Mirror Plane

Mirror this cut feature about the Mirror Plane

Feature to be mirrored about the mirror plane.

⇥ **NOTE:** *You cannot mirror edge rounds because the reference edge is no longer the same on the mirrored geometry. If rounds are selected as features to be mirrored, they will fail regeneration. You can, however, redefine them to the new edges of the mirrored geometry and the rounds will be created.*

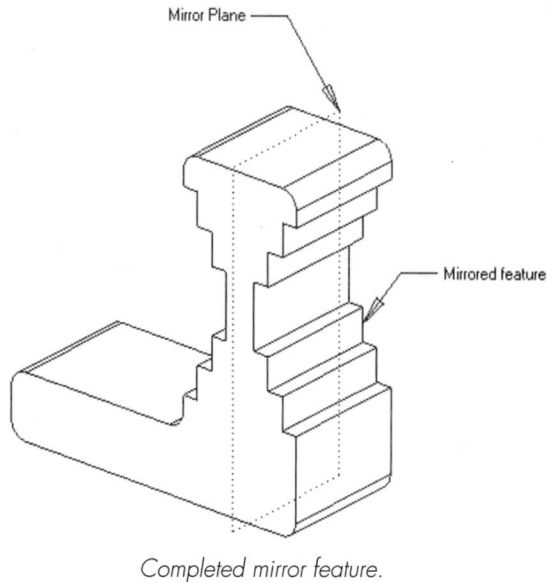

Completed mirror feature.

Exercise 10-7: Mirroring All Features

This exercise demonstrates the differences between mirroring single features and using the All Features option.

1. Make the necessary changes to the model used in the previous exercise to match the model shown in the illustration at right.

2. Select **Feature** ➥ **Copy** ➥ **Mirror** ➥ **All Feat** ➥ **Done**.

3. The next prompt is to select the mirror plane or to use

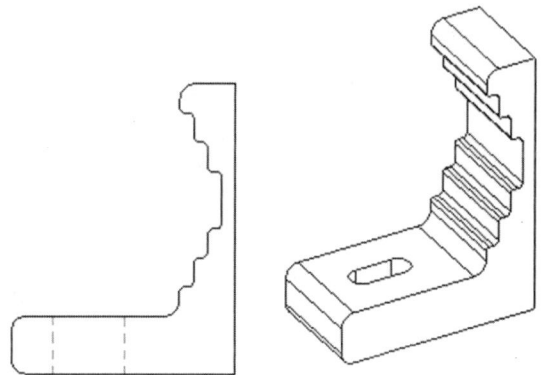

Model for All Features mirroring.

the Make Datum option. The mirror plane is shown in the first of the following illustrations. Select the datum plane and all features will be created about the mirror plane. Notice in the second of the following illustrations that *all* features have been mirrored, including the edge rounds, and that all of the datums have been duplicated.

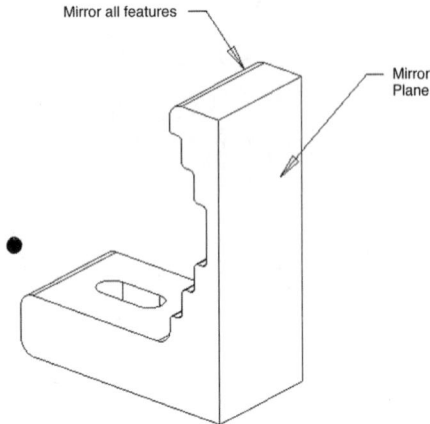

Mirror plane for All Features mirror exercise.

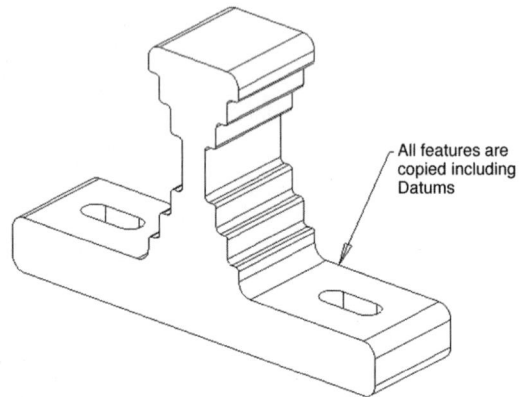

Completed All Features mirror. All features, including edge rounds and datums, were copied.

User-defined Features

User-defined features (UDFs) are similar to local groups in many ways. Both local groups and UDFs must be created from sequential features, and both can be used to treat several features as a single feature when patterning or suppressing and resuming. However, the UDF has the capability to go far beyond these capabilities. Where a local group can only be used in the current model, a UDF can be used to recreate features from one model to another. To be able to do this, the UDF requires more information up front, which is discussed later in the chapter.

UDFs are an excellent means of creating a library of standard features. When creating geometry to be used as a UDF, be very careful about creating unnecessary parent/child relationships. Create as few references as possible, and select stable geometry when a reference is required. Use of the default datum planes for sketching and for sketch orientation planes will simplify the prompts that must be entered when the UDF is created. The same holds true for the dimensioning of the sketches for the UDF features. Be aware of every reference created during this stage of feature creation.

In exercise 10-8 you will create a UDF of a fastener drive, then place the UDF into the model of a second type of fastener. This will require you to create the fasteners shown in the following illustrations.

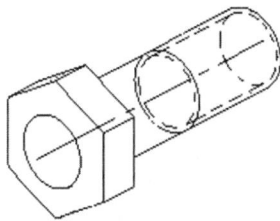

This fastener will receive the UDF in the second part of the exercise.

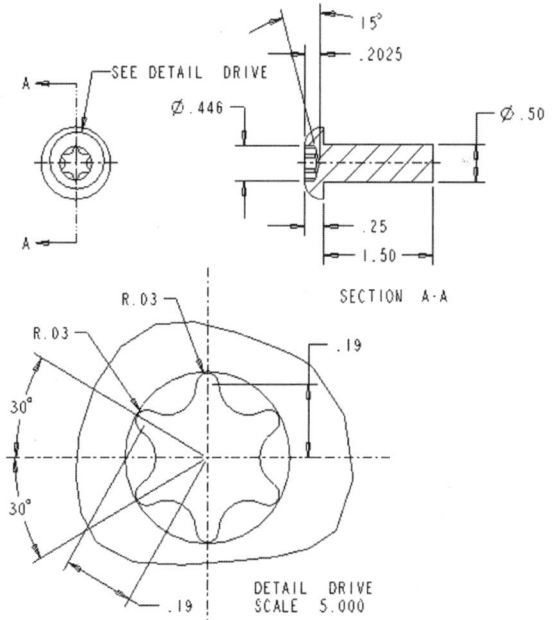

First fastener needed for the UDF exercise with the drive geometry.

Exercise 10-8: Creating a Standard Feature Using a UDF

In this exercise, you will copy the fastener drive geometry as a UDF. In a subsequent exercise, you will place it on another part.

1. To create the UDF, select **Feature** ➡ **UDF Library** ➡ **Create**. Pro/ENGINEER will then ask for the name of the UDF. Enter a name that will be obvious to the next or other users of the UDF as to what the geometry will be. If possible, follow a company CAD or drafting standard when a name is required for any

Pro/ENGINEER geometry. For this exercise, enter *Drive* for the name of the UDF. If there is no standard in place to help guide you, bring it to someone's attention in the department responsible for standards and let them know that a standard is needed.

2. After a name has been entered, there are some options you need to set for the UDF. The first is whether the UDF should be Stand Alone or Subordinate to the parent geometry. The Stand Alone option copies all of the information required to recreate the geometry to the UDF. Therefore, you would not be required to bring the original UDF geometry into session at the time future copies were created. Selecting Subordinate means that the UDF will get the information from the original model geometry whenever the UDF is brought into session. When a subordinate UDF is copied into another part, the original UDF model must also be brought into session so that the UDF is created to the latest geometry. For this exercise, select **Stand Alone**.

3. Select the features that will make up the UDF. Features in a UDF must be sequential and cannot include the base feature or shell features. Using the model tree to select the features can be very helpful; it makes it easy to see just what features are sequential. Pick all of the features required

Features that are required to create the fastener drive geometry are selected to create the User Defined Feature (UDF)

Drive geometry to be selected for the UDF.

to create the drive geometry (as shown in the illustration above), then select **Done Sel ➡ Done ➡ Done/Return**.

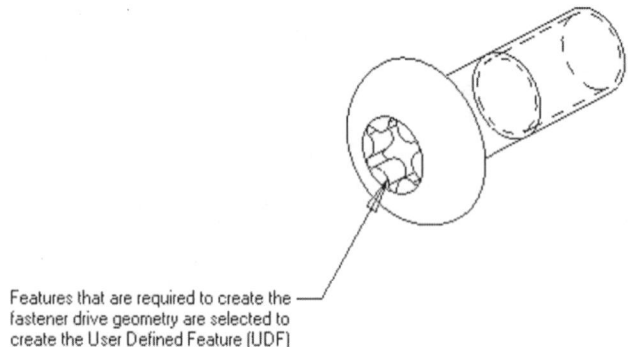

Exercise 10-9: Establishing Reference Features

One of the most critical steps in creating an efficient UDF is creating intuitive prompts for reference features. It is very likely that the same reference will be used more than once during the creation of the features included in the UDF. Therefore, Pro/ENGINEER will ask if you want to use single or multiple prompts for these references.

1. The Single prompt option will allow you to enter a single prompt that will appear only once, and then be applied to all of the instances for which the reference is used. The Multiple option will allow you to enter a different prompt for the same reference each time it is required to create the UDF. Select either **Single** or **Multiple**.

2. Once you have selected Single or Multiple option, you need to enter the prompts themselves. This is where paying attention to references while creating UDF features pays off. It makes writing prompts very easy if you use default datum planes as sketcher and dimensioning references. Simply enter *Select the default datum plane DTM_Y* as the prompt. This works especially well if start parts are used, where there is a consistent naming of the default datums.

Unfortunately, there are no real guidelines on creating good prompts, other than trying to keep them concise and intuitive. If you are having trouble coming up with wording you think will work, ask someone who is not involved in your project what they would call this feature or that feature. Because they are not involved with the part directly, their answer very well may be the most intuitive.

How you created the drive geometry will determine how many and what references will be required. The drive geometry for the fastener used four references. The following table shows the references and the prompts used.

Reference	Prompt
Default Datum Plane DTM_X	The default datum plane DTM_X
Default Datum Plane DTM_Y	The default datum plane DTM_Y
The top surface of the fastener	The top surface of the fastener head
The center axis of the fastener	The center axis of the fastener

3. Pro/ENGINEER will place the word *Select* in front of each prompt when the UDF is being placed. After all prompts for the reference features are entered, Pro/ENGINEER will then allow you to step through each prompt using the **Next** and **Previous** picks. You can review and modify each prompt. A prompt itself can be changed, or the Single/Multiple option can be switched. After you complete the prompt review, select **Done/Return** ➥ **OK**. Look in the message window for the message "Group 'DRIVE' has been stored" and you will know that the UDF is finished.

Exercise 10-10: Placing a UDF in a Model

There are two ways to start the UDF placement operation. The first is to select Feature ➡ Create ➡ User Defined ➡ Retrieve. At this point, the Open dialog box will open allowing you to navigate through the directories or folders to find the name of the UDF you want to place. The second is to select Feature ➡ Group ➡ Create ➡ From UDF Lib. This leaves you at the same point as the first method, and the creation from this point on is the same.

1. Using one of the previously mentioned methods, locate the UDF "DRIVE" and select it from the list.

Before placing the UDF on the model, you first need to set a few options. Independent or UDF Driven is the first option to be determined. The Independent option will copy the required information to the model, and will allow the features of the UDF to be modified. This geometry will no longer be tied to the original UDF, and any modifications to the UDF will not be reflected in the model. When the Independent option is selected, the additional option of Scale will be asked.

The scale options are Same Size, Same Dims, and User Scale. Same Size will create the UDF at the same size as the original, Same Dims will create the UDF with the same dimensional values regardless of units, and User Scale will prompt you for a scale factor to rescale the UDF without any change of units. The UDF Driven option ties the copy to the original UDF. Therefore, any modifications to the original will be made to the copy when the Update option in the GROUP menu is selected.

2. Select **UDF Driven** ➡ **Same Dims** ➡ **Done** for this exercise.

The three options available for the display of the dimension of the UDF are Normal, Read Only, and Blank. Selecting the Normal option means that dimensions will be displayed in any mode (Drawing, Part, or Assembly) and can be selected for modification. The Read Only option dimensions can be displayed but cannot be selected for modification. When the Blank option is selected, the dimensions cannot be displayed or modified in the model or drawing. If the Blank option is used, and the dimensions end up being needed on the drawing, the only way to display them is to delete the UDF and replace it with a new one using the Normal or Read Only option.

3. For this exercise, select **Read Only** ➡ **Done**. The references Pro/ENGINEER will prompt you for are shown in the following illustration.

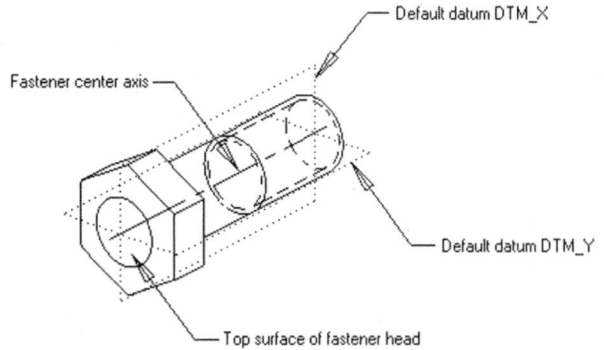

References Pro/ENGINEER will prompt for to place the UDF.

Default datum DTM_X
Fastener center axis
Default datum DTM_Y
Top surface of fastener head

4. Pro/ENGINEER now displays the prompts entered during the creation of the UDF for each of the reference features. Select the geometry as prompted (see the previous illustration). You have the option of skipping a reference, but it will have to be redefined before the UDF can be placed.

5. After the references have been selected, the viewing direction for each of the sketching planes must be determined by selecting Okay or Flip. Select **Okay** for all of the references. The UDF can now be placed and you will be given the options of Showing the Placement, Redefine, or Info. Once the placement of the UDF is correct, select **Done** to accept the UDF location and the UDF will be created, as shown in the following illustration.

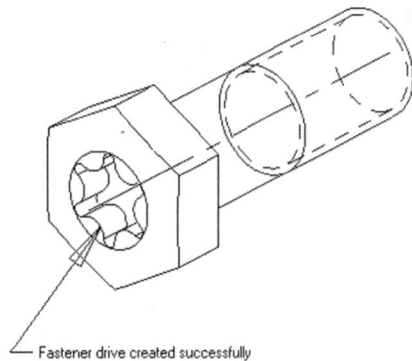

Final placement of the UDF on the new style fastener.

Fastener drive created successfully

Summary

In this chapter, you have learned how to copy features. You first learned about local groups and pattern types, and worked through exercises involving rotational, varying, general, and table patterns. You were then introduced to the concept of mirroring geometry to copy features. You were exposed to exercises in which you mirrored single features of a model and all features of a model. Along the way, you learned about independent and dependent copies. In the last series of exercises, you learned about user-defined features, creating a standard feature using a UDF, establishing UDF reference features, and placing a UDF in a model.

11

Patterning Complex Sweeps

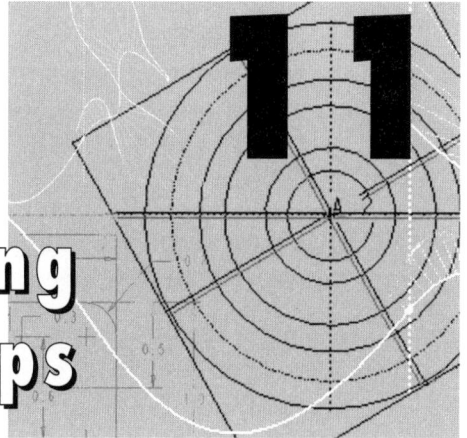

To this point, all of the patterns used in the exercises have been fairly simple 2D patterns. However, patterns can be used with much more complex geometry and in more complex ways. This chapter explores some of these advanced patterning techniques. By patterning a bump feature along a 3D curve, you will experience the use of datum points on a curve, datum curve splits, variable section sweeps, local groups, parameters, and relationships. Gaining a fundamental knowledge of these items will allow you to create virtually any sweep along a complex path.

Design Intent of the Pattern

The result of exercises in this chapter is a part to be used as a flexible protective covering for a machine cooling line. To meet this goal, the following design intent points were determined.

- The length of the tube must be easily modifiable.

- The shape of the curve path must be modifiable.

- The tube's centerline must be 3D.

- The shape of the individual bumps must be modifiable.

- The number of bumps must be able to be patterned.

Exercise 11-1: Creating a Tube Centerline

The design intent calls for a 3D tube centerline. To accomplish this goal, you will create the centerline in phases. The first phase is to create a 2D spline; the second is to create an extruded datum plane. The final phase is to project the spline onto the datum plane. This projected datum curve will be the 3D, flexible centerline of the tube.

1. You first need to create a 2D datum curve. This feature will be a datum curve sketched on the default datum plane DTM_Y. This datum curve will be the feature that will control the length of the tube. To create the curve, select **Feature ➮ Create ➮ Datum ➮ Curve ➮ Sketch ➮ Done**. As previously stated, the curve will be sketched on DTM_Y. Therefore, select it as the sketch plane, select **Okay** for the feature creation direction, select **Bottom** from the SKET VIEW menu, and pick DTM_Z as the sketcher orientation plane.

2. Once in Sketcher mode, sketch two horizontal centerlines. These will be used as the ends of the datum curve. They also provide an entity to dimension tangency of the curve ends to. Sketch a spline by selecting **Sketch ➮ Adv Geometry ➮ Spline**. Use the **Sketch Points** option and set the tangency to **Both**. Pick four points to define the spline, and dimension per the illustration at right.

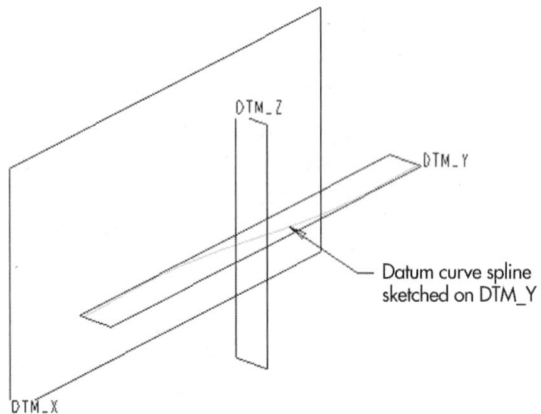

2D datum curve sketched on DTM_Y that will form centerline of tube.

3. Create the extruded surface using a sketch similar to that used to create the 2D curve in step 1 and extruding it perpendicular to the step 1 curve. Select **Feature ➮ Create ➮ Surface ➮ New ➮ Extrude ➮ Done** and use the **One Side** and **Open Ends** options. Select **Done**. Pick DTM_X as the sketching plane and **Flip** the feature creation direction arrow so that it points toward the datum curve created in step 1. Pick DTM_Y as the **Top** sketcher orientation plane.

4. Select the curve end points as sketcher references and sketch a vertical center-line through each of these points. Sketch the spline that will define the shape of the surface by selecting **Sketch ➡ Adv Geometry ➡ Spline**. Use the **Sketch Points** option and set the tangency to **Both**. Pick four points to define the spline and dimension as shown in the first of the following illustrations (the datum curve that will drive the extruded surface). The second of the following illustrations shows the completed surface.

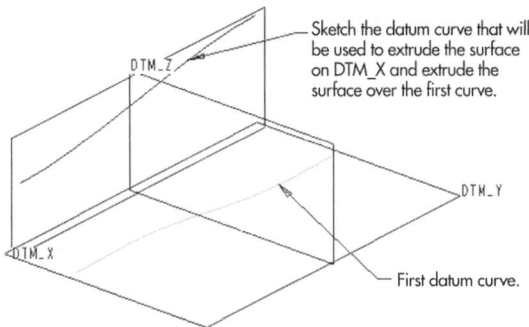

Datum curve that will drive the extruded surface.

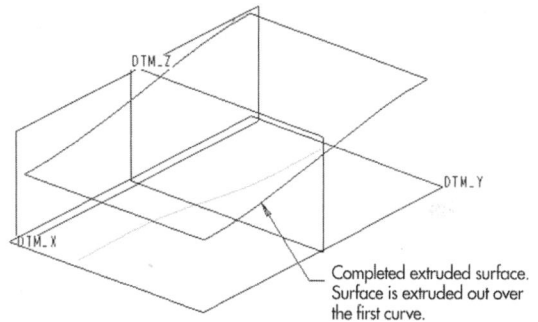

Completed surface.

5. Project the curve onto the surface. To get the 3D centerline curve required by the design intent, you must project the 2D datum curve onto the surface feature. To do this, select **Feature ➡ Create ➡ Datum ➡ Curve ➡ Projected ➡ Done**, and then select the **Select** option and **Done**. The first item you need to select is the datum curve to be projected. Pick the datum curve created in step 1 as the curve to be projected, and then select **Done Sel ➡ Done**. The next item to be selected is the surface the curve will be projected onto. Select the extruded surface and **Done**. Select the projection type **Along Dir**. Select **Done**, then select the **Plane** option and pick DTM_Y as the plane the projection direction will be normal to. The direction arrow needs to point in the direction for the surface (it should be correct as is). When it is, select **Okay** and **OK** to project the curve. You should now have a 3D curve that lies on a surface, as shown in the following illustration. This projected curve is the centerline of the flexible tube.

*First datum curve projected
onto the surface to form the
3D centerline of the tube.*

Projected datum curve that
will be used as the 3D
centerline of the tube.

DTM_Z

DTM_Y

DTM_X

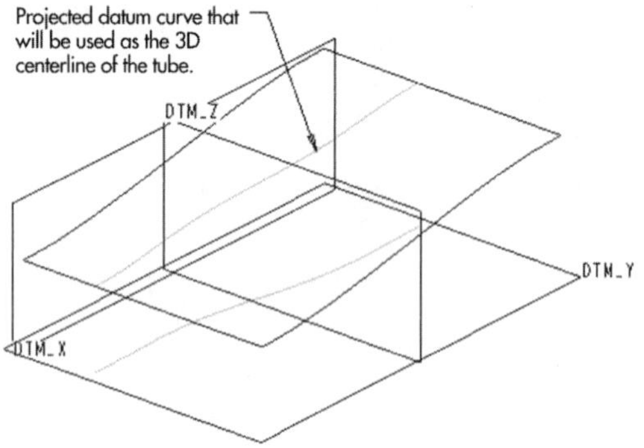

Exercise 11-2: Preparing the Centerline for Bump Creation

Before the bump geometry can be created, three steps need to be completed to prepare the centerline for the bumps. For this exercise, the datum curve needs to be split so that it can be more easily patterned. To do this, datum points need to be placed on the curve to determine where the splits will take place. The first phase of the exercise is to create those datum points. The second phase is to split the curve itself. The final phase is to test the modifiability of the split curve.

1. Create datum points to split the curve. This is the critical phase of the exercise because the way datum points are created and the way the curve splits allow the curve to be patterned so that it will follow the 3D centerline as well as allow for a varying number of splits in the curve. The method required for creating the datum points is the Length Ratio option. It is the length ratio dimensions that will be patterned later on. Create the datum points and see how it fits together later. Select **Feature** ➟ **Create** ➟ **Datum** ➟ **Point** ➟ **On Curve** ➟ **Length Ratio** and pick the centerline curve and then **Done Sel**. Pro/ENGINEER now prompts for the ratio. The ratio for the place where the curve was selected is the default value. Enter *.100*. To create the second point, select **On Curve** ➟ **Length Ratio** and pick the curve a second time, entering a ratio of *.200*. Select **Done** ➟ **Quit** to exit the DATUM POINT menu. The following illustration shows the two datum points created on the centerline curve.

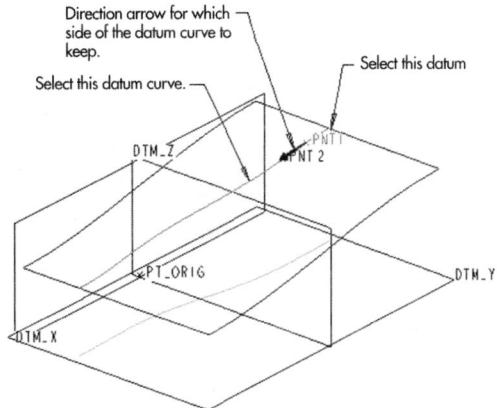

Datum points that will be used to split the centerline curve.

2. Split the curve. The bump feature will be created between the two datum points created in the previous step, and will only require the curve segment between the two points. To split the curve, select **Feature ➥ Create ➥ Datum ➥ Curve ➥ Split ➥ Done**. Then pick the 3D-tube centerline curve and one of the datum points (see the following illustration). An arrow will appear and Pro/ENGINEER will prompt you to choose which part of the datum curve to keep. Use the **Flip** option to get the arrow to point in the direction of the other datum point, then select **Okay ➥ OK**. The curve will be "trimmed" at the datum point.

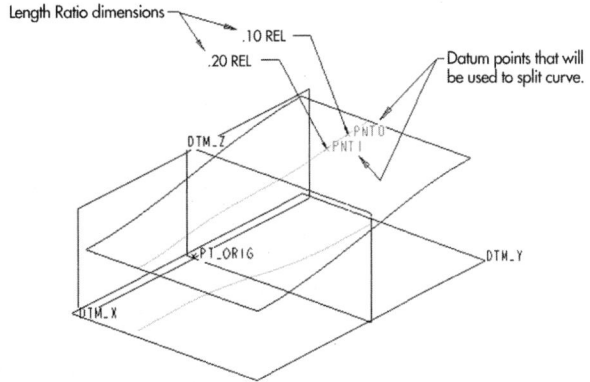

Direction in which the arrow must point to properly split the curve.

3. Repeat the procedure to split the curve at the second point. Select **Feature ➥ Create ➥ Datum ➥ Curve ➥ Split ➥ Done**. Then pick the 3D tube centerline curve again and the second datum point. Use the **Flip** option to force the arrow to point toward the first datum point and select **Okay** ➥ **OK**. After splitting the datum curve, only the short section between the datum point will remain, as shown in the following illustration.

Centerline curve after the splits have been made.

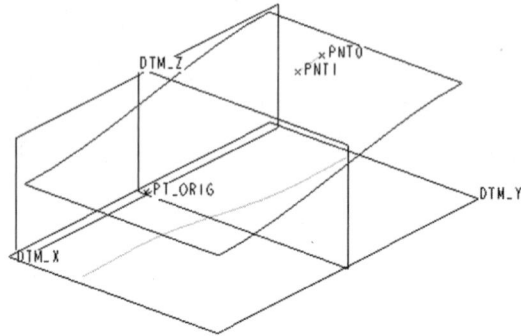

4. Test the split. When the bumps are created, they will be tied to the datum points. The length ratio dimensions of the datum points are the dimensions that will be placed in the pattern. This is how the bumps will be patterned along the 3D centerline curve. To ensure the pattern will work successfully, you can modify the length ratio dimensions of the points to guarantee that they, and the section of centerline curve between them, move up and down the entire length on the centerline curve.

5. Modify the length ratio values by selecting **Modify**, and pick the point shown in the following illustration. Modify the value from .100 to *0*. Do the same to the second point but modify its value from .200 to *.100*. The distance between the points remains the same but the position of the points and the shape of the curve changes. Now modify the values from 0 and .100 to *.5* and *.6*, respectively. Modify the points to the far end of the curve by changing the values to *.900* and *1.000*. Now that the curve has been tested and proved, return the values to *0* and *.100* so that you can move on to the next phase, which is creating the bump geometry.

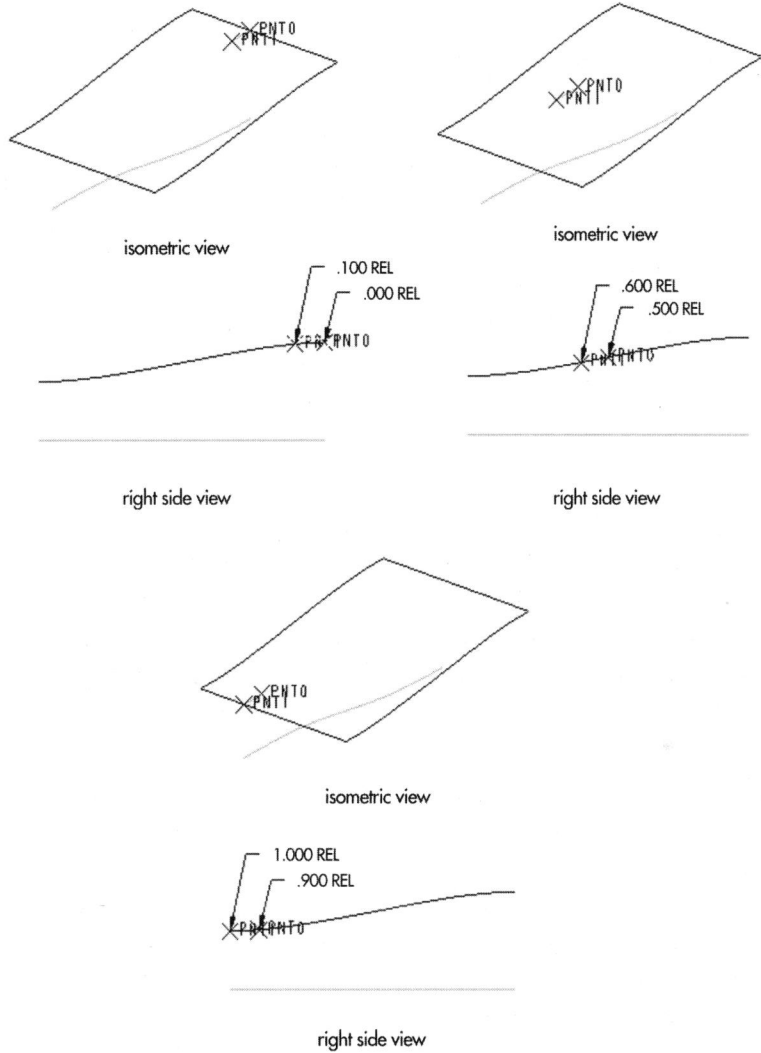

isometric view

isometric view

.100 REL
.000 REL

.600 REL
.500 REL

right side view

right side view

isometric view

1.000 REL
.900 REL

right side view

Position of the two datum points changing but following the original 3D centerline trajectory as values are modified.

Exercise 11-3: Creating the Bump To Be Patterned

In this exercise, you will create the bump of the flexible tube that will be patterned. The bump protrusion will be created with a variable section sweep. This type of feature gives you control of the bump shape by varying the diameter of the section as it moves along the 3D tube centerline. However, the length of the bump changes depending on how many bumps are patterned along the centerline.

How can you vary the length of the sweep to move with the number of bumps? You accomplish this by driving the sweep with a graph feature and then tying the graph feature to the sweep with a relation. This relation will equate the length of the sweep with the length of the graph (X coordinates). The relation also sets the diameter of the bump by reading the height of the graph (Y coordinates) as it moves along the length of the graph. Because the graph feature is used by the sweep, the graph must be created before the sweep or reordered to come before it. Therefore, this exercise involves two phases. The first is to create the graph feature, and the second is to create the variable section sweep.

1. Create the graph feature that will drive the bump shape. Select **Feature** ➡ **Create** ➡ **Datum** ➡ **Graph** and enter the name *Bumps* for the graph. You will now be in Sketcher mode, in which you must first create a Sketcher coordinate system. Because the graph works from X and Y coordinates, it requires the coordinate system. Next, sketch a series of horizontal and vertical centerlines, as shown in the following illustration. These centerlines aid in defining the symmetry of the sketch and provide an entity to dimension tangency of the spline to.

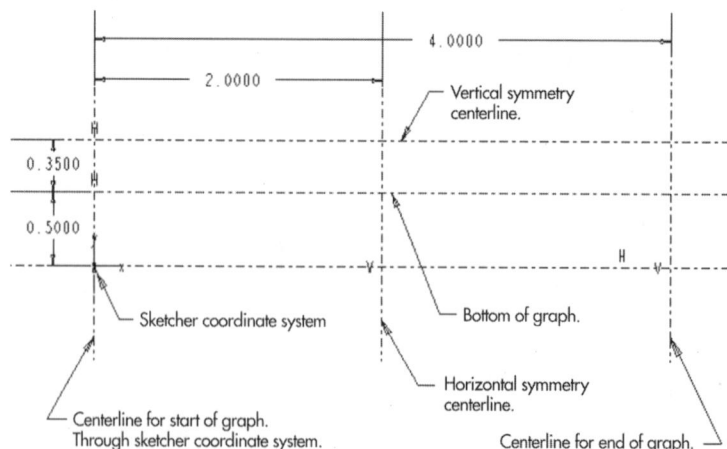

Construction geometry needed for the graph feature sketch.

4.0000

2.0000

Vertical symmetry centerline.

0.3500

0.5000

Sketcher coordinate system.

Bottom of graph.

Horizontal symmetry centerline.

Centerline for start of graph. Through sketcher coordinate system.

Centerline for end of graph.

2. Sketch the spline with tangency set to **Both** ends and dimension it as show in the following illustration. Select **Adv Geometry** ➥ **Spline** ➥ **Both**. The spline will consist of three points, with the center point on the horizontal symmetry centerline.

Completed graph feature sketched section.

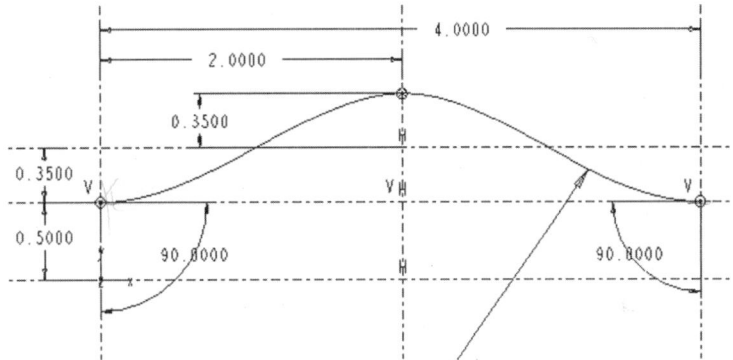

Sketched spline Y coordinate value will drive the diameter of the Variable Section Sweep protrusion. Diameter will start at .500 and grow to 1.200 at the center and back down to .500 at the end.

3. With the graph feature complete, you will need to know the symbolic dimension name for the overall length dimension of the graph. This is so that you can use the symbolic dimension name in the relation that will tie the graph to the variable section sweep protrusion. To do this, select the **Modify** option from the PART menu. A separate window will open, with the graph feature displayed. Select the INFO menu from the pull-down menu bar and select the **Switch Dims** option, shown in the following illustration. Make a note of the symbolic dimension name of the 4.00 overall length of the graph dimension. An example of a symbolic name is d20.

Symbolic dimension name for the overall length will be needed for relations later on.

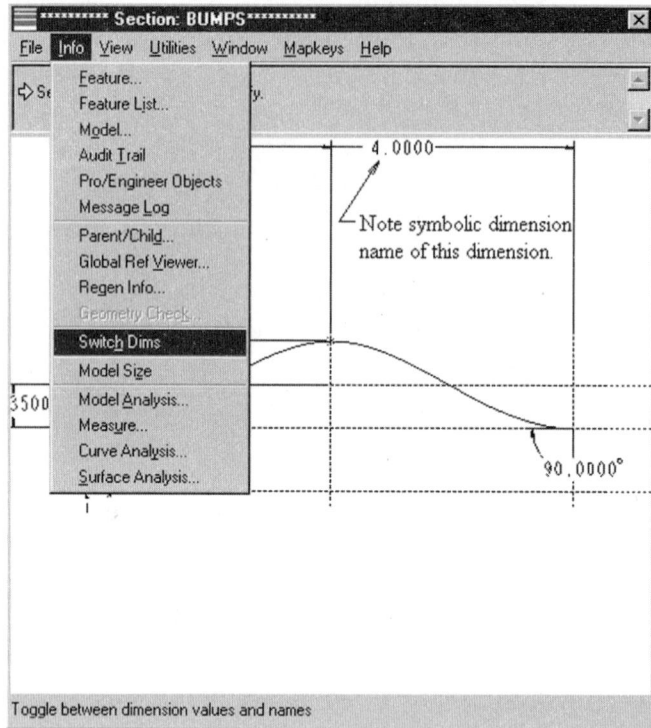

✗ **WARNING:** *Do not use* **File** ➥ **Exit** *from the pull-down menu at the top of this separate window. It will exit Pro/ENGINEER completely. Use* **Done** *from the MODIFY menu.*

4. Create the bump. To create the variable section sweep, select **Feature** ➥ **Create** ➥ **Protrusion** ➥ **Advanced** ➥ **Done** ➥ **Var Sect Swp** ➥ **Done** ➥ **NrnTo-OriginTraj** ➥ **Done** ➥ **Select Traj**. Pick the small section of the 3D centerline curve left after the split (see the following illustration) and select **Done Sel**. The start point of the sweep needs to be at PNT 0. Select **Done** when the start point is set. The next item Pro/ENGINEER requires is the trajectory for the horizontal vector.

➥ **NOTE:** *See Chapter 9 for an explanation of how this trajectory works.*

5. Select the **Select Traj** option and pick the original 2D curve for the horizontal vector trajectory; then select **Done Sel** ➥ **Done** ➥ **Done** ➥ **Origin Start**.

6. The view will reorient itself and place you in Sketcher mode. Sketch a circle and modify its value to *.50*. Because this tube needs to be hollow, sketch a second circle inside the first and dimension it with a thickness dimension (shown in the relation that follows). Dimensioning the sketch this way will keep a constant wall thickness as the diameter of the bump varies. Now that the sketch is complete, write a Sketcher relation that will tie the graph feature BUMPS to the swept protrusion. Select **Relation** ➥ **Add** and enter the following relation.

```
sd4=evalgraph ("BUMPS", trajpar*d20)
```

In this relation, *sd4* is the symbolic dimension name for the diameter of the circle in the variable section sweep. *Evalgraph* means to evaluate the Y coordinate of the spline sketched in the graph feature. "BUMPS" tells the relation what graph to evaluate. *Trajpar*d20* determines the X coordinate value along the graph for which the relation is to evaluate the Y coordinate. *Trajpar* is a value between 0 and 1.00; that is, it is a percentage, if you will, of the total length of the graph when multiplied by the length of the graph dimension *d20*.

In the end, the relation works as follows: as the section of the sweep moves along the trajectory, it goes back to the graph feature *BUMPS*, where it uses *trajpar*d20* to determine where the section is along the graph and evaluates the Y coordinate value at that location. It then applies that Y coordinate value to the dimension that controls the diameter of the circle in the sweep section. The first of the following illustrations shows the sweep sketch section. The second of the following illustrations shows the completed bump protrusion.

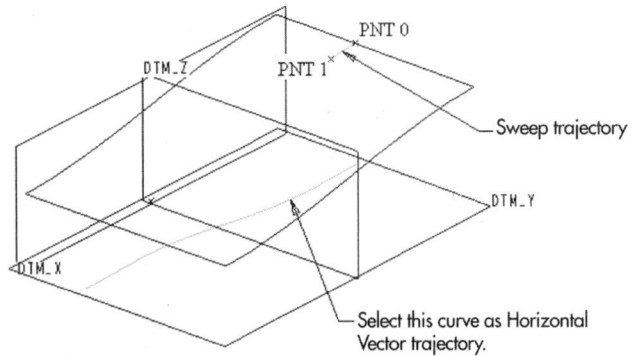

Variable section
sweep sketch
section.

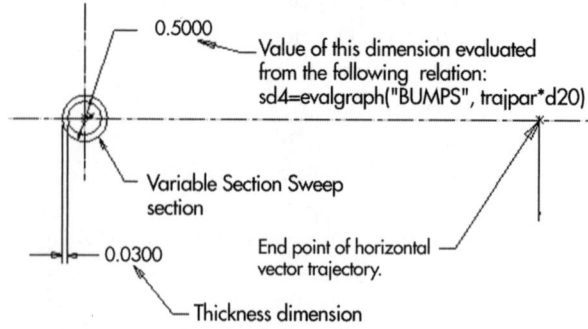

0.5000 — Value of this dimension evaluated
from the following relation:
sd4=evalgraph("BUMPS", trajpar*d20)

Variable Section Sweep
section

End point of horizontal
vector trajectory.

0.0300

Thickness dimension

⊷ **NOTE:** *The sd4 and d20 symbolic dimension names apply only to the model used in this section. Yours will likely be different.*

Completed bump
protrusion feature.

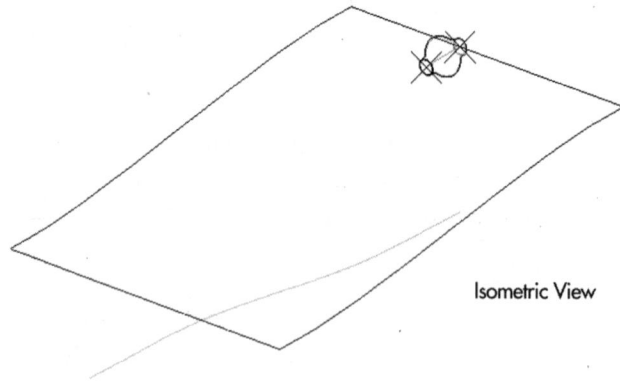

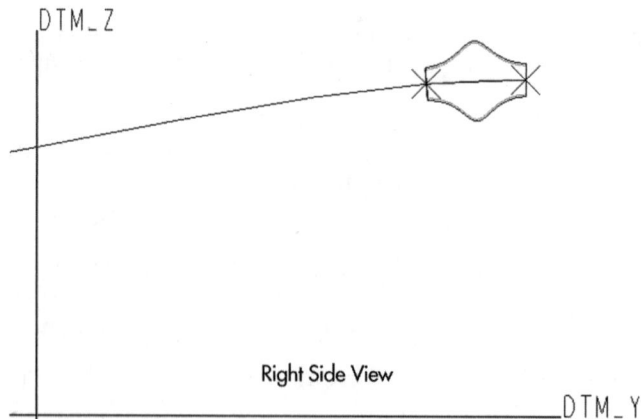

Isometric View

DTM_Z

Right Side View

DTM_Y

Exercise 11-4: Creating the Pattern of Bumps

In this exercise, you will create the pattern for the bump feature. This pattern is more complex than the pattern exercises in Chapter 10 in that several features need to be selected in order for the geometry pattern to work. Because you cannot pick more than one feature to pattern (you can pick more than one dimension, but not more than one feature), the features needed to create the pattern must be made into a "local group" so that they can be selected as a single feature. The first phase of the exercise is to create a local group of the features in the pattern. The second is to create the actual pattern. In exercise 11-5, you will set up some relations that will help drive the pattern in a predictable way.

1. Create a local group of the features to be patterned. The features need to be grouped so that they can be selected as a single feature that can be patterned. To create a local group, select **Feature ➥ Group ➥ Local Group** and enter the name *BUMP_PATTERN* for the group. You will then be prompted to select the features of the group. Select the datum points used to split the curve, the two split

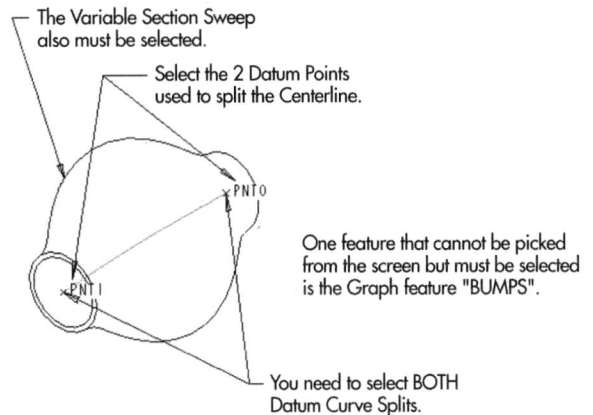

The Variable Section Sweep also must be selected.

Select the 2 Datum Points used to split the Centerline.

PNT0

PNT1

One feature that cannot be picked from the screen but must be selected is the Graph feature "BUMPS".

You need to select BOTH Datum Curve Splits.

Features that need to be selected to create the local group BUMP_PATTERN.

curve features, the graph feature BUMPS, and the bump protrusion (see the illustration at right). Then select **Done Sel ➥ Done** from the FEATURE SEL menu and **Done/Return ➥ Done**.

2. Pattern the group. In this step, you will pattern the group created in the previous step, which will make the flexible tube look like one. Select **Feature ➥ Group ➥ Pattern** and pick the local group BUMP_PATTERN. Next, select the dimensions that will be incremented by the pattern. These dimensions will be the two Length Ratio dimensions used to create the datum points that split the curve (see the following illustration).

3. Enter *.100* for the increment amount of each of the dimensions and then select **Done**. Pro/ENGINEER will prompt you to enter the number of total instances. Enter *10* and select **Done**. Select **Done** for the dimensions in the second direction prompt. The pattern will be created, which should look like that shown in the following illustration.

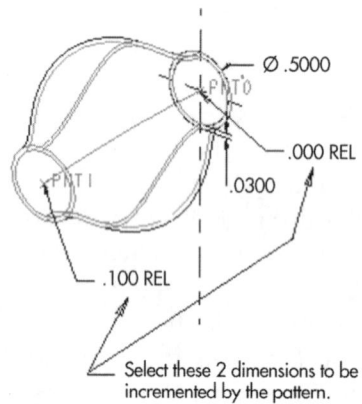

Dimensions to be incremented by the pattern.

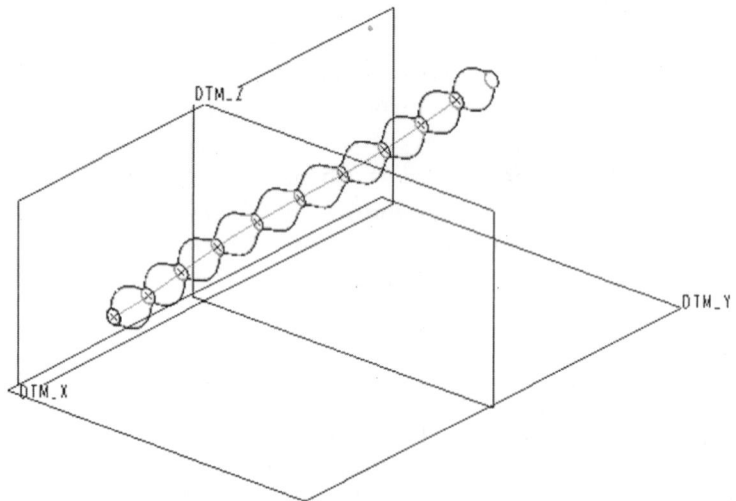

Completed pattern.

The pattern at this stage looks pretty good, and if the design intent were to change you would no doubt be able to make the necessary changes to match the new design intent. However, what if it was some time down the road or you were to hand off this model to someone else to make the changes? What do you think the chances are that all of the required changes will be made the first time? Exercise 11-5 is going to help guarantee they will.

This is not to say that the values will be guaranteed to work, only that all of the required values will be changed. The way you will do this is to create some parameters that you can assign values to and then write some relations that will tie the necessary dimensions together so that the pattern will have the best chance to regenerate successfully.

Exercise 11-5: Creating the Parameters

Two parameters need to be created: one for the number of bumps and one for the length ratio for the second datum point used to split the curve.

1. Create the parameters. Select **Set Up** ➡ **Parameters** ➡ **Part** ➡ **Create** ➡ **Real Number** and enter the name *NUMBER_BUMPS* for the first parameter. Enter *10* for the value. Now select **Real Number** from the ADD PARAM menu, enter the name *BUMP_RATIO* for the second parameter, and enter a value of *.100*. The value of the BUMP_RATIO parameter is 1/"the total number of bumps."

2. Create the relations that will drive the pattern. Create and enter the following five part relations by selecting **Relations** ➡ **Part Rel** ➡ **Edit Rel**. **Save** and **Exit** the relations edit file.

- number_bumps = p0

- bump_ratio = 1/number_bumps

- d14 = bump_ratio

- d28 = bump_ratio

- d29 = bump_ratio

Here, *p0* is the symbolic dimension name for the number of pattern instances. It sets the value for the parameter you created earlier. The value for the parameter BUMP_RATIO is set by the relation number_bumps = p0 above it. With the value for BUMP_RATIO defined, the values for the symbolic dimensions d14, d28, and d29 can be set. *D14* is the length ratio dimension used to create the datum point PNT 1. *D28* and *d29* are the increment dimensions for the pattern. These three dimensions guarantee that the curve will be split into the proper length segment (d14) and that each instance of the pattern will be the proper length (d28 and d29) for any number of instances entered (p0). The dimensions used in the relations are highlighted in the following illustration.

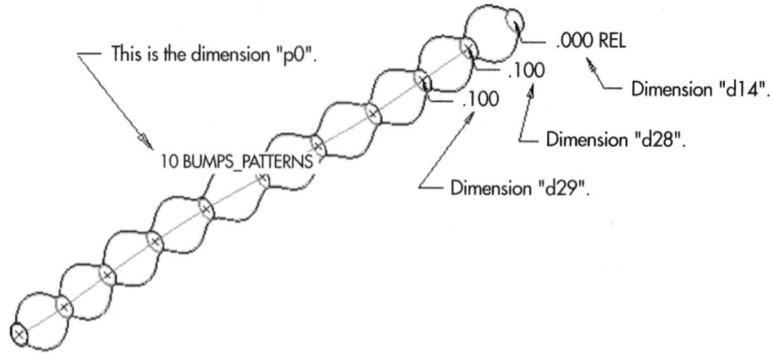

Dimensions that need to be selected for the relations that control pattern flexibility.

Modifying Design Intent

The model is now complete and meets the design intent, but changes will inevitably have to be made. This section discusses some of the changes you will likely see. The first is the need to change the number of bumps. To modify the number of bumps, you would select one of the bumps and modify the value for the number of bumps and regenerate.

You will notice that the bumps change shape as the number of bumps is changed. They become more stretched out with fewer pattern instances, and more compressed as the number of instances goes up. This is because the diameters of the bumps remain the same but the bump protrusion feature is created over a longer length when the pattern instance number is reduced. The following illustration shows various design intent models for the tube.

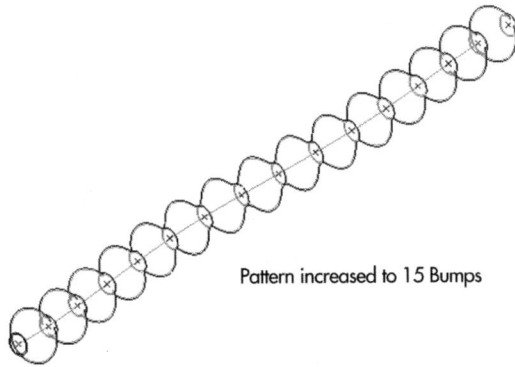

Pattern increased to 15 Bumps

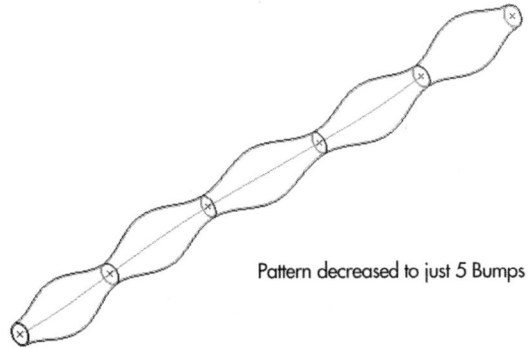

Pattern decreased to just 5 Bumps

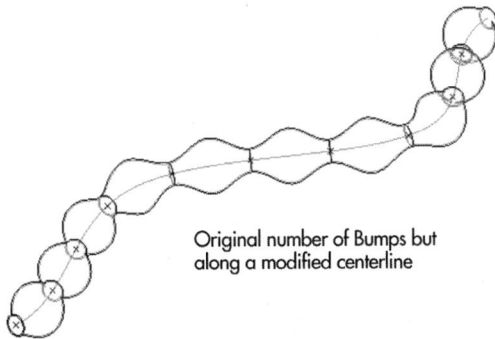

Original number of Bumps but
along a modified centerline

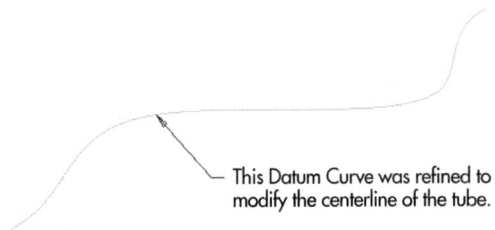

This Datum Curve was refined to
modify the centerline of the tube.

Various design intent models of the flexible tube.

The 3D centerline can be changed two ways. The first is to redefine the first
2D datum curve created, and the second is to modify the shape of the curve
that creates the extruded surface the 2D curve is projected onto. The following illustration shows which features need to be modified to change the 3D
centerline of the tube. In addition to being able to change the centerline of
the tube, the first 2D curve also controls the length of the centerline.

These curves can be redefined to modify the 3D centerline of the tube.

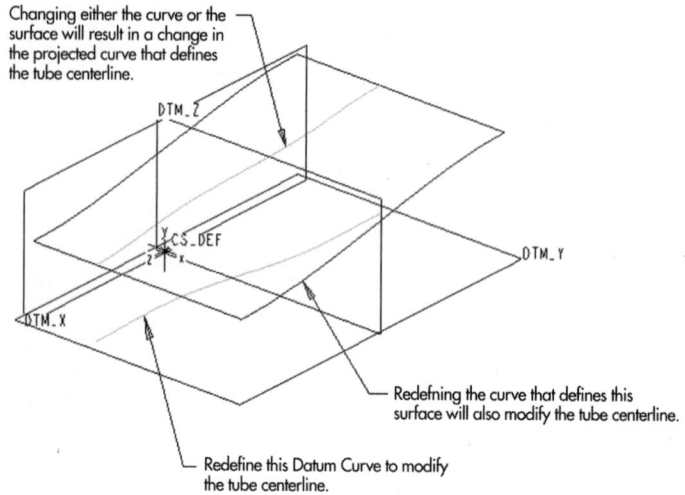

Changing either the curve or the surface will result in a change in the projected curve that defines the tube centerline.

DTM_Z

CS_DEF

DTM_Y

DTM_X

Redefining the curve that defines this surface will also modify the tube centerline.

Redefine this Datum Curve to modify the tube centerline.

Two features are used to control the shape of the bumps. The graph feature (shown in the following illustration) controls how the bump "flows," or how quickly the bump rises to the maximum diameter. It also determines the value of the maximum diameter. The other is the variable section sweep protrusion that creates the bump. The minimum diameter and the material thickness are controlled by this swept protrusion.

The graph feature BUMPS and the variable section sweep control the shape of the bumps.

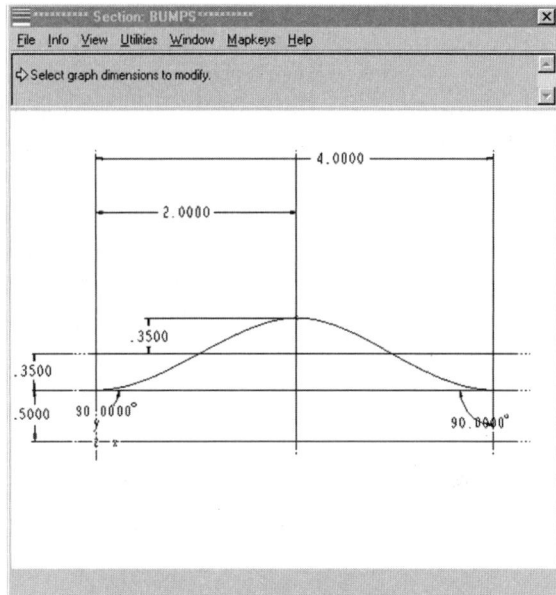

This is the window that will open when the Graph feature BUMPS is selected to be modified.

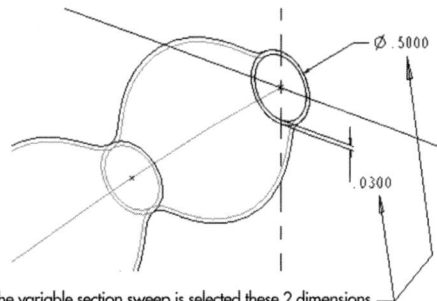

When the variable section sweep is selected these 2 dimensions will appear. Select the .5000 dim to change the start and finish diameters and the .030 dimension to change the material thickness.

Summary

This chapter demonstrated how to pattern a complex shape that required several features to be grouped in a local group to make the pattern successful. In addition, you created datum curves, used them to create surfaces, and projected them onto a surface. You also created datum curve splits. You used a variable section sweep to create a bump, and used a graph feature to drive the shape of the bump. You also used parameters and relations to help control the pattern itself. Perhaps the most important thing you did in this chapter was to create a model that was flexible and robust enough to accept a wide range of changes in design intent.

Datum Curves

12

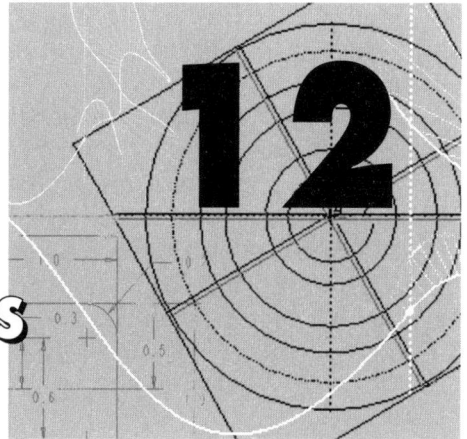

Datum curves serve several valuable purposes during the part design process in Pro/ENGINEER. Curves can be used as a tool for capturing design intent, a handle for feature modification, and a support structure for model geometry creation. Datum curves can, however, be overused, which results in models cluttered with excess geometry. Datum curves are what used to be the "2.5D" CAD of the not too distant past. Those were the days of wireframe and surfaces only.

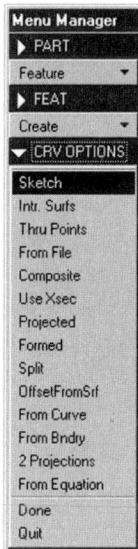

Datum Curve Basics

In the solid modeling world of today, some of the reasons for creating wireframes are still valid. Not too long ago, in wireframe modeling you would create several cross-sectional curves and then stretch surfaces over the wires. With solid modeling, the user now employs a datum curve to capture specific section information as a map of design intent. Throughout this chapter you will see examples and complete exercises that will develop your skills in the use of the Datum Curve feature. Pro/ENGINEER has provided the designer with a wide variety of datum curve types, which are listed in the CRV OPTIONS menu, shown in the illustration at left.

Menu Manager
- ▶ PART
- Feature ▼
- ▶ FEAT
- Create ▼
- ▼ CRV OPTIONS
- Sketch
- Intr. Surfs
- Thru Points
- From File
- Composite
- Use Xsec
- Projected
- Formed
- Split
- OffsetFromSrf
- From Curve
- From Bndry
- 2 Projections
- From Equation
- Done
- Quit

CRV OPTIONS menu.

➥ **NOTE:** *For an explanation of each menu option, and some additional examples of datum curves, see the* Pro/ENGINEER Part Modeling User's Guide, *Release 20.0, Chapter 3, pages 17 through 52.*

In this chapter you will be focusing on the Sketch, Projected, and From Equation options. The exercises that follow take you through hands-on examples of using these three options. Exercise 12-1 deals with sketched curves. Exercise 12-2 explores projected curves and patterns. Exercise 12-3 covers curves created with the From Equation option, as well as sine curves and wave springs.

Exercise 12–1: Using Sketched Curves

The basis of all geometry, except that imported into Pro/ENGINEER, is the 2D sketch. Previous chapters have familiarized you with sketcher functionality and have shown you how to set up design intent and create splines and conics. In this exercise, you will be working through the process of capturing the design intent of a model in which the external geometry is driven as a wall thickness. This is typically the case when working with engine castings and flow path critical geometry. The following illustration shows an image of a part that, because of internal clearance criteria, has internal geometry that drives the design.

For this exercise, you will be building the housing cover shown in the first illustration on the following page. The cover's internal perimeter and flange must align to a mating housing, and all geometry for the cover is defined by wall thickness.

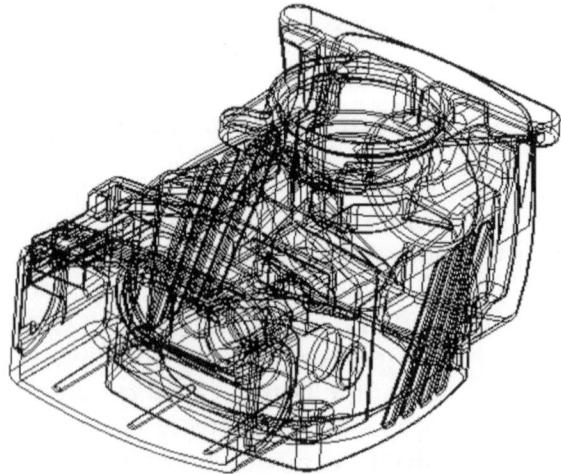

Internal critical geometry.

The drawing views shown in the illustration below are representative of the type of information desired at the end of this part modeling process.

The cover part you will be building could be easily built using the Shell feature within Pro/ENGINEER. However, here you will be working with a technique that comes in very handy on more complex casting models.

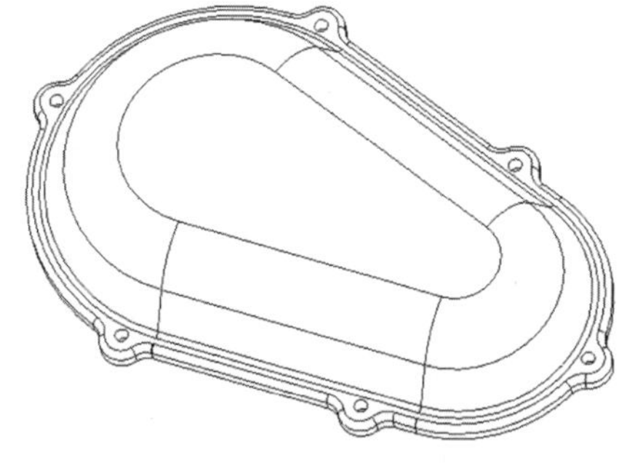

Housing cover model.

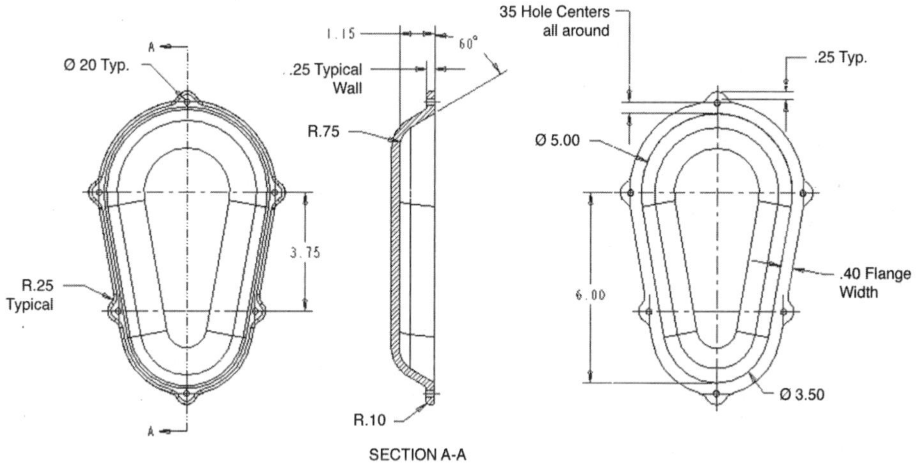

Cover drawing data.

1. As usual, start this model with **Default Datum Planes**. Next you will be sketching a datum curve that will contain the dimensions you need to show in the detailed drawing. The curve will also be used as a reference and parent for the flange and mounting holes, capturing the design intent. From the PART menu, select **Feature ➥ Create ➥ Datum ➥ Curve ➥ Sketch**; then set up a sketching plane and create the section shown in the following illustration.

2. Once the sketch is complete, select **Done** and then pick **OK** from the Curve dialog box. Sketch the datum curve to be used to locate the flange mounting bosses and holes. Again, select **Feature** ➡ **Create** ➡ **Datum** ➡ **Curve** ➡ **Sketch** from the PART menu, pick as your sketch plane the plane you used for the previous curve, and create the six 0.20-diameter circles in the section shown in the following illustration.

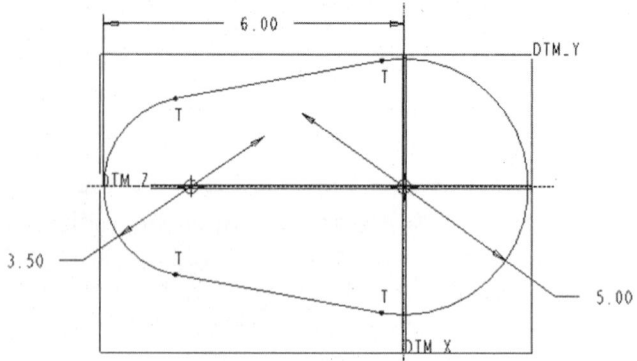

Base curve section.

Flange curve section.

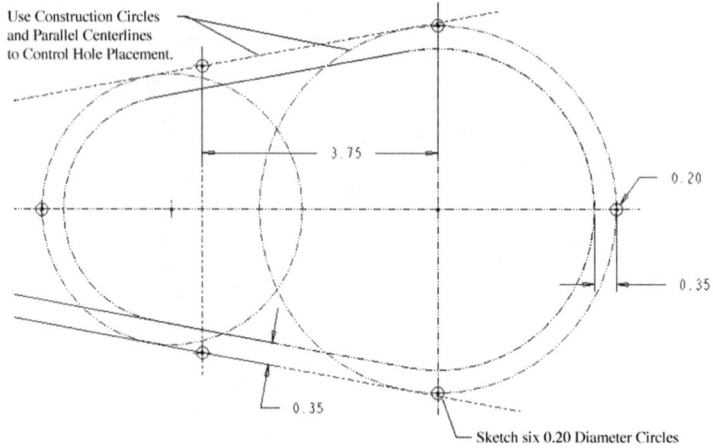

3. When you have completed the sketch, select **Done** and then pick **OK** from the Curve dialog box. You now have two 2D datum curve features in space. The two curve features should look similar to those shown in the following illustration.

4. Because you have created these curves for the purpose of driving part geometry, you should name them. By naming these two features, you are telling the next user of this part that these two features are important. From the PART menu, select **Set Up** ➡ **Name** ➡ **Feature**, pick the first curve you created and name it Base_Curve, and pick the six-circle curve and name it Flange_Curve.

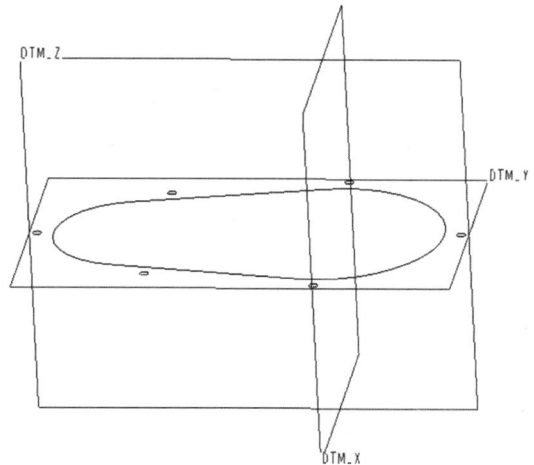

Base and flange curves.

5. Create the main cover body as an **AddInnFcs Sweep**. You will be using the Base_Curve as the trajectory for this feature. Select **Feature** ➡ **Create** ➡ **Sweep** ➡ **Solid** ➡ **Done** ➡ **Select Traj** ➡ **Curve Chain**, pick the Base_Curve, and select **Select All** ➡ **Done** and then **AddInnFcs** ➡ **Done**. The system will now orient the model to the sketching plane view. Sketch the section shown in the illustration at right.

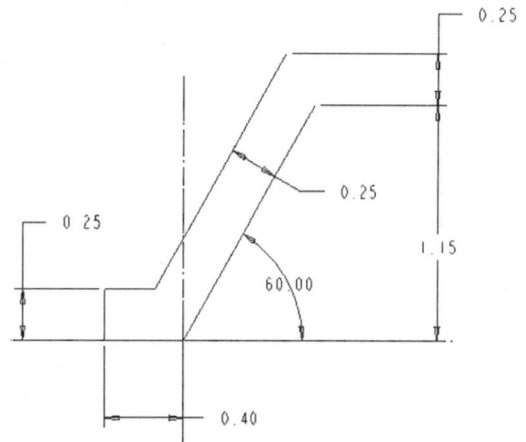

Sweep section sketch.

6. When the section is complete, pick **OK** from the Protrusion dialog box to finish the feature.

The part should look like that shown in the following illustration.

7. Use the Flange_Curve feature to add the geometry for the six mounting bosses. The mounting bosses will be a straight protrusion consisting of six circles. Enter a **Blind** depth of *2* to complete the feature. The section for the boss feature is shown in the second illustration at right. You will be controlling the boss size as a tangent dimension from one of the Flange_Curve circles.

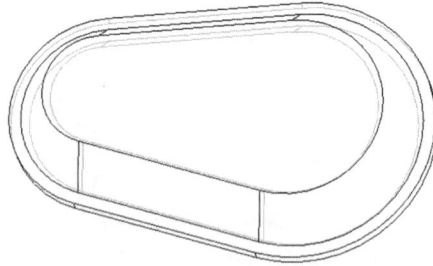

Cover with 0.40 flange.

8. When the section is finished, pick **OK** from the Protrusion dialog box. The cover should look as it does in the third illustration at right.

9. To finish the cover, add the mounting holes as a single cut feature. In the section sketch, you will select the **Use Edge** option from the GEOM TOOLS menu and pick each of the six circles in the Flange_Curve to complete the geometry. The section will look like that shown in the following illustration. Use the **Thru All** option for the feature's depth.

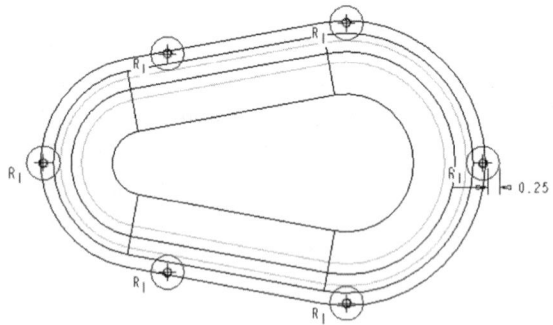

Flange boss section.

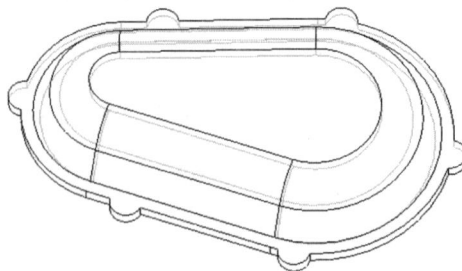

Cover part with flange bosses.

10. Add a .1-inch radius round to the flange edge and a .75- and 1.0-inch round to the housing cover interior and exterior edges. The completed part should look like that shown in the second illustration at right.

The design intent during the creation of this part was to have sketched datum curves control the part's interior size and shape, as well as the mounting boss locations. To make modifications to these features, you would need to select the Base_Curve or Flange_Curve. This method of geometry creation can be very useful when creating geometry to enclose a void.

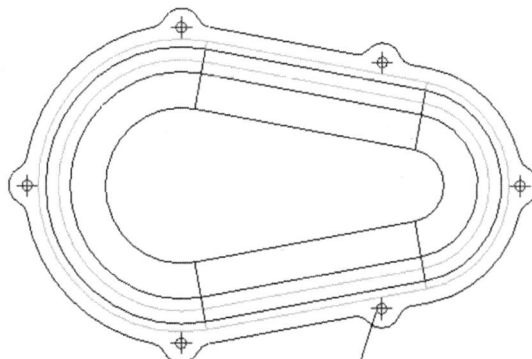

Use Edge to Create the 6 Holes

Mounting hole section.

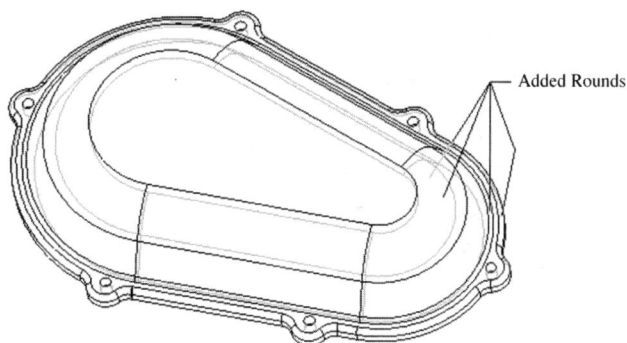

Added Rounds

Completed cover.

Exercise 12-2: Projecting Curves and Patterns

The ability to project a datum curve onto existing part surfaces can be very useful in the creation of relatively complex geometry. In this exercise, you will be creating an impeller similar to the one shown in the illustration at right.

The first vane will be created as a sweep, its trajectory being a projected curve on the center hub. You will then copy and rotate the first vane and create the remaining vanes as a pattern from the copied vane.

Impeller sample part.

1. Start your impeller part using the default datum planes and then create the base hub as a 360-degree revolved protrusion. The section for the base hub is shown in the following illustration.

Base hub section.

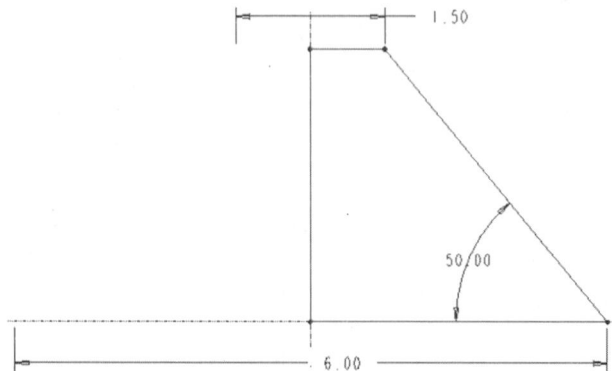

2. After you have finished the base hub feature, create a **Shell** feature and pick the top and bottom flat surfaces to be removed. Enter *.20* for the shell thickness. The base hub should look like that shown in the following illustration.

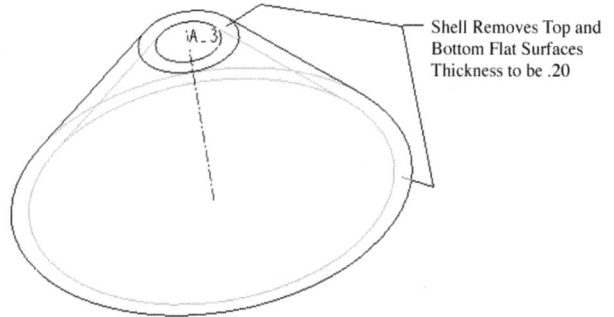

Base hub after shell feature added.

Shell Removes Top and Bottom Flat Surfaces Thickness to be .20

3. Now it is time to create the projected datum curve to be used as the sweep trajectory of the first vane. From the PART menu, select **Feature ➟ Create ➟ Datum ➟ Curve ➟ Projected ➟ Done**. You will now be given the options to Sketch or Select geometry to project. You will be selecting **Sketch**, but keep in mind that if you had an existing datum curve or geometry edge in your model, you could select either one to project onto a surface of your model. Select **Sketch ➟ Done**.

You will now need to set up a sketching plane on which to draw your curve section. This sketching plane will also be used as the origin plane to project the curve from. Select the bottom flat surface as your sketching plane. You will then be prompted: "Select direction of viewing the sketching plane." For this exercise, you will want the arrow to be pointing from the small diameter top surface to the larger diameter bottom surface. Use the **Flip** option to get the arrow pointing in the right direction, and then select **Okay**.

You will next be prompted: "Arrow shows direction of feature creation. Pick FLIP or OKAY." For this exercise, you will want the arrow to point from the large diameter bottom surface to the small diameter top surface. Use the **Flip** option to get the arrow pointing in the right direction, and then select **Okay**. Finish the sketch plane orientation setup by selecting a horizontal or vertical reference; then sketch the section shown in the following illustration.

Curve section.

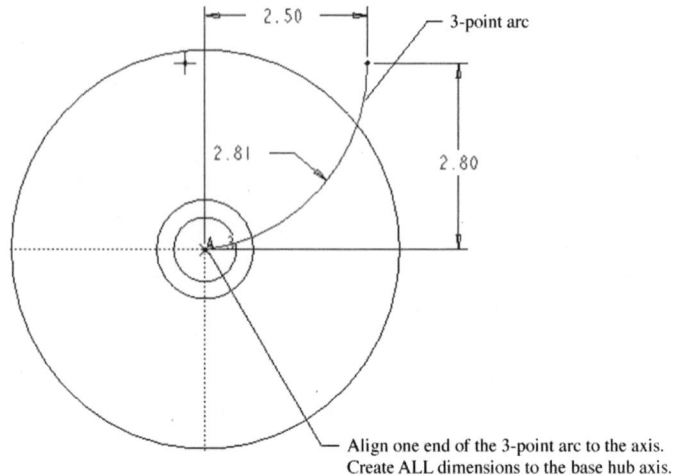

2.50   3-point arc

2.81

2.80

Align one end of the 3-point arc to the axis.
Create ALL dimensions to the base hub axis.

4. Once the section is complete, select **Done**. You will now need to select the surfaces to project the curve onto. Pick both halves of the outer tapered surface of the base hub, and select **Done** and **Norm to Sket** ➡ **Done**. Complete the feature by selecting the **OK** button from the Curve dialog box. Your part should now look like that shown in the following illustration.

5. To create the first vane for the impeller, select **Feature** ➡ **Create** ➡ **Protrusion** ➡ **Sweep** ➡ **Done** ➡ **Select Traj** ➡ **One By One**, and then pick the projected curve. For this exercise, make the **Start Point** of the trajectory the top small diameter end of the curve; then select **Done**. You will then be prompted: "Select upward direction of horizontal plane for sweep section." Use the **Flip** option to set the upward direction toward the outside of the part; then select **Okay**. Create the vane section shown in the following illustration. Because you will be

Projected curve.

unable to reference the existing edges of the hub, sketch the section so that its lower segment is clearly buried in the hub.

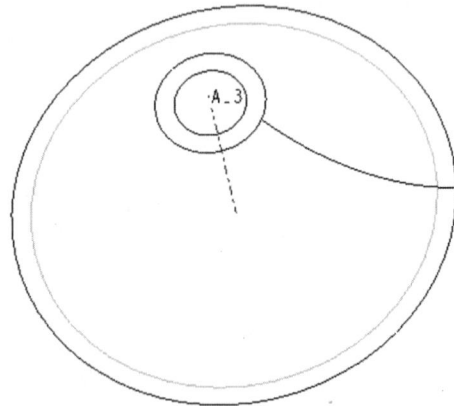

6. When you have successfully regenerated the section, select **Done**; then select the **OK** button from the Protrusion dialog box. The part with the single vane should look like that shown in the illustration below. Notice that the vane sticks through the lower flat surface at the bottom of the base hub. This excess geometry can be cleaned up later using a cut. In this exercise, that is your design intent, so you built the base hub with enough extra material to allow for the cut.

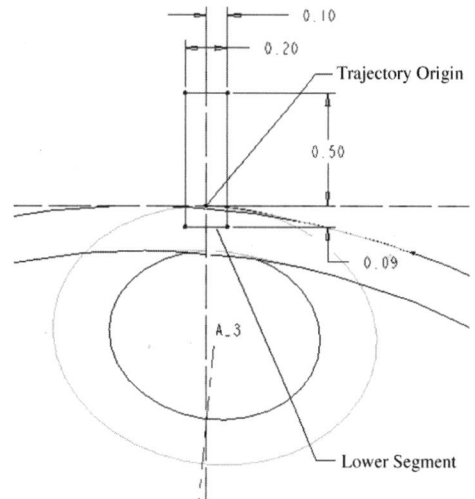

Vane section.

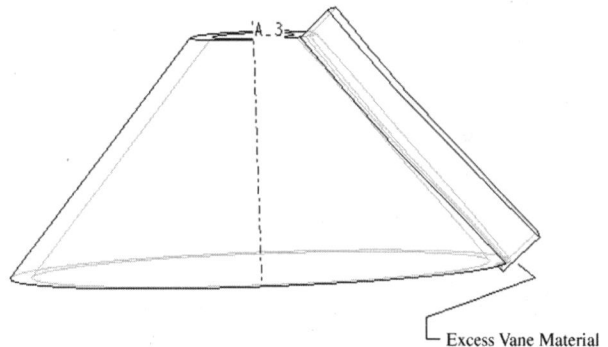

First vane feature.

7. Add a round to the top surface of the first vane. Select **Feature** ➡ **Create** ➡ **Round** ➡ **General Rnd** ➡ **Done** ➡ **Simple** ➡ **Done** ➡ **Full Round** ➡ **Edge Pair** ➡ **Done,** and pick the top two edges of the vane, as shown in the following illustration.

Round edge selection.

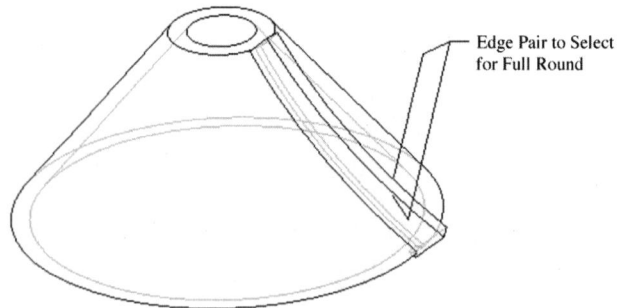

Edge Pair to Select
for Full Round

8. To complete the round, select **OK** from the Round dialog box. The completed round is shown in the illustration at right.

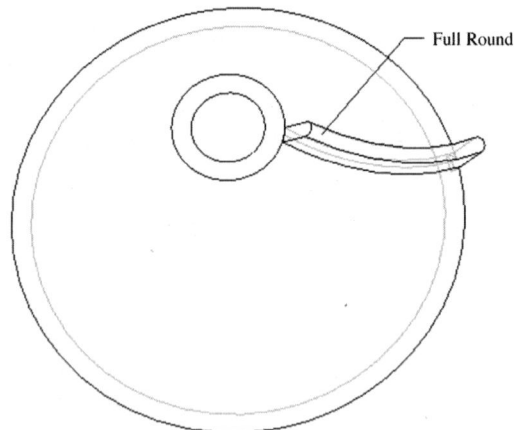

The technique for adding the remaining seven impeller vanes uses a combination of the Copy and Pattern functionality. During the creation of the first vane, you did not use an angular dimension. An angular dimension, however, is critical to the creation of a radial pattern. Under the functionality of the Copy command, you can generate a copy of the first vane and its full round and rotate that copy radially about the part's axis, thus creating a patternable angular dimension.

Full Round

Completed vane round.

9. To generate the copy of the first vane and its round, select **Feature** ➥ **Copy** ➥ **Move** ➥ **Select** ➥ **Dependent** ➥ **Done** ➥ **Select**. Pick the vane protrusion and the full round features, select **Done** ➥ **Rotate** ➥ **Crv/Edg/Axis**, and select the part's center axis. An arrow will appear on the axis. Use the **Flip** option to make the arrow point toward the top, small diameter surface, and select **Okay**. You will be prompted: "Enter Rotation Angle." Enter *45*, and select **Done Move** from the MOVE FEATURE menu. Now a Group dialog box and GP VAR DIMS menu will appear. Because you will not need any variable dimensions for this part, select **Done**; then select **OK** from the Group Elements dialog box. The completed vane copy should look like that shown in the following illustration.

Copy of first vane and round.

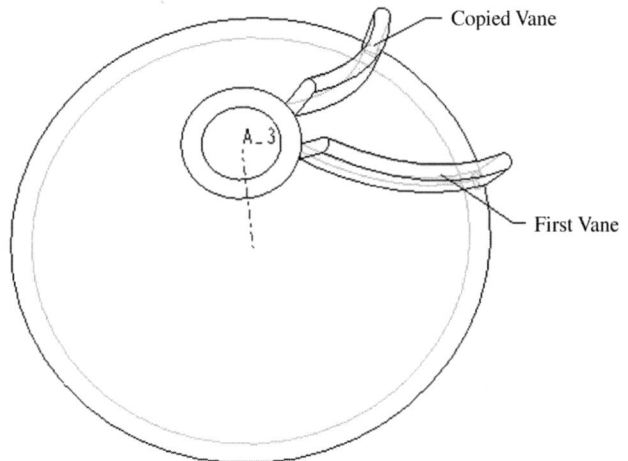

Copied Vane

A_3

First Vane

10. Create the additional six vanes as a pattern of the copied vane. When you copied the vane and its round, you generated a group of those two features. Therefore, to pattern them you need to use the pattern functionality under the GROUP menu. Select **Feature** ➡ **Group** ➡ **Pattern** ➡ **Select**, and then pick the copied vane. The dimensions for the copy are now displayed. Pick the 45-degree dimension as the value to be patterned and enter *45*

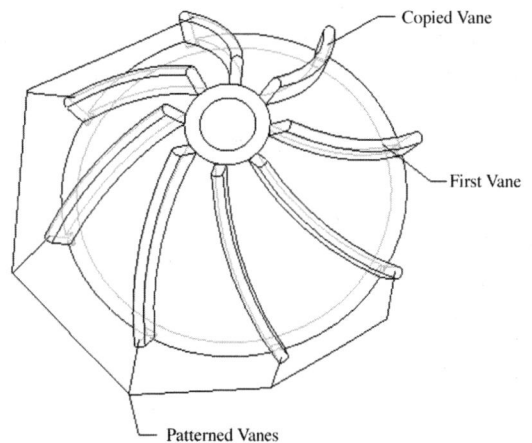

Copied Vane

First Vane

Patterned Vanes

Completed pattern.

for the amount by which to increment the pattern. Select **Done**, enter *7* for the number of instances, and select **Done** again to complete the pattern feature. Your part should now look like that shown in the illustration at right, showing a total of eight vanes spaced 45 degrees apart.

➥ **NOTE:** *More information on copying features and patterns can be found in Chapter 10.*

11. Clean up the lower edge of the impeller by cutting away the bottom .45 inch of impeller. The finished part should look similar to that shown in the illustration at right.

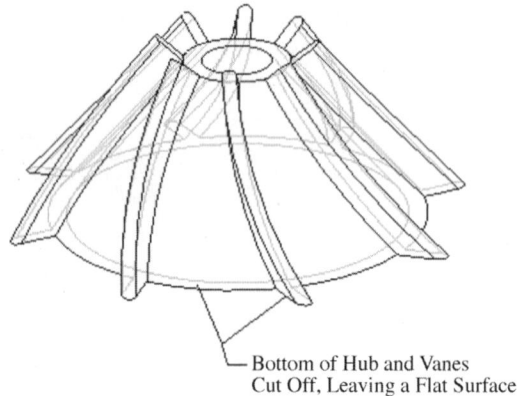

Bottom of Hub and Vanes
Cut Off, Leaving a Flat Surface

Finished impeller part.

Exercise 12-3: Creating a Datum Curve Using From Equation, Sine Curves, and Wave Springs

In this exercise, you will be creating a datum curve using the From Equation option. The ability to specify or write an equation to control the shape of a datum curve allows you to very quickly and simply control all internal point locations and dimensions for a datum curve without having to sketch a single entity.

1. Start this exercise by creating the default datum planes and default coordinate system. To help clarify the directions of the X, Y, and Z coordinates, rename your datum planes as shown in the illustration at right.

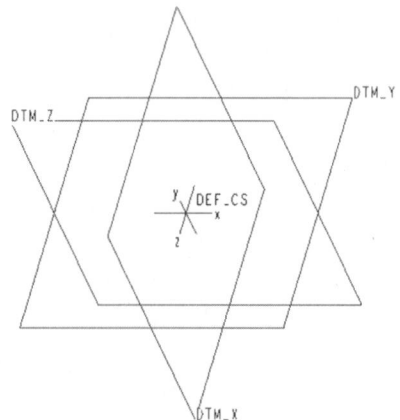

This is the Default View
in Trimetric Orientation

Renamed datum planes.

2. From the PART menu, select **Feature** ⇢ **Create** ⇢ **Datum** ⇢ **Curve** ⇢ **From Equation** ⇢ **Done**, select the default coordinate system, and select **Cartesian** as the coordinate type. At this point, a new window will open. Depending on what your *config.pro* setting is for PRO_EDITOR_COMMAND, you will see either a system window or the Pro/TABLE window. The following illustration shows the Pro/TABLE editor window.

Pro/TABLE Editor window.

	C1	C2
R1	/* For cartesian coordinate system, enter parametric equation	
R2	/* in terms of t (which will vary from 0 to 1) for x, y and z	
R3	/* For example: for a circle in x-y plane, centered at origin	
R4	/* and radius = 4, the parametric equations will be:	
R5	/* x = 4 * cos (t * 360)	
R6	/* y = 4 * sin (t * 360)	
R7	/* z = 0	
R8	/*_____	
R9		
R10		
R11		
R12		
R13		
R14		
R15		

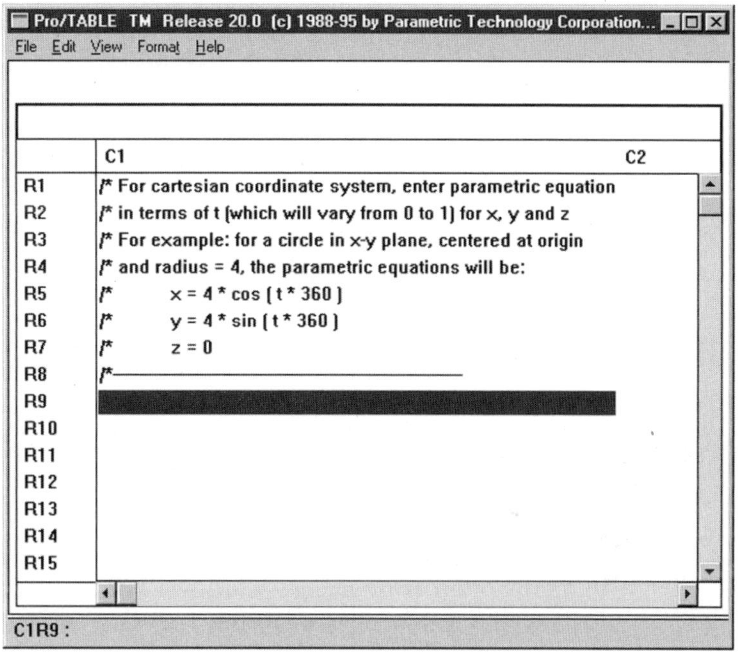

`Pro/TABLE  TM  Release 20.0  (c) 1988-95 by Parametric Technology Corporation...`
`File  Edit  View  Format  Help`
`C1R9 :`

The directions in the Editor window show how to create a 2D circle with a 4-inch radius, centered at the origin and oriented on the *x-y* plane. This is pretty straightforward, because as the value of *t* varies from 0 to 1, the angle for the sine and cosine vary from 0 to 360 degrees. You will most likely never use an equation to create a circle. For this exercise, you will be creating the curve shown in the illustration at left.

3. In that you already have an Editor window open, enter the following equations for the values of X, Y, and Z.

Sine curve.

```
X = sin ((2*t) * 360)
Y = 4 * t
Z = 0
```

This equation will generate a 2D sine curve 4 inches long in the Y direction, and the total travel in the X direction will be 2 inches. To make this a 3D curve, you need to add an equation for the Z direction as it relates to the *t* value. The following illustration shows an example of the 3D sine curve you will be generating by redefining the equation you have already created. In the last portion of this exercise, you will use this curve to create a wave spring.

3D sine curve.

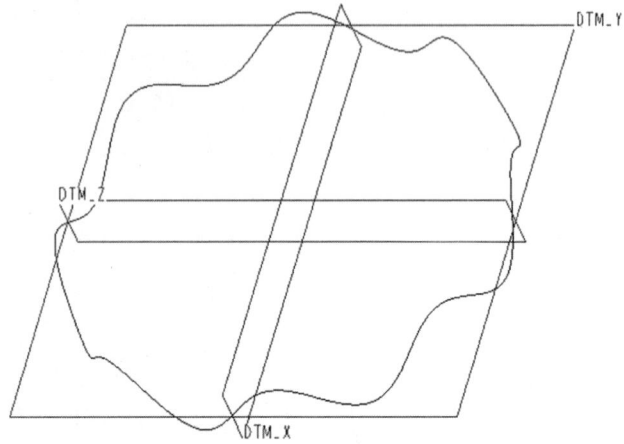

4. Because all of the dimensions for a From Equation curve are contained within the feature's equation, you must use the Redefine option to change the equation. From the FEATURE menu, select **Redefine**, and then pick the sine curve. In the Curve dialog box, highlight **Equation** and pick the **Define** button. This will again open your Editor window. The current equation and window are shown in the following illustration.

Current equation values.

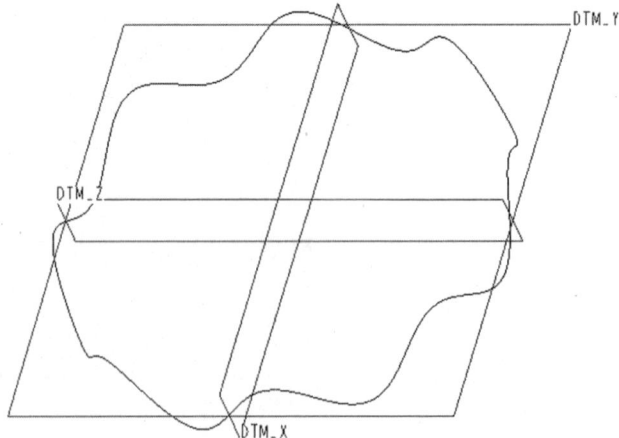

5. The wave spring base curve you will be creating has a 4-inch radius in the *x-z* plane, 8 peaks, and 8 valleys, and the total height from valley to peak is 1 inch in the *y* direction. One of the first things you must do with the design constraints is break them up into the equations you will need. Because the basic circle you will be creating is in the *x-z* plane, you can use the circle equation shown in the Editor window. The following equations will set up the 4-inch-radius circle. Enter the equations for X and Z into the Equation Editor window.

```
X = 4 * cos(t * 360)
Z = 4 * sin(t * 360)
```

6. Solve for Y. You know that the spring needs 8 peaks and 8 valleys, which means that the *t* value must cycle from 0 to 1 eight times. Therefore, part of the equation must contain the following expression.

```
Y = Sin((8 * t) * 360)
```

7. The last thing to add to the equation that solves for Y is the height value. Because the sine wave cycle starts at the 0 value and the total travel you want is 1 inch from peak to valley, you will need to multiply the Y equation by 0.5. The complete equation for Y follows. Enter the equation shown below into the Editor window.

```
Y = .5 * sin((8 * t) * 360)
```

The complete set of equations as they appear in the Pro/TABLE Editor window are shown in the following illustration.

Wave spring equations.

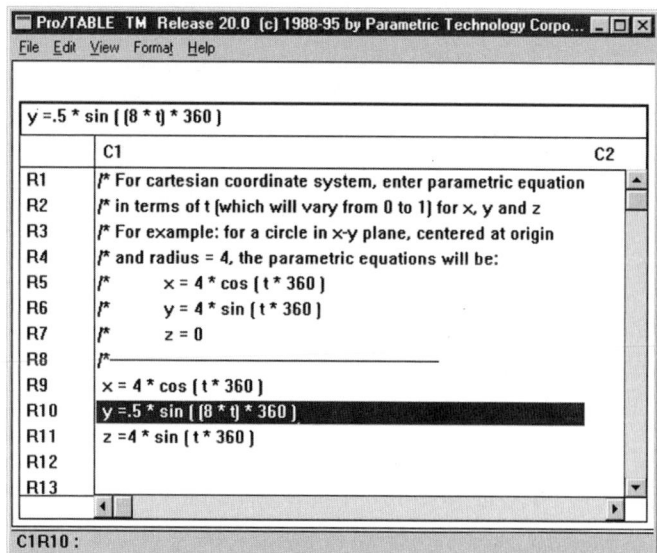

8. Once you have entered the equations for the wave spring, save them and exit the Editor window. Select **OK** from the Curve dialog box and the system will generate the curve shown in the following illustration.

Completed wave spring curve.

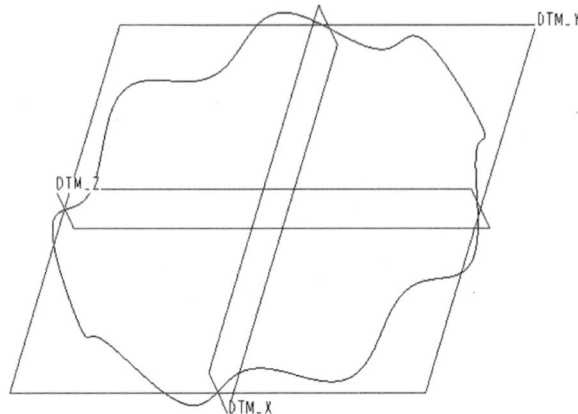

9. Your From Equation curve currently contains only one segment, with its start and end point at angle 0. To create the wave spring, you will need at least one other point on this curve. To complete the wave spring geometry, you must first Split the wave spring curve into two segments. To accomplish this, first orient the part to the default view. Select **Feature ➡ Create ➡ Datum ➡ Curve ➡ Split**, pick the wave spring curve near the 180-degree area, and then pick DTM_Z as the divider, shown in the following illustration.

Split curve picks.

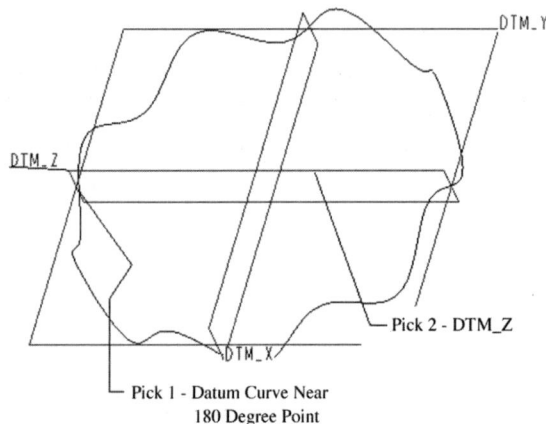

Pick 2 - DTM_Z

Pick 1 - Datum Curve Near 180 Degree Point

10. Once you have selected the datum plane as a divider, an arrow will appear and you will be prompted: "Select Part of Curve to Keep." From the TRIM CURVE menu, select **Both** and **Okay**. To finish the curve Split feature, select **OK** from the Curve dialog box. Your wave spring base curve will now have two segments, and you can now create the spring protrusion.

11. From the PART menu, select **Feature** ➥ **Create** ➥ **Protrusion** ➥ **Advanced** ➥ **Done** ➥ **Swept Blend** ➥ **Done** ➥ **Sketch Sec** ➥ **NrmToOriginTraj** ➥ **Done** ➥ **Select Traj** ➥ **Curve Chain**, pick the curve, and select **Select All** ➥ **Done**.

12. From the SEC ORIENT menu, select **Automatic**. You will then need to set the upward direction of the horizontal plane for the sweep section. Use the **Flip** and **Okay** options to set the upward direction, as shown in the following illustration.

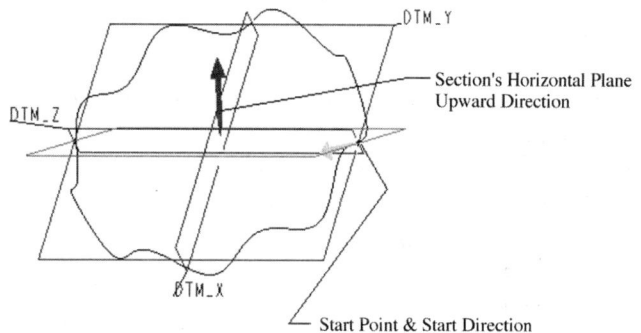

Upward direction.

13. You will next be prompted: "Accept points to sketch more sections (end points of Origin Traj are in green)." Here you must select **Accept** or the feature will not have enough sections to blend. The following illustration shows the origin point and the point you created by splitting the curve.

Two section point locations.

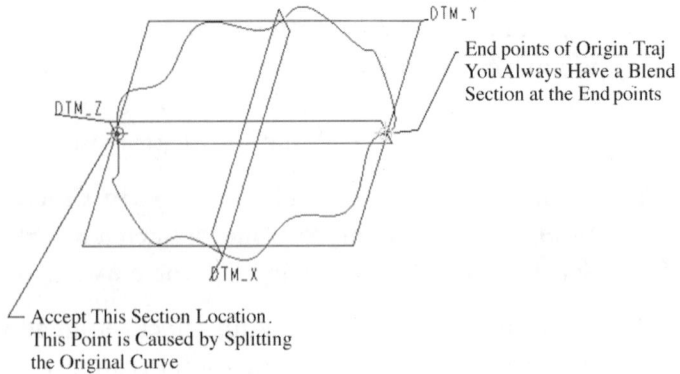

DTM_Y

End points of Origin Traj
You Always Have a Blend
Section at the End points

DTM_Z

DTM_X

Accept This Section Location.
This Point is Caused by Splitting
the Original Curve

14. To complete the wave spring, you will need to sketch the section shown in the following illustration at both sketch locations. Accept default values for *all* section orientation or rotation prompts.

Wave spring section.

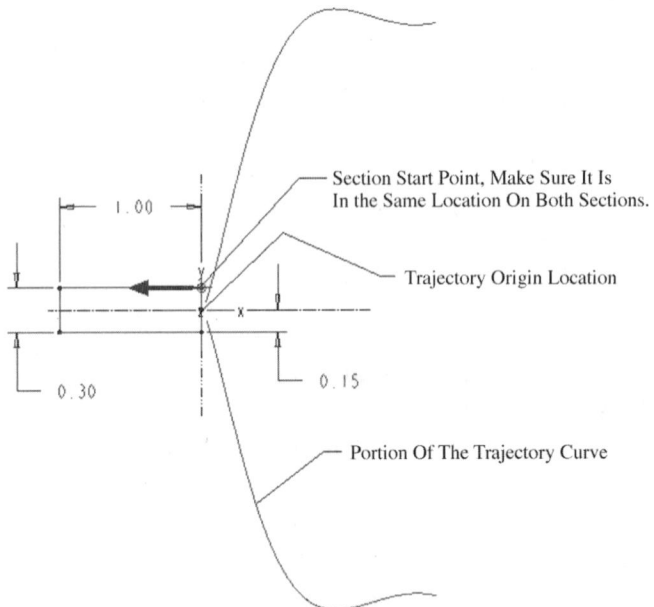

1.00

Section Start Point, Make Sure It Is
In the Same Location On Both Sections.

Trajectory Origin Location

0.30

0.15

Portion Of The Trajectory Curve

15. Once the two sections are complete, select **OK** from the Protrusion dialog box. The completed wave spring should look like that shown in the following illustration.

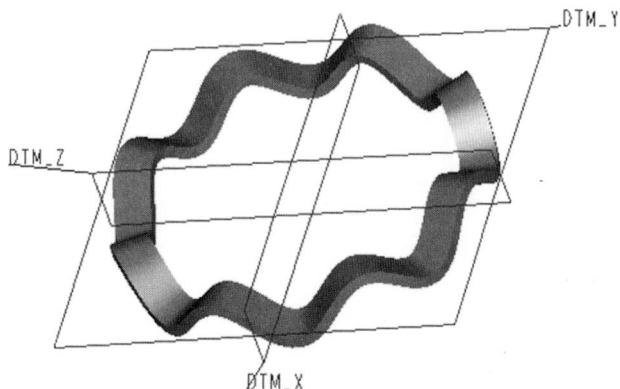

Completed wave spring.

Summary

As demonstrated in this chapter, datum curves have a variety of uses. They can be used when creating castings, as a footprint for external geometry references, as a sweep trajectory, or as the basis for an entire part. In some cases it is helpful to create datum curves for later use in a model. These curves serve as "space holders" for maintaining a spatial reference for later model geometry and features. The following illustrations show a pot handle, the solid geometry of which is the portion that attaches to a pot. The curves shown have two uses. The first is as a size reference for the loop of the handle; the other use is as a selectable trajectory and selectable sections for the Swept Blend feature, which will create the loop portion of the handle.

Selectable Blend Sections

Selectable Trajectory

Solid Portion Attaches to Pot

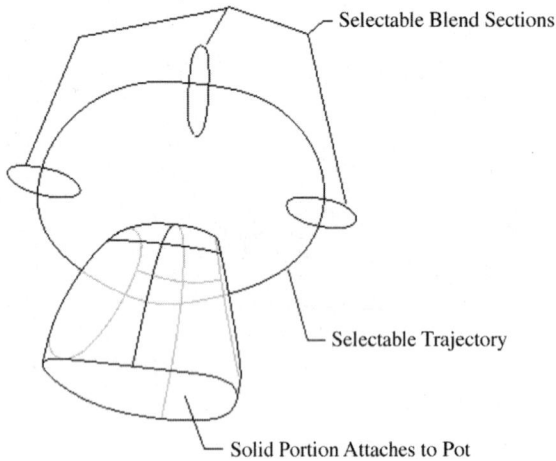

Pot handle solid and curves.   *Completed pot handle.*

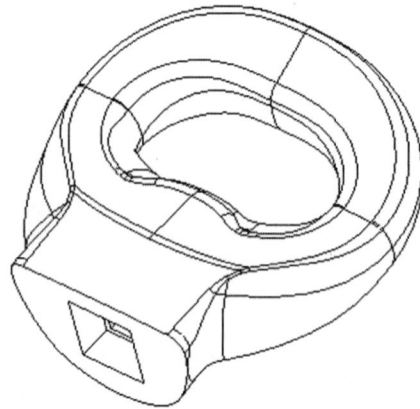

⊶ **NOTE:** *When you use datum curves as trajectory, section, or bound-ary references, you will need to modify the curves to change the solid geometry.*

Although it is true that you are working in a solid modeling world, the datum curve can be very helpful in capturing design intent. In exercise 12-3, the pro-trusion feature contained only the spring's material thickness and width, whereas the datum curve contained and controlled the internal diameter, the number of waves, and the overall wave height.

The exercises in this chapter explored some common uses of the Datum Curve option, but there are many more. In Chapter 13, you will work through a few more uses for curves, including an exercise on importing curves for use in a modifiable model.

13

Surfacing

The use of Pro/ENGINEER surfaces allows the designer and engineer more freedom in model design. This is because surfaces do not need to be constrained as tightly as solid geometry, making surfaces quite handy in solving many complex design problems. With the freedoms surfacing grants the designer, it is important to do a good job of capturing design intent and to leave a trail for future model modification.

In general, surfaces are defined using many of the same methods as are used in working with solid geometry. They can be created as extrusions, revolves, sweeps, blends, or any of the advanced feature forms, such as Swept Blend and Var Sec Swp. In addition, they can be created as a flat sketch or defined by a series of boundaries. The one absolute rule about using surfaces is that you need to locate surface sketches to the current model, just as you would in any geometry sketch.

Exercises

Through the exercises in this chapter, you will gain an understanding of surfacing basics such as the surface merge, extend, surface by boundary, and replace operations. In each case, after you have gone through the basic operation, you will undertake more difficult design problems, such as creating models from imported geometry and making modifiable models from imported curves. Keep in mind that Pro/ENGINEER is a solid modeling tool. Therefore, use surfacing when necessary, not just because you can.

Excessive use of surfacing can result in overly complex models that are difficult to modify and that lose their design intent.

Also before you begin the exercises, note that when creating surfaces in Pro/ENGINEER you will tend to create numerous support features used in the surface definition. The support features you create will be used to modify the model geometry. In many cases, surfaces and merged surfaces used to create solid geometry will become part of the solid feature, making them difficult to locate and select for modification later in the modeling process. Because of this, datum point, curve, and surface geometry should be named, making them easier to locate and modify later.

Exercise 13-1: Learning Surfacing Basics

In this exercise you will be creating the part shown in the following illustration.

The solid model shown in the illustration at right consists of an extruded surface, two flat surfaces, a surface extend, two surface merges, a protrusion using a closed quilt, a surface by boundaries, and a replace surface. To begin this exercise, start a part using the default datum planes. The first geometry you will be creating is a surface containing the two concentric circles.

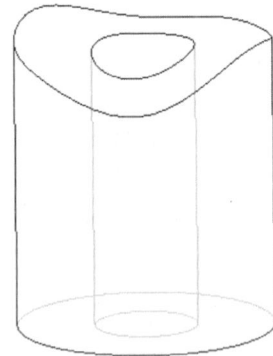

Exercise 13-1 model.

1. From the PART menu, select **Feature** ➡ **Create**. From the FEAT CLASS menu, select **Surface**.

Notice that the SRF OPTS menu contains many of the options available to you in solid modeling. Each of the basic options (extrude, revolve, blend, and sweep) works the same as its solid counterpart, with one exception: you have the ability to create your surfaces with open ends or capped ends. A capped surface is one that can be used to create a solid protrusion (with use of the Use Quilt option), an operation in which an open surface is closed before it is converted to a solid.

2. From the SRF OPTS menu, select **Extrude** ➡ **Done** ➡ **One Side** ➡ **Open Ends** ➡ **Done**. Pick DTM2 as the sketching plane. Use the **Flip** and **Okay** options to set the feature creation direction, as shown in the following illustration.

3. Pick DTM3 as the **Bottom** option for the section's horizontal reference. Sketch the section shown in the second illustration at right.

4. After you have completed the section, select **Done** and then **Blind** from the SPEC FROM menu, and enter *6.0* for the depth; then select **OK** from the Surface dialog box.

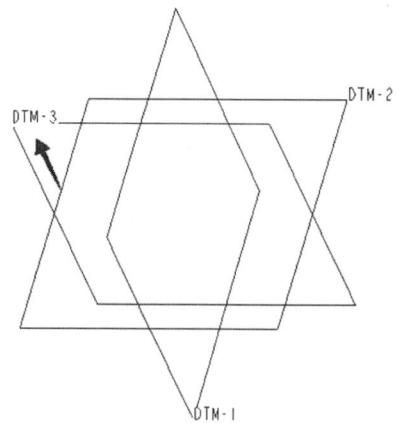

Feature creation direction.

You should now have two concentric cylinders on your screen. Both cylinders will be purple with yellow edges. When creating protrusions from surface quilts, all surface edges should be purple. Purple edges represent a closed quilt. Yellow edges represent an open quilt. Looking for yellow edges will be very important if you are importing surface geometry or another designer's model based on surface geometry. Open surface quilts cause problems for prototyping and tool path generation. The only time you can use an open surface for a protrusion is when using the Thin option.

5. Create the two Flat surfaces to close the extruded surface. Select **Feature ➥ Create ➥ Surface**. When you select Surface, the QUILT SURF menu will pop up. This menu is shown in the illustration at right.

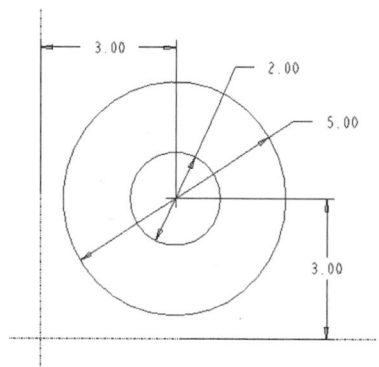

Extruded surface section.

The following list describes the options under the QUILT SURF menu.

New	Creates a new surface.
Merge	Intersects or joins two surfaces, resulting in a single surface quilt.
Trim	Removes a portion of a surface or quilt.
Extend	Adds length to a surface or quilt.

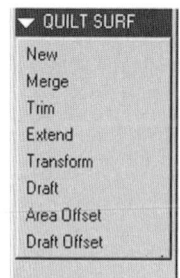

QUILT SURF menu.

Transform	Creates a copy of datum curves, surfaces, or quilts. Location can be translated, rotated, or mirrored. This option is very handy when the Copy option under the FEATURE menu fails to copy a surface quilt properly.
Draft	Adds draft to a closed quilt's surfaces.
Area Offset	Creates an offset of selected surfaces from a quilt.
Draft Offset	Creates an offset of selected draft surfaces from a quilt.

6. Select **New** from the QUILT SURF menu, then **Flat ➥ Done**. Select **Make Datum** from the SETUP PLANE menu and create an **Offset** plane from DTM2 offset toward the **Yellow** side 2 inches. Select DTM3 as the **Bottom** option for the section's horizontal reference. Sketch the section shown in the following illustration. When you are done, select **OK** from the Surface dialog box.

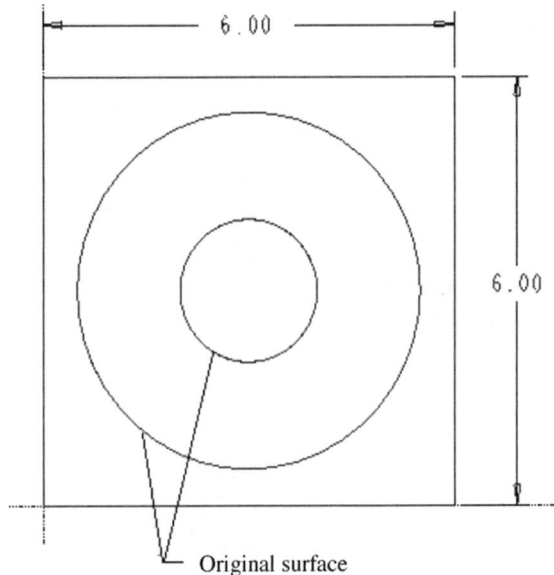

Flat surface section.

7. Create the second flat surface, this time using the **Transform** option from the QUILT SURF menu. Select **Feature ➥ Create ➥ Surface ➥ Transform**. In the OPTIONS menu that pops up, you will see the options Move, Mirror, and Copy/No Copy. The Copy/No Copy options allow you to make a copy or move the original surface. From the OPTIONS menu, select **Move ➥ Copy ➥ Done**. Select the flat surface you created; then select **Translate** and **Plane**. Pick DTM2 when prompted "Select a Plane the direction will be perpendicular to." Use the **Flip** and **Okay** options to set the translation direction, as shown in the following illustration.

Translation direction.

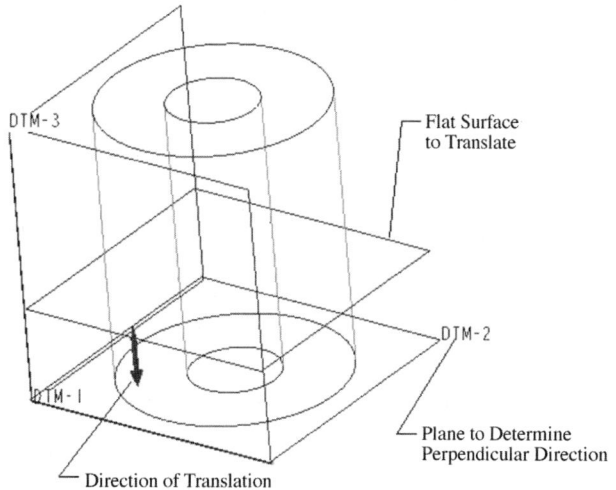

DTM-3

Flat Surface
to Translate

DTM-2

DTM-1

Plane to Determine
Perpendicular Direction

Direction of Translation

8. Once you set the translation direction, you will be prompted for a distance value. The distance is taken from the original flat surface. Enter *5.0* and select **Done Move**. Your surface model should look like that shown in the illustration at right.

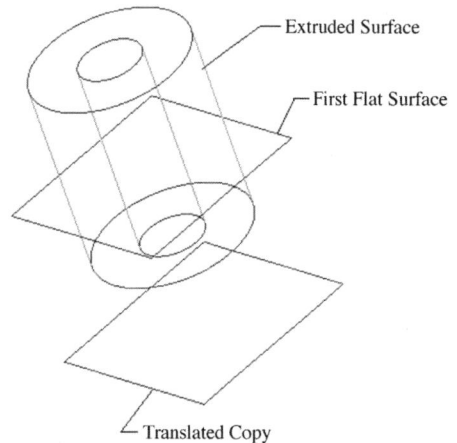

Extruded Surface

First Flat Surface

Translated Copy

Extrusion and flat surfaces.

9. Now that you have a few surface features containing your design information, it is a good time to start naming them. From the PART menu, select **Set Up ➥ Name ➥ Other**, pick the extruded cylinder and name it *Cyl_Surf*, pick the original flat surface and name it *Flat_Surf*, and pick the copied surface and name it *Trans_Surf*.

10. Start closing the quilt. From the PART menu, select **Feature ➥ Create ➥ Surface ➥ Merge**. You will be given the options **Join** and **Intersect**. A quick rule of thumb is: If the surfaces do not share a boundary, use Intersect. In this exercise, the Flat_Surf edges are well outside the Cyl_Surf boundary. Therefore, select **Intersect**. Pick Cyl_Surf as the first surface to merge and then Flat_Surf as the surface to add to the merge.

You will now be presented with a set of arrows and the prompt "Arrows indicate portion of quilt to keep. Pick FLIP or OKAY." The highlighted boundary belongs to the quilt the arrows are applied to, as shown in the first illustration at right.

11. Set the arrows for the portion to be kept, as shown in the previous illustration, and then select **Okay**. You will again be prompted "Arrows indicate portion of quilt to keep. Pick FLIP or OKAY." Use the **Flip** and **Okay** options to set the direction to be kept, as shown in the second illustration at right.

12. Once you have selected Okay, you may or may not be prompted to add another surface. This seems to be a random occurring prompt; about the time you expect the menu to allow additional merges, it does not appear. If the feature did not already complete, select **Done**. The model should now look like that shown in the illustration at right on the following page. Note that the edge where the merge occurred is now purple (on your screen), signifying that this end of the quilt is closed.

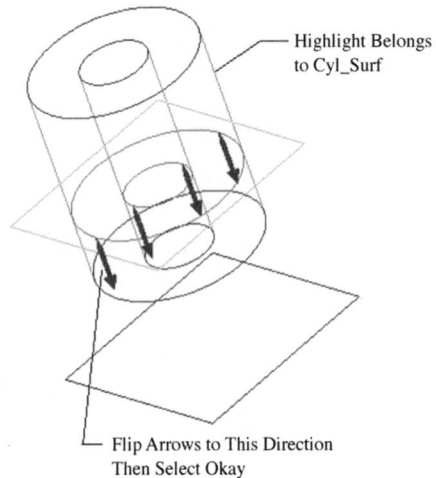

Highlight Belongs to Cyl_Surf

Flip Arrows to This Direction Then Select Okay

Portion of Cyl_Surf quilt to be kept.

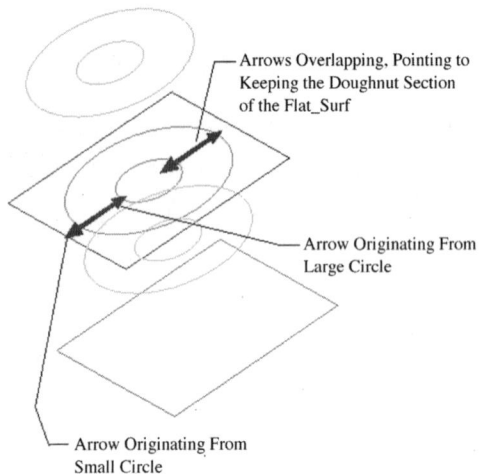

Arrows Overlapping, Pointing to Keeping the Doughnut Section of the Flat_Surf

Arrow Originating From Large Circle

Arrow Originating From Small Circle

Portion of Flat_Surf to be kept.

Menu Manager
▶ PART
Feature ▼
▶ FEAT
Create ▼
▼ OPTIONS
Same Srf
Approx Srf
Along Dir
Tangent Srf
Single Dst
Variable
Dist On Srf
Dist In Pln
Up To Plane
Done
Quit

*Surface
extend
OPTIONS
menu.*

13. Merge the lower end of the Cyl_Surf merge to the Trans_Surf. To do this, first make the lower end touch or intersect the Trans_Surf. Lengthening the open end of a surface is accomplished using the **Extend** option from the QUILT SURF menu. In general, it is a good practice to make sure any surfaces to be merged clearly intersect each other along all edges. Excess surface length will be removed during the merge.

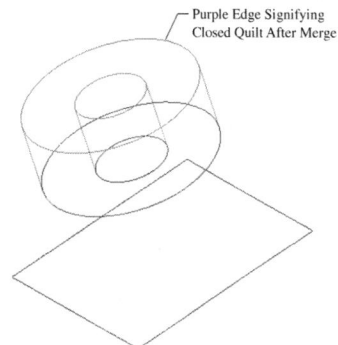

On your merged Cyl_Surf, you have two open yellow edges to be extended. Each edge will need to be extended separately because the Extend option uses adjoining surface edges or closed loops. In this exercise, you have two closed loops; therefore, each will need to be extended.

Purple Edge Signifying Closed Quilt After Merge

Model with Flat_Surf and Cyl_Surf merged.

14. From the PART menu, select **Feature** ➥ **Create** ➥ **Surface** ➥ **Extend**, which will bring up the OPTIONS menu, shown in the illustration at left.

The following list describes the surface extension options.

Same Srf	Extends a surface as the same type of feature as the original surface: Flat Surface Plane, Cylinder, Cone, or Spline. Of these surface types, the one that may extend in an odd fashion is the spline. This is because the extend will approximate the flow direction and curvature of the existing surface, potentially resulting in a curling of the surface edge.
Approx Srf	Extends the surface and holds the edge boundary. When extending spline edges, this may cause the extended portion of the surface to deform slightly.
Along Dir	Selects or creates a datum plane to determine the extend length and direction. This option must be used in combination with the Up To Plane option. The directional plane and termination plane can be different planes.
Tangent Srf	Extends a surface tangent to the surface end along the selected edge.
Single Dst	Enters a single value for all surface edges to be extended.
Variable	Allows you to extend selected surface edges by different values.
Dist On Srf	Determines the extension length from the surface edge.

Dist In Pln	Selects the plane from which to determine the extension length.
Up To Plane	Must be used in combination with the Along Dir option.

15. From the OPTIONS menu, select **Same Srf ➡ Single Dst ➡ Dist On Srf ➡ Done**. Pick both yellow surface edges belonging to the large diameter of the Cyl_Surf merge, as shown in the first illustration at right. Notice when using the One By One option under the CHAIN menu that you can pick each 180-degree arc of the large diameter. This means that you could Extend one or both portions of the cylinder.

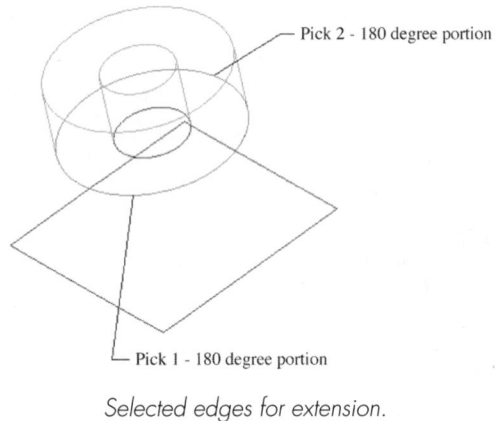

Pick 2 - 180 degree portion

Pick 1 - 180 degree portion

Selected edges for extension.

16. After you have picked the two edges of the surface to be extended, select **Done**. Pick one of the two surface edges again, from which to measure the extend distance. Enter *5.0* for the extend distance value and select **Done Extend** from the SURF EXTEND menu. The extended surface should look like that shown in the second illustration at right.

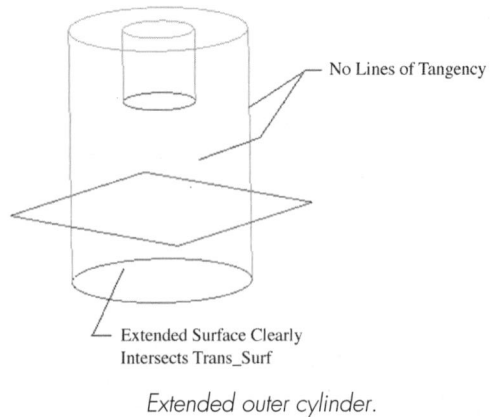

No Lines of Tangency

Extended Surface Clearly Intersects Trans_Surf

Extended outer cylinder.

Notice that no lines of tangency exist along any edges of the extend feature. If you had selected the Approx Srf or Tangent Srf option, the system would generate lines of tangency at the start of the extend feature and at the ends of the 180-degree arc sections, as shown in the following illustration. On straightforward geometrical shapes such as the one you have created, the lines of tangency will not be a problem. However, if you were merging spline surfaces with multiple tangency lines, you could have difficulty with rounding those edges.

17. Extend the small diameter surface by selecting **Feature** ➡ **Create** ➡ **Surface** ➡ **Extend** ➡ **Same Srf** ➡ **Single Dst** ➡ **Dist On Srf** ➡ **Done**. Select **Tangnt Chain** and then pick the yellow surface edge belonging to the small diameter of the Cyl_Surf merge and the edge from which to measure the extend distance, as indicated in the illustration below right.

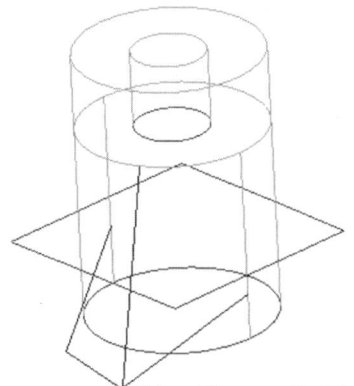

18. Select **Done**, enter *5.0* for the extend distance, and select **Done Ext**. The surface model should now look like that shown in the illustration below left.

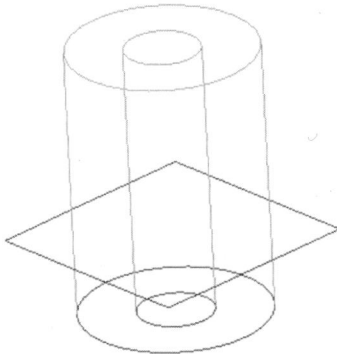

Lines of Tangency Created When Using the Approx Srf and Tangent Srf Options

Lines of tangency.

Pick 1 - Small Diameter

Pick 2 - Edge to Measure Extend Distance

Extended cylinder model.

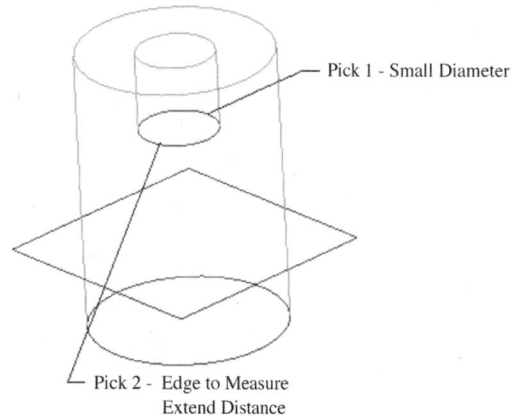

Inner cylinder extend.

19. Merge the extended Cyl_Surf and the Trans_Surf to close the surface quilt. Select **Feature** ➡ **Create** ➡ **Surface** ➡ **Merge** ➡ **Intersect**; then pick the cylinder and then the flat surface. You will then need to use the **Flip** and **Okay** options to select the portion of each quilt to be kept. Set the arrows to the directions shown in the following two illustrations.

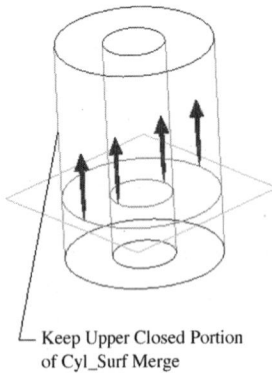

Keep Upper Closed Portion
of Cyl_Surf Merge

Cylinder portion to be kept.

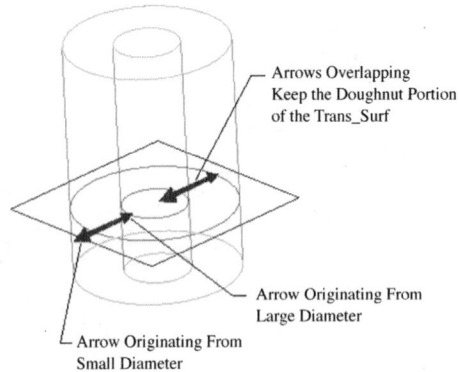

Arrows Overlapping
Keep the Doughnut Portion
of the Trans_Surf

Arrow Originating From
Large Diameter

Arrow Originating From
Small Diameter

Trans_Surf portion to be kept.

The completed closed surface quilt should now look like that shown in the following illustration below right.

20. To create a protrusion from this closed quilt, select **Feature** ➝ **Create** ➝ **Solid** ➝ **Protrusion** ➝ **Use Quilt** ➝ **Solid** ➝ **Done**, pick the quilted surface feature, and select the **OK** button from the Protrusion dialog box to complete the feature.

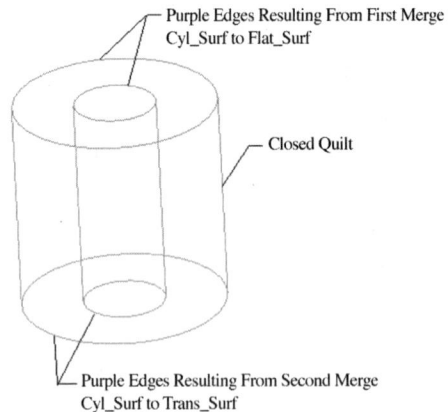

Purple Edges Resulting From First Merge
Cyl_Surf to Flat_Surf

Closed Quilt

Purple Edges Resulting From Second Merge
Cyl_Surf to Trans_Surf

Closed quilt.

Now you will have a solid model in which all of the purple edges of the quilt have turned white, or your system's default solid geometry color.

> ➝ **NOTE:** *Whenever you use or create surface geometry for use in a solid model, your feature count expands dramatically. This exercise has required eight features, excluding datum planes and coordinate systems. In addition, if you want to make modifications to the solid cylinder part, you must select the surface features, which are no longer visible.*

Exercise 13-2: Surfacing with Boundaries and Replacement

In this exercise, you will be building a surface by boundary, which will then be used to reform the top surface of the cylinder part created in the previous exercise. To create a surface from boundaries, you must first have boundaries to select. The boundaries can be existing model geometry edges, datum curves, two vertices, or a single point. Because there are no existing edges to use, you will be creating datum curves. The most important rule about the boundaries is that they must form a closed loop to create a surface.

→ **NOTE:** *If you are not familiar with spline and datum curve creation, see chapters 5 and 12 for background information and exercises.*

The surface you will be creating in this exercise will need to be larger than the top face of the solid cylinder because you will be using it to replace that top surface.

1. Create a sketched datum curve, Datum Curve 1, on DTM1 as shown in the illustration at right. The exact positions of interior spline points are not critical.

2. Create a second sketched datum curve, Datum Curve 2, on a plane parallel to DTM1 and offset 6.0 inches in the yellow side direction. Sketch the section shown in the following illustration. Again, the internal point locations are not critical.

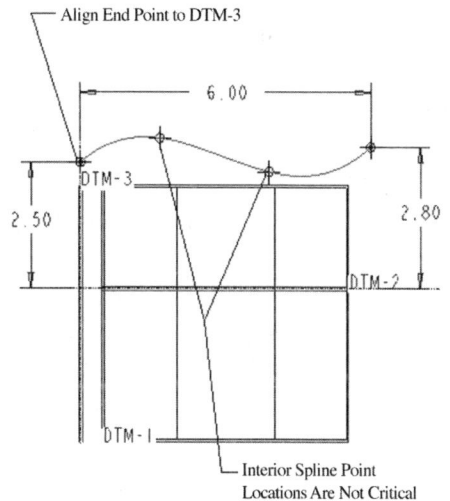

Datum curve 1 section.

3. Create a feature containing two datum points: one at the end of Datum Curve 1 and at the end of Datum Curve 2 where they intersect with DTM3. To do this, select **Feature ➡ Create ➡ Datum ➡ Point ➡ Add New ➡ Curve X Srf**, pick Datum Curve 1 and DTM3, select **Add New** again, and pick Datum Curve 2 and DTM3. Select **Done** after selecting the second datum point's location. The illustrations that follow show the screen picks for placing the datum points, as well as the model features created so far.

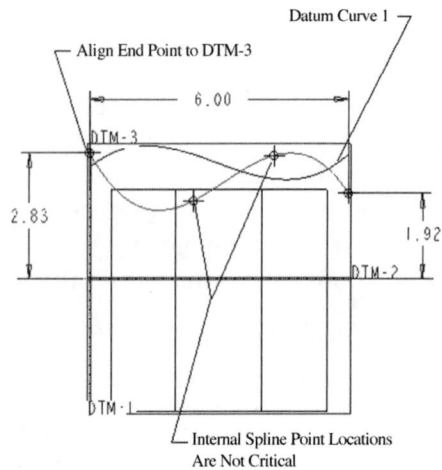

Datum curve 2 section.

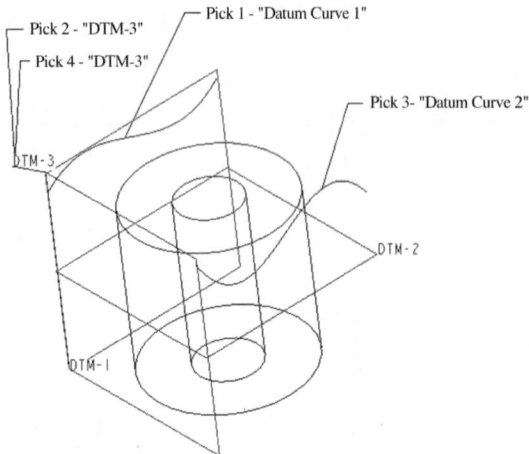

Datum point screen picks.

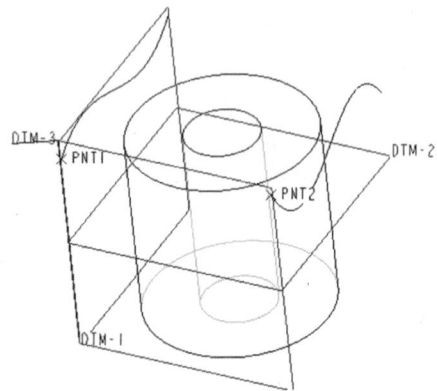

Model so far.

4. The next datum curve you will be creating will be on DTM3, and its end points will be aligned to PNT1 and PNT2. Sketch the datum curve, Datum Curve 3, shown in the following illustration.

5. Make sure you explicitly align the spline's end points to the datum points.

↝ **NOTE:** *Whenever you are sketching a spline the end or internal points of which are to be aligned to existing datum points, always explicitly pick the datum point and the spline point for alignment. Do not simply align by selecting the datum point and then pressing the middle mouse button. This can result in point-to-curve misalignments, which will cause the surface creation "from" boundary to fail.*

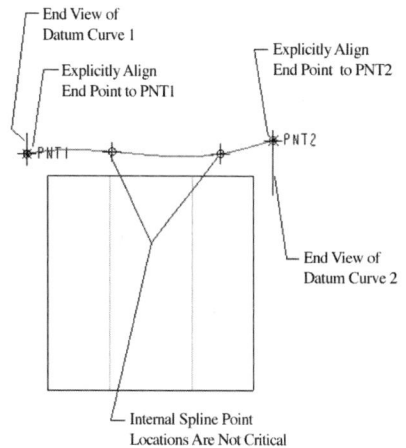

Datum curve 3 section.

At this point, your model should consist of the solid cylinder, three datum curves, and two datum points. You are now ready to create the surface "from" boundaries.

6. From the PART menu, select **Feature** ↝ **Create** ↝ **Surface** ↝ **New** ↝ **Advanced** ↝ **Done** ↝ **Boundaries** ↝ **Done**. The BNDRS OPTS menu will pop up. The list that follows describes the options contained in this menu.

Blended Surf	Blends a surface between selected curves, edges, and datum points. The blend can be controlled in one or two directions.
Conic Surf	Generates a conic surface between two boundary curves along a control curve.
Approx Blend	Generates a surface by approximating the selected curves. Allows the user to enter from selected curves the amount of acceptable deviation.
N-Sided Surf	Allows the user to create a surface with more than four bounding curves.

7. From the BNDRS OPTS menu, select **Blended Surf** ↝ **Done Blended Surf**. The CRV_OPTS menu, shown in the illustration at right, will pop up.

By default, the First Dir option is selected and you can start picking the bounding curves that define the first direction. Notice that under the SELECT ITEM menu you have Curve, Point/Vertex, and

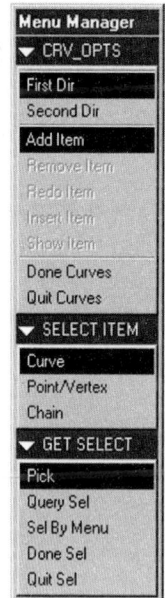

CRV_OPTS menu.

Chain. These options allow you to pick different types of entities. Curve will allow you to select existing datum curves. Point/Vertex allows selection of existing datum points or end points of datum curves.

The Chain option allows you to select existing geometry edges or datum curves and to trim or extend them if necessary. This is very helpful if you are creating surface patches where one boundary may be defined by the outer profile of a part and you need only a small portion of it to define the patch feature. You can select the long boundary by chain and then trim it back to the desired length.

8. To set up the **First Dir** of your surface, pick Datum Curve 1 and then Datum Curve 2, as shown in the illustration at right.

9. The first direction is complete. Select **Second Dir** and then pick Datum Curve 3. This is the first curve of the second direction. Because you do not have an existing curve defining a closed boundary, you will need to use the **Chain** option. At the bottom of the CHAIN menu, select the option **Two Points**; then pick the two curve end points opposite DTM3, as shown in the following illustration.

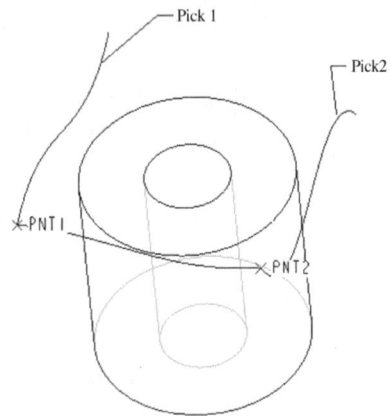

First direction curve selection.

Two-point chain selection.

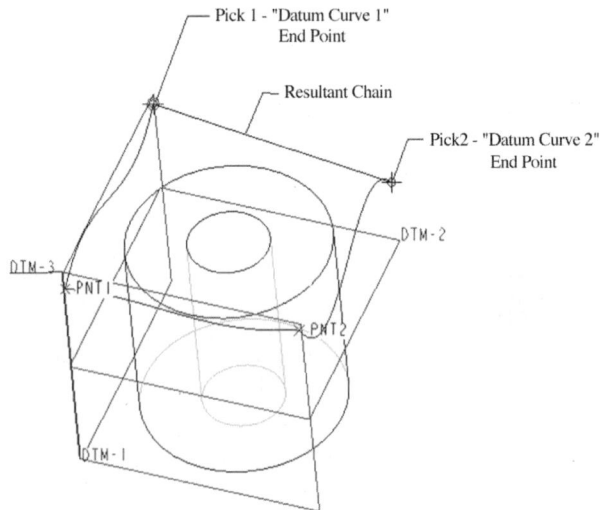

10. After picking the two curve end points, select **Done/Return** from the DEF
TAN menu and **Done** from the bottom of the CHAIN menu. These two picks
will complete the Two Point curve and the surface boundary. To finish the surface, select **Done Curves** from the CRV_OPTS menu and then pick the **OK**
button from the Quilt dialog box. The completed surface should look like that
shown in the first illustration at right.

Inside the Quilt dialog box are some
additional elements that will be useful
from time to time. The second illustration at right shows a screen shot of the
expanded dialog box.

The first element in the dialog box tells
you how many curves in each direction
make up this surface. In this exercise, it is
two by two. The second element, Bndry
Conds, allows you to select the boundary
edges of your surface and make them
Free, the default, or make them Tangent,
Normal, or Crvtr Cont to adjoining surfaces. The third element, Control Pts, lets
you determine how the curves in each
direction will blend. It does this by high-
lighting entity point locations along each
selected boundary and allowing you to
select which points blend to which,
between curves.

Completed boundary surface.

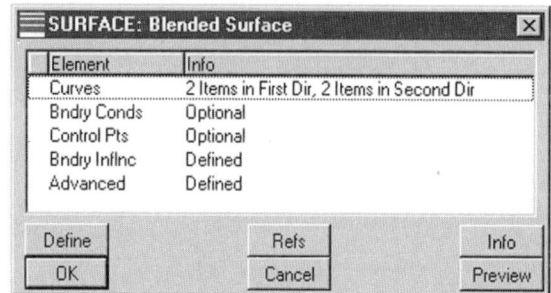

When selecting control points, if you
cause the surface to fold over or intersect
itself, the feature will fail. Exercise 13-3

Quilt dialog box.

contains an example of control point selection. Element four, Bndry Inflnc, determines
how much control each boundary curve has over the shape of the resulting surface. The
last element, Advanced, lets you control the surface's tangency to its inner edges in both
directions.

The final step in this exercise is using the boundary surface to shape the top flat surface
of the cylinder. Within the TWEAK menu is the option Replace, which allows you to

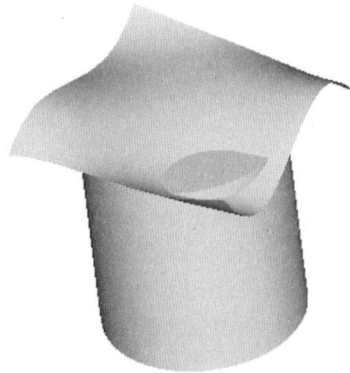

replace an existing solid face with a new surface, trimming or extending the geometry where needed.

11. From the PART menu, select **Feature ➥ Create ➥ Solid ➥ Tweak ➥ Replace**. Pick the flat cylinder top surface as the surface to replace, and then pick the boundary surface as the new surface contour, as shown in the first illustration at right.

12. Select **OK** from the Replace Surface dialog box and the boundary surface will become part of the solid model, as shown in the second illustration at right.

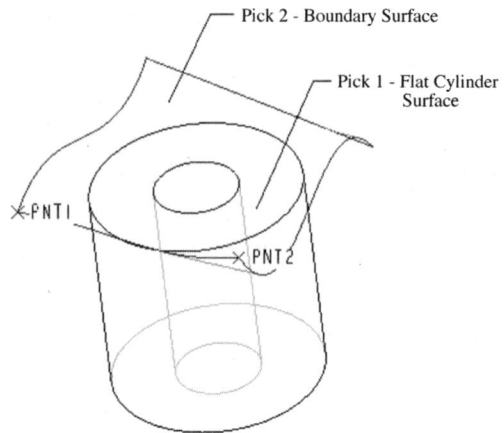

Replace surface screen picks.

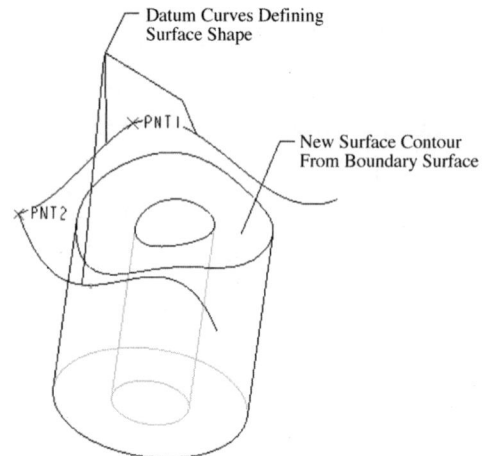

Completed Exercise 13-1 model.

In exercises 13-2 and 13-3 you have worked through some of the basic yet mandatory tools of surface modeling. You have created Pro/ENGINEER native surface geometry, modified surface boundaries, and generated solid geometry from a closed surface quilt. Sometimes your product and part design is passed from software other than Pro/ENGINEER. In these cases, you will need to gain some familiarity with the Import and Export tools. In the following exercises, you will be creating a surface model and then using it as the basis for an IGES file export. Following that, you will be working through some of the problems that arise during model generation based on an IGES file import.

Exercise 13-3: Importing Curves for Surfacing

From time to time during your career as a solid model designer or engineer you will need to work with design information developed on CAD systems other than Pro/ENGI-NEER. In some cases you will need to import curve and surface data to define the boundaries of the model you will be constructing. In the example shown in the following illustration, the part boundaries for the model were determined by an imported curve and surface file that constitute a rough outline of a dashboard and vent area near the fly window on the driver's side of a truck. The goal is to develop an electronic mirror control panel, shown in the illustration at right, that fits along the A-pillar of the truck but does not interfere with the airflow from the vent.

Because surface and curve data existed for the dash and vent area, there was a boundary this model could not violate. The control panel itself had to be located above the vent and dash area within the boundary defined by the windshield, and the A-pillar cover had to be tapered away from the vent but around the outside of the pillar itself. One of the most important parts of a project such as this is to gain an understanding of the import data, simplify or delete excess information, and then break the data into boundaries and plan your model within the boundary constraints.

In this exercise, you will be creating a set of curves, exporting them as a datum curve IGES file, and then importing them as the basis for your model. You will then be creating surfaces through your imported curves, and in the process learning a little about controlling surface appearance.

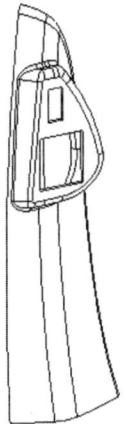

Mirror control panel.

1. Create the default datum planes and the default coordinate system.

2. From the PART menu, select **Feature** ➥ **Create** ➥ **Solid** ➥ **Protrusion** ➥ **Blend** ➥ **Solid** ➥ **Done** ➥ **Parallel** ➥ **Regular Sec** ➥ **Sketch Sec** ➥ **Done** ➥ **Smooth** ➥ **Done**. Pick DTM3 as your sketch plane, create the feature in the yellow direction, and then pick DTM2 as the top for the horizontal reference. Sketch the sections shown in the following illustration. Remember to toggle between the two sections, and make your start points on the same entity segment, as shown in the illustration below.

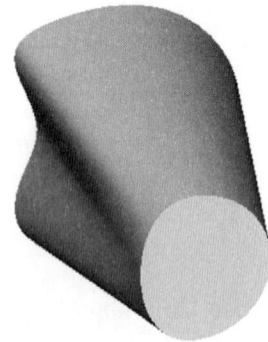

Base geometry.

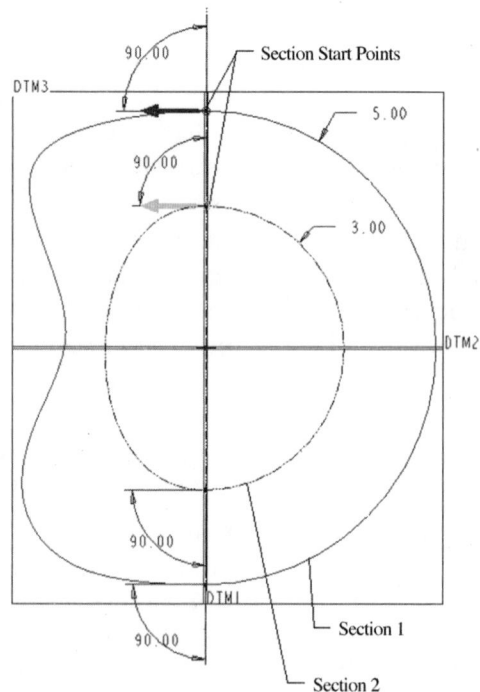

Blend sections.

3. Once the section is complete, select **Done** ➥ **Blind** ➥ **Done** and enter *12.0* for the protrusion depth. Select **OK** from the Protrusion dialog box to complete the feature.

4. Set up the curves you will be exporting by creating datum curves at the intersection of two surfaces. First create a new datum plane offset from DTM3 by 2.0 inches; then create a datum curve at the intersection of the plane and the solid model surfaces.

5. From the PART menu, select **Feature ➤ Create ➤ Datum ➤ Curve ➤ Intr Surfs ➤ Done ➤ Whole**; then pick the newly created datum plane. Select **Whole** again, and pick the solid model. Select **Done** and the system will generate a datum curve around the entire outside of the model at that intersection, as shown in the first illustration at right.

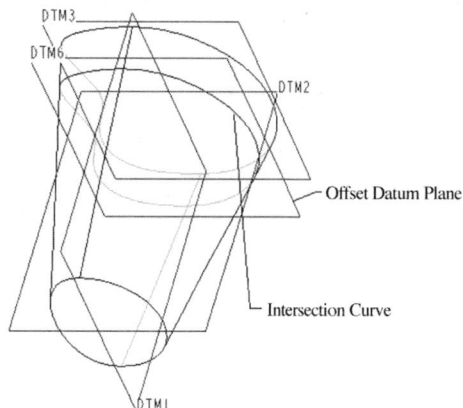

Intersection curve.

6. Create a pattern of the datum plane and curve by selecting **Feature ➤ Create ➤ Group ➤ Create ➤ Local Group**. When prompted for a group name, enter *Export* and press the Return key. Pick the offset datum plane and the intersection curve to be grouped, and select **Done** from the SELECT FEAT menu. This will complete the local group creation. Select **Pattern** from the LOCAL GROUP menu and pick the offset datum plane. You will now be prompted "Select pattern dimensions for FIRST direction, or increment type." Pick the offset dimension **2.0** and then **Done**. Enter a total number of instances of 6 and select **Done**. The complete group pattern should look like that shown in the second illustration at right.

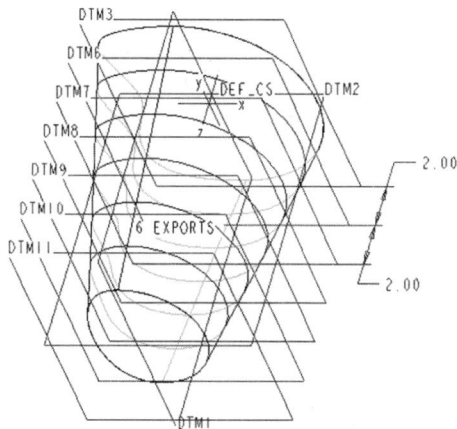

Datum plane pattern.

7. Create a datum curve at the intersection of DTM3 and the solid model. This time, create the curve as a composite. Select **Feature** ➡ **Create** ➡ **Datum** ➡ **Curve** ➡ **Composite** ➡ **Done** ➡ **Exact** ➡ **Done** ➡ **Tangent Chain**. Pick the edge along DTM3 and accept the start point location. Select **Done** and then the **OK** button from the Curve dialog box to complete the curve feature.

8. Export the model as a set of datum curves to use for your new surface model. From the **File** pull-down menu at the top of the Pro/ENGI-NEER window, select **Export** ➡ **Model**. From the pop-up EXPORT menu, select **IGES**. When prompted "Enter export file name," enter *Curves* and press the Return key. The Export IGES dialog box, shown in the illustration at right, will pop up.

Export IGES dialog box.

9. The Export IGES dialog box allows you to select what types of information you want to include in your IGES file. For this exercise, pick the **Datum Curves and Points** box and deselect the **Surfaces** box. This dialog box also lets you select part layers for export, in addition to specific surfaces for export instead of a blanket ALL Surfaces, which is the default.

10. Another area of interest in the Export IGES dialog box is the Coordinate System selection box, which lets you pick any coordinate system on the part. For this exercise, select **Default**. The other important area on this box is the **Options** button. This button allows you to read in specific *iges_config.pro* files. The *iges_config* file allows you to set how Pro/ENGINEER will import or export specific entities. When sending IGES data to other systems, it is important to know how that system translates entities such as splines and surfaces.

11. After setting the Export IGES dialog box settings, as shown in the previous illustration, pick the **OK** button to generate the IGES file. **Save** the model; you will want to use it as a reference later in this exercise.

12. Create a new part consisting of the default datum planes and coordinate system. Select **File** from the pull-down menu across the top of your window, and then pick **Import ➡ Append to Model**. From the Append Model pop-up window, select the *Curves.igs* file to add to your new model. An INFORMATION window will open. It shows data about IGES entity types, numbers, and conversion totals from the file you are importing. Select the **Close** button at the bottom of this window. In the meantime, a second dialog will have popped up. This dialog allows you to select the coordinate system to

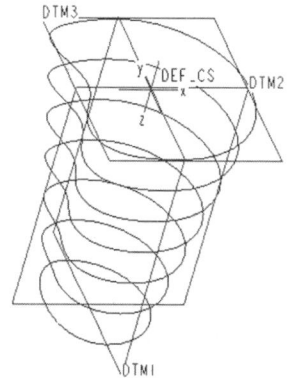

Imported IGES file.

locate the import IGES file to. For this exercise, select the **Default** setting. Select the **OK** button to complete the file import. The new model should look like that shown in the illustration at right.

At this point, the set of seven datum curves is usable, but not modifiable. It is also a single feature. Therefore, if you wanted to delete one of the curves by simply using the Delete option, all of the curves would be deleted and you would be left with only the default datum planes and coordinate system.

To modify the entities of an imported feature, you need to use the Redefine option. The options under the REDEFINE import menu allow you to read in a different file, delete entities in the current import, modify imported curve line styles, and modify and fix imported surfaces. Keep in mind that like almost all other functionality in Pro/ENGI-NEER, there is no undo command under this redefine menu. If you delete entities, you need you will need to re-read in the import file.

> ➡ **NOTE:** *It is always a good idea to save often when redefining imported geometry. Modify or delete a small amount of the imported feature, save it, and then redefine and modify some more.*

13. To see some of the control issues that arise when building surfaces from imported curves, create a surface by boundaries through the import feature. From the PART menu, select **Feature ➡ Create ➡ Surface ➡ New ➡ Advanced ➡ Done ➡ Boundaries ➡ Done ➡ Blended Surf ➡ Done**. For this surface, use the default menu selection from the CRV_OPTS menu. Pick each curve from the imported feature, starting with the largest curve and moving to the smallest, as shown in the following illustration.

14. After you have picked the seven curves, select the **Done Curves** option and select the **OK** button from the Surface dialog box.

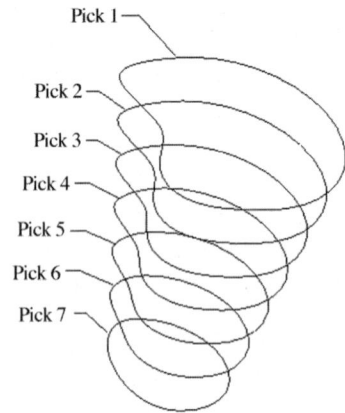

Pick 1
Pick 2
Pick 3
Pick 4
Pick 5
Pick 6
Pick 7

Order of curve selections.

You should now be presented with a Resolve button in place of the OK button on the dialog box. This is telling you that Pro/ENGINEER needs more information to complete the surface. In effect, what is happening is that because each imported curve consists of two entities, the system is trying to blend a surface between curves whose end points are flipping between top and bottom locations, twisting the surface. This is similar to the effects shown in exercise 8-1, Creating Parallel Blends, but in this case the surface is intersecting itself and failing.

15. A very quick way of solving this problem is to select the **Control Pts** option in the dialog box and then the **Define** button. This option will highlight each curve you select and show each curve end point, starting with the first curve you select, as shown in the second illustration at right.

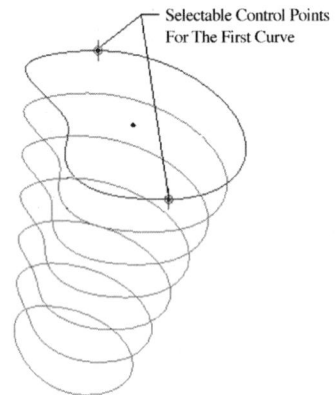

Selectable Control Points For The First Curve

Highlighted control points.

16. Pick the upper point as the **Control Point**. After selecting the control point on the first curve, two points will appear on the second curve. Pick the upper point as the control for each curve; then select **Done**. Select **OK** from the dialog box.

Now that you have told the system how to blend the surface, it can generate the feature, shown in the third illustration at right. Notice the variance in this model's surface shape from the model you originally generated the curves from.

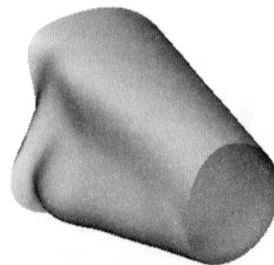

Boundary surface.

The blended surface shown in the previous illustration is allowed to pucker and bulge as it flows between curves, and represents a surface that does not have enough constraints. The issue of how many control curves to add to a surface, and whether or not to have control curves in one or both directions, is a very difficult one to solve. For the example shown in the previous illustration, you could smooth out the surface by redefining it and removing some of the curves, but you would then need to experiment until you got the appearance you desired. Another way of controlling the surface shape is to add curves to the second direction, which will give the surface more definition by constraining its shape in both directions.

17. Because this first surface model is unacceptable, delete it. Create the control curves for the second direction of the surface. For the first curve, select **Feature ➥ Create ➥ Datum ➥ Curve ➥ Thru Points ➥ Done**. From the CONNECTION TYPE menu, select **Spline ➥ Single Point ➥ Add Point**, and then pick each of the seven imported curves' upper end vertices, as shown in the following illustration. You may want to use the **Query Sel** option to locate and pick the vertices.

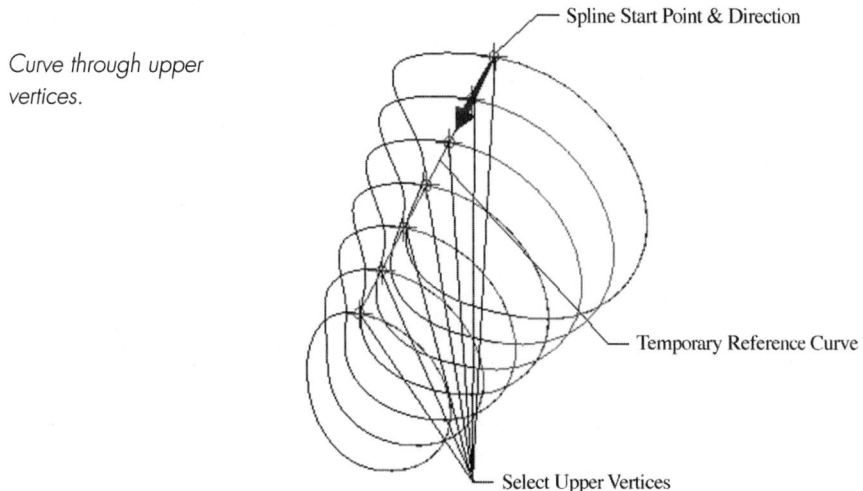

Curve through upper vertices.

Spline Start Point & Direction

Temporary Reference Curve

Select Upper Vertices

18. Once you have selected the vertices, select **Done**; then select the **OK** button from the Datum Curve dialog box. This will complete the first control curve. Repeat this process to create a datum curve through the lower end vertices. The vertex selections are shown in the following illustration.

The curve model should now look like that shown in the second illustration at right. At this point, if you were to create the surface again through the curve you have just created, the resulting surface would be very similar to the wavy model surface you deleted. You still need more control over the surface. Therefore, you need to create two more curves along the sides of the imported curves. These curves will take out the rest of the surface's undulations. Unfortunately, the imported curves do not have vertices along their sides. Therefore, here you will have to make a choice.

Under the datum curve CRV OPTIONS menu you have the option to Split existing curves into segments. This option will work very well for this case, but will create fourteen additional features once you have split the curves on both sides. Another option is to create a series of datum points at the intersection of DTM2 and the imported-curves feature. This option will create one additional feature by ganging all points into one array. To keep the feature count down, the datum point option is the one to use here.

19. From the PART menu, select **Feature ➡ Create ➡ Datum ➡ Point ➡ Add New ➡ Curve X Srf**. Pick the largest imported curve along one side, then DTM2, as shown in the third illustration at right.

20. Continue to **Add New** points using the **Curve X Srf** option and selecting each curve, down one side then down the other, until all curves have two points, one on each side.

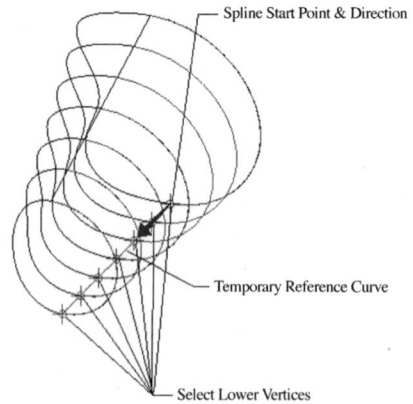

Curve through lower vertices.

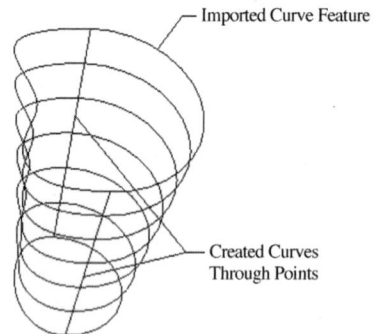

Curve model.

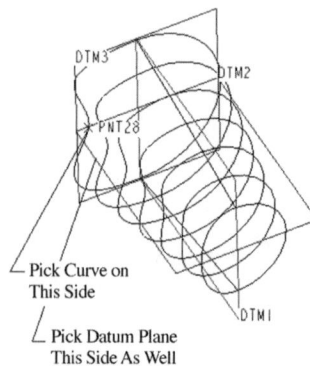

Curve X Srf picks.

After the datum point feature containing all fourteen points has been created, it should look like that shown in the following illustration.

Completed datum point feature.

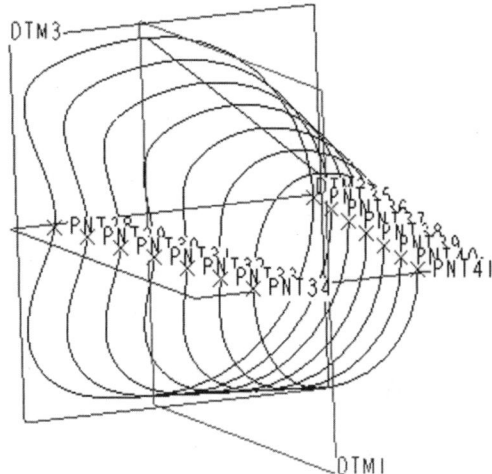

21. Now that you have points created, create the last two datum curves for controlling the shape of the boundary surface. From the PART menu, select **Feature ➥ Create ➥ Datum ➥ Curve ➥ Thru Points ➥ Done**; then select **Spline ➥ Single Point ➥ Add Point**.

The reason you use the Single Point option instead of Whole Array in this case is that the Single Point option allows you to pick each point individually, skipping some if you desire. If you had used the Whole Array option, you would only need to pick one of the points you created and the entire feature would have been picked. The resulting curve would wrap around both sides of the model, which is not what you want in this case.

22. For this curve feature, select each point along the left side of the model, which are PNT28 through PNT34 in the previous illustration. After picking the last point, select **Done**; then select **OK** from the Datum Curve dialog box. Repeat the last feature creation procedure to make a datum curve through PNT35 to PNT41. Once this curve is complete, the model should look like that shown in the following illustration.

Curves through datum points.

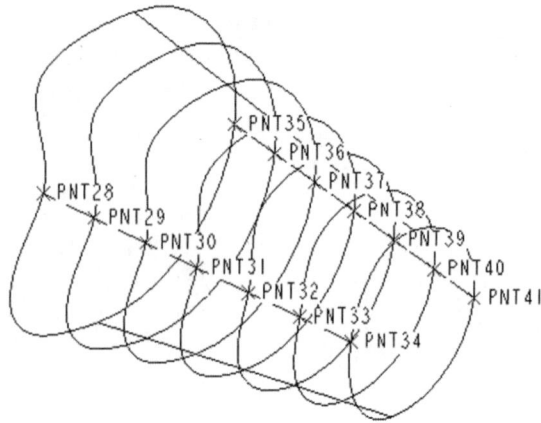

23. Again using boundaries, create a surface through the imported curves, this time adding a second direction of control by using the newly created datum curves. Select **Feature** ➙ **Create** ➙ **Surface** ➙ **New** ➙ **Advanced** ➙ **Done** ➙ **Boundaries** ➙ **Done** ➙ **Blended Surf** ➙ **Done**. For the **First Dir** curves, pick the imported curves, starting with the largest. Pick the curves in order from the largest to the smallest, and then select the **Second Dir** option. The curves for the second direction are the straight curves along the top, bottom, and sides. Pick each of the four **Second Dir** curves. When you have finished, select the **Done Curves** option, and then select the **OK** button from the Surface dialog box. The new surface is shown in the following illustration.

Surface controlled in two directions.

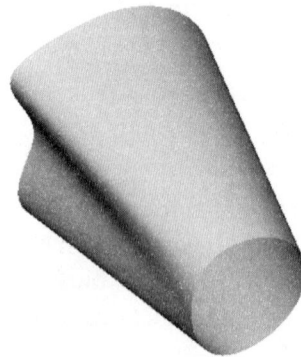

This new model should more closely represent the original model from which the curves were exported. In this section of the exercise you could have used the Control Pts option to define the blending of the surface along the top and bottom because the imported curves had end points present. However, this illustrated two types of curve "through-point" selection types: implicit points/vertices selection and explicit point selection.

Keep in mind that surfacing in Pro/ENGINEER is not an exact science. The number of control points you need for defining a surface depends greatly on the amount of detail and curvature you need to create. In this area, there is no substitute for experience. Use surfaces only when and where necessary in your daily project modeling.

As of now you have successfully created a model based on imported IGES curves. The model looks very much like the original part on which the curves were based. This is a good thing. The only problem is that the model is not modifiable. If it is a design that will never change, or if it is strictly for archival purposes, this might be alright. However, because this is seldom the case, you need to find a method to make the model more user- and design-friendly. In the following exercise you will do just that. Using an imported IGES curve file as the basis of your model design, you will create a modifiable Pro/ENGINEER part.

Exercise 13-4: Making Imports Modifiable

In this exercise, you will be redefining an imported datum curve file (to delete some of the excess curves) and then using those curves as the basis for modifiable surface geometry. The imported IGES file you will be using is the same one you used in exercise 13-3.

1. Create a new part consisting of the default datum planes and coordinate system. Select **File** from the pull-down menu across the top of your window; then select **Import** ➡ **Append to Model**. From the Append Model pop-up window, select the *Curves.igs* file to add to your new model. As before, an INFORMATION window will open. It shows data about IGES entity types, numbers, and conversion totals from the file you are importing. Click on the **Close** button at the bottom of this window. In the meantime, a second dialog will have popped up. This dialog allows you to select the coordinate system for locating the import IGES file to. For this exercise, select **Default**. Select the **OK** button to complete the file import. The new model should look like that shown in the first of the following illustrations.

Imported IGES file.

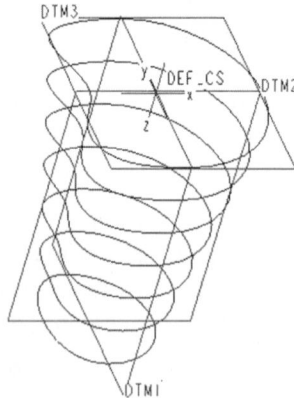

REDEF IMPT
(Redefine Import)
menu.

2. Redefine the imported geometry. Select **Feature ➡ Redefine** and then pick the imported curve set. This will bring up the REDEF IMPT menu, shown in the illustration above right.

3. The following list describes the options within the REDEF IMPT menu.

File	Allows you to re-read in the current import file or select a new file to read into Pro/ENGINEER.
Delete	Allows you to delete points, datum planes, curves, and surfaces. It also opens a secondary menu, allowing you to select exactly what type of entity you wish to delete. By default, all entity types are selected.
Line Style	Allows you to change the line style of selected wireframe and datum curve entities. In some cases, it is very handy to modify the line style or color of imported curves to assist in differentiating them from Pro/ENGINEER native geometry.
Attributes	Is used when importing surface geometry. When selected, this option opens another menu containing the options Make Solid and Join Surfs. Before an imported surface can be converted to solid geometry, you must first Join Surfs. If both options are selected, Pro/ENGINEER will join all surface boundaries and then convert the joined quilts into solid geometry. One thing to be aware of: Not all IGES files are complete; some contain small gaps, duplicate surfaces, and/or nontrimmed surfaces. The geometry created after joining and solidifying some files may be unexpected and incomplete.
Exclude Srfs	Allows you to deselect surfaces being joined under the Attributes option. For situations in which extra surfaces or nontrimmed surfaces may cause problems in joining a set of surfaces, you can exclude the problem surfaces from the quilt.
Fix Bndries	Is useful for correcting problems in imported surface geometry. In many cases, IGES import surfaces have small gaps and misalignments. This option allows you to use Zip Gaps and to edit surface boundaries.

For this exercise, select **Delete** and then pick both entities of the five interior curves for deletion. The curves you need to pick are shown in the first illustration at right.

4. After picking the curve entities shown in the previous illustration, select **Done** from the DELETE ENT menu. This will complete the deletion process, leaving your file looking like that shown in the second illustration at right.

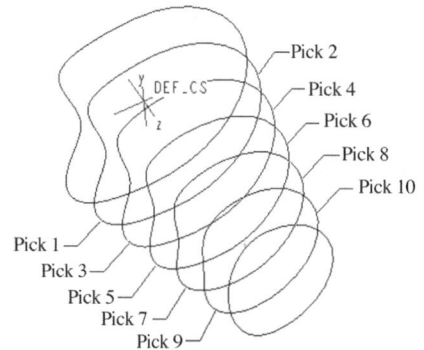

Curves to be deleted from import.

5. Select **Done/Return** from the REDEF IMPT menu. This will complete the redefine process and put you back into the feature creation mode.

Now that you have cleaned up the import file, you are ready to make two modifiable and Pro/ENGINEER-native datum curves from which to create a new surface.

6. Select **Feature** ➡ **Create** ➡ **Datum** ➡ **Curve** ➡ **Sketch** ➡ **Done**. Pick DTM3 as the sketch plane and DTM2 as the top horizontal reference.

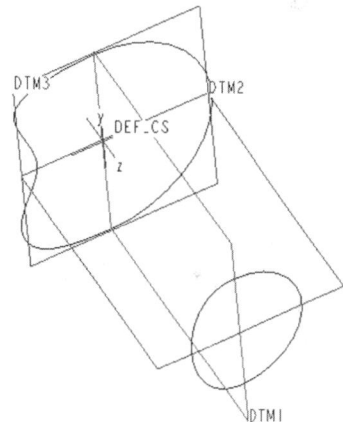

Import file after curve deletions.

7. In Sketch mode, select **Geom Tools** ➡ **Use Edge** ➡ **Sel Loop**; then pick the larger curve to use for the entity creation. You now have an exact copy of the imported curve as the new curve section. However, the section is not yet modifiable. Select **Alignment** from the SKETCHER menu, then **Unalign All** ➡ **Confirm**. This will give you an exact duplicate of the original import curve, which you now need to dimension to your datum planes. You can also add, delete, and modify all internal spline points to change the shape if necessary, but this does result in a curve that no longer references the import. Add tangency to the end points of the spline segment and align the curve end points to DTM1, then dimension the curve section as shown in the following illustration.

*Curve section
dimensioning.*

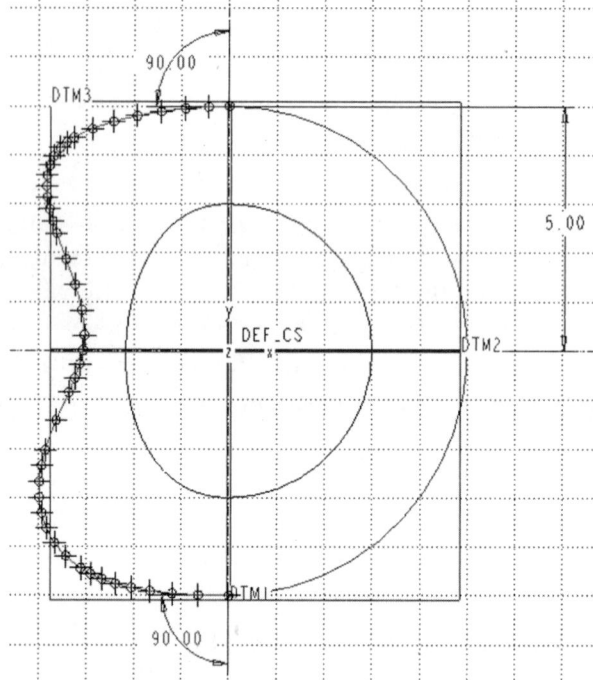

8. Once dimensioning is complete, select **Done**, then select **OK** from the Curve
dialog box to finish the datum curve creation.

9. Create a new datum curve for the smaller import curve. In this case, you know
the offset distance to the small curve is 12. However, if you did not know this,
you could use the Info pull-down menu and select the Measure option to
determine that value. Create the smaller curve's duplicate curve by repeating
the procedure used for the large curve. This time, offset the sketch plane from
DTM3 by 12. Create the curve using the existing curve edge and dimension
the section as shown in the following illustration.

Small curve section
dimensioning.

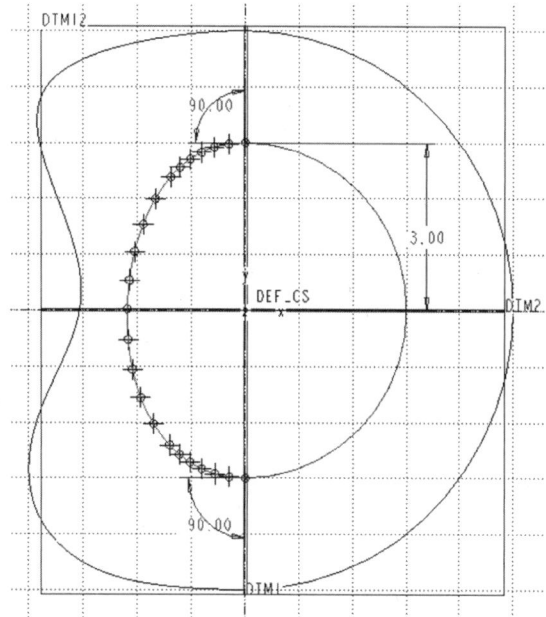

DTMI2

90.00

3.00

DEF_CS

DIM2

90.00

DTMI

10. Because you have just created two new datum curves that are identical to the import curve but no longer reference the import geometry, **Delete** the import feature. Use the **Query Sel** option or model tree to ensure you select the correct feature for deletion.

11. The final step is to create a surface between the two modifiable datum curves. Create a **Boundary** surface between your two datum curves. Select **Feature** ➥ **Create** ➥ **Surface** ➥ **New** ➥ **Advanced** ➥ **Done** ➥ **Boundaries** ➥ **Done** ➥ **Blended Surf** ➥ **Done**. Pick the larger curve as the first curve boundary, then the small curve. To complete the surface, select **Done Curves**; then select **OK** from the Surface dialog box. The surface you have just created can now be modified, by changing the base curves. This method of using imported curve data as a temporary reference for Pro/ENGINEER native geometry is a good method of making modifiable models from imported data.

Summary

Throughout this chapter you have become familiar with the basic and most commonly used tools for surface creation in Pro/ENGINEER. The ability to create modifiable solid model geometry while capturing design intent is critical, and never more so than when you start using surfaces in your models. Remember that those who subsequently work on your models may not have your skill level with surfaces and datum curves. Therefore, they will need a very clear trail to follow. While using surfaces for solid model features, the quilts you build disappear when used as a protrusion, which makes it less than obvious that a surface was used. By naming important features, you leave behind that necessary trail through surface models.

In many cases, surfaces allow you to build geometry you otherwise might have a difficult time creating. This is the main reason for using surfaces. You should not use them just because you can. Surface geometry will always cause your model feature count to balloon. This is because surface geometry will always need to be trimmed, extended, merged, or constrained before it can be used as a protrusion or cut. When using imported geometry, this becomes even more apparent.

An import can sometimes be a very messy file, containing a great deal of extemporaneous geometry. In most cases, you should analyze the geometry on your screen and then plan what should be kept and what deleted. Then delete the unnecessary curves, points, and surfaces to make visualizing your model easier. You can always re-read in the original import file if you need it, or break the original import into multiple features by importing it several times and deleting varying data.

For example, you might be faced with a situation in which you import a file containing an entire car, and need to break the four doors out into separate import features. You could import the file the first time, then delete all of the geometry except for the driver side front door. You would then import the file again as a new feature and delete all geometry except the passenger side front door. You would then have two features referencing the same file but displaying different geometry, making model maintenance and visualization a bit easier.

In the final exercise of this chapter, you had the opportunity to use an imported feature as the basis for native Pro/ENGINEER geometry. The method employed in this exercise of temporarily referencing imported curve data, then disassociating your native Pro/ENGINEER entities from that imported feature, can be very useful. It results in modifiable geometry, and allows you to delete the excess import feature to lower feature count and ease model visualization.

14

Draft

In the early days of solid modeling, draft was one of the more difficult things to model. Discussion centered on whether draft had to be modeled at all, with the possibility of it being covered in a standard note on the drawing. Things have changed. More often than not these days, a part is already well on its way to being made before the drawing is even started. With time to market as critical as it is, there is no time to wait for a drawing when the mold, die, or pattern can be, and often must be, started right away.

The technologies that make this type of scenario possible, such as rapid prototyping and numerically controlled machining, require that draft be added to models. This is a must, and luckily for designers the techniques for creating draft have improved over the years to handle more and more complex situations.

In this chapter, you will learn about the draft feature and its options. The two types of draft, neutral plane and neutral curve, will be covered, as well as the various methods of creating split drafts. There is also an example of creating draft without using the draft feature, in which making a surface at the desired draft angle using a Variable Section sweep creates the draft surface. However, first you need to explore the basics of draft.

Draft Feature Basics

The first thing to learn about the draft feature is where it is located in the Pro/ENGINEER menu structure. It is located under the TWEAK menu with 12 other options used to modify existing model geometry or to create very specialized geometry. A draft surface is a surface at an angle to the parting line of a tool (e.g., mold or die) that will allow for the part to be removed from the tool after the part is cast (see the following illustration). Even a surface perfectly perpendicular to the parting line will cause too much drag on the part to release it from the tool.

What draft is and why it is important.

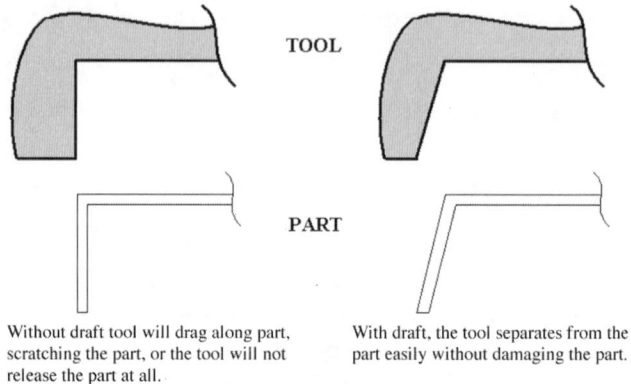

TOOL

PART

Without draft tool will drag along part, scratching the part, or the tool will not release the part at all.

With draft, the tool separates from the part easily without damaging the part.

The fact that your model is likely going to be used directly or indirectly to create the tool for the part is why creating draft in the model becomes so important. Pro/ENGINEER creates a draft surface by pivoting the model surface at the intersection of the model surface and the neutral plane or about a neutral curve. The following illustration shows how Pro/ENGINEER determines a simple neutral plane draft surface.

The basic idea behind how Pro/ENGINEER creates a draft surface.

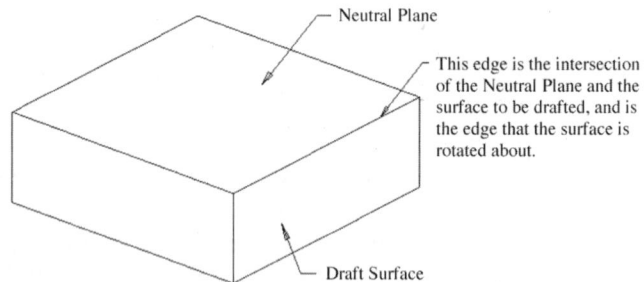

— Neutral Plane

This edge is the intersection of the Neutral Plane and the surface to be drafted, and is the edge that the surface is rotated about.

Draft Surface

The following is the information you need to provide Pro/ENGINEER for it to be able to create a draft surface.

- Model surfaces to be drafted

- Neutral plane or curve

- Angle at which the surface is to be rotated

Neutral Plane Drafts

The neutral plane draft requires that a plane be selected to determine where to pivot the model surfaces in creating the draft. The selection of the neutral plane is critical to the end result of the draft surface. The following illustration shows two protrusions of equal height and diameter.

Identical protrusions before any draft has been added.

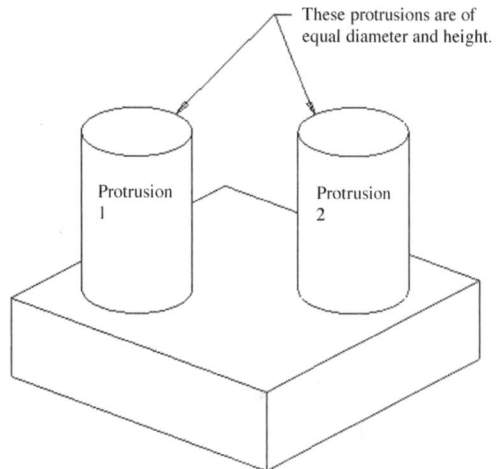

These protrusions are of equal diameter and height.

Protrusion 1

Protrusion 2

In the previous illustration, the same angle draft has been added to each of the protrusions, but different neutral planes were selected to create the two drafts. The draft for protrusion 1 was created selecting the plane at the base of the protrusion. Because the neutral plane determines the pivot edge for the draft, the diameter at the base of protrusion 1 remains the same, and the boss tapers from there.

For protrusion 2, the bottom of the block was selected as the neutral plane. This moves the pivot edge down to the bottom of the block and is where the taper of the draft begins. Because the draft starts at the bottom of the block, and at the original diameter of the boss, by the time it reaches the top of the block the boss diameter has already been reduced by the draft and continues to get even smaller. Clearly this is not the design intent of the boss. As indicated in the following illustration, extreme care needs to be taken when selecting the neutral plane when creating draft.

This diameter is the same as the
original protrusion before any
draft was added.

The draft angle is the same, but
the angle starts at the different
Neutral Planes. This causes the
diameters at the base and top of
the protrusions to be different.

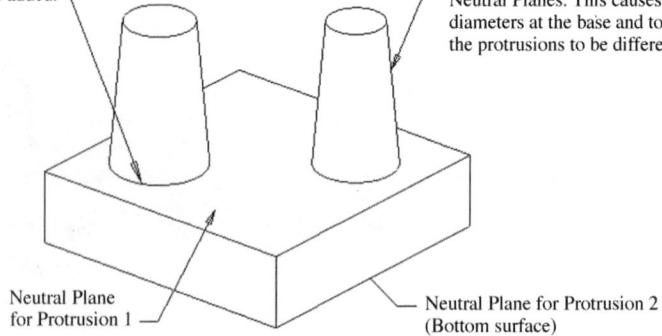

*How selection of the neutral
plane can affect the final
geometry of a part.*

Neutral Plane
for Protrusion 1

Neutral Plane for Protrusion 2
(Bottom surface)

In this first exercise, you will create a very simple Neutral Plane draft. This is
the most common type of draft feature used.

Exercise 14-1: Creating a Neutral Plane Draft

In this exercise, you will be creating a simple neutral plane draft.

1. Select **Feature** ➡ **Create** ➡ **Tweak** ➡ **Draft**. Select **Neutral Pln** ➡ **Done** ➡ **No
Split** ➡ **Constant** ➡ **Done**. The No Split option means that the entire surface
that was selected will be drafted as one piece. The Constant option tells you
that the draft surface selected will have the same draft angle across the entire
surface. Select the surfaces shown in the following illustration and select **Done
Sel** ➡ **Done**. You now have a choice of either selecting the neutral plane or
making a datum that will be used as the neutral plane. Select the neutral plane
shown in the illustration.

This top surface will be
used as the Neutral plane.

*Items that need to be
selected to create a
neutral plane draft.*

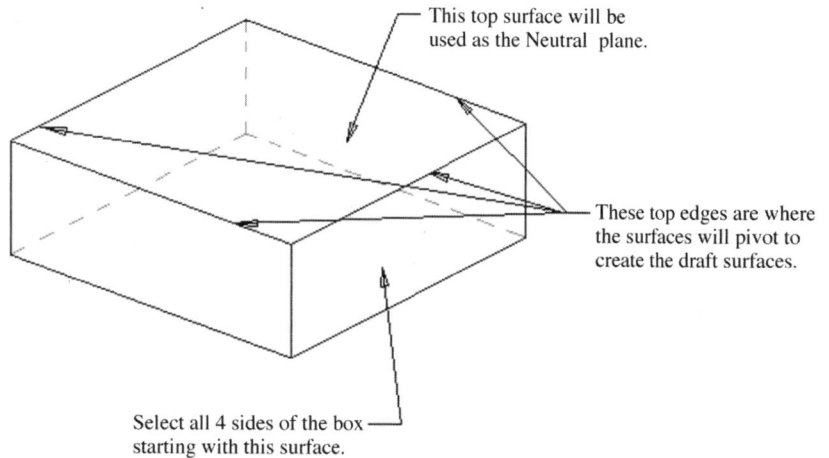

These top edges are where
the surfaces will pivot to
create the draft surfaces.

Select all 4 sides of the box
starting with this surface.

2. The next prompt you will receive is to set the reference direction. The reference direction is most often perpendicular to the surfaces being drafted. You
have several ways of selecting the reference direction entity. The first is to use
the Use Neut Pln option and not have to select anything. The second is to use
the Select option. The Select option allows you to pick the reference direction
by using one of three methods. The Plane option allows you to select an existing plane or create a new datum; the Crv/Edg/Axis option allows for a curve,
edge, or axis to be selected; and the Csys option allows you to select the axis of
a coordinate system. For this exercise, use the **Use Neut Pln** option and select
Done.

3. The last prompt is to enter the draft angle. There is a range of valid draft
angles. Before Release 20, the range was +15 degrees to –15 degrees, but has
now been increased to +30 degrees to –30 degrees. The first of the following
illustrations indicates that two arrows will appear and one of the draft surfaces
selected will highlight in yellow. The red arrow indicates the draft direction
and the yellow arrow shows the positive direction for the draft rotation. For a
No Split draft, a positive draft angle will remove material and a negative draft
angle will add material. You need to add material in this case; therefore, enter
a –5-degree angle and then select the **OK** button to complete the draft, shown
in the second of the following illustrations.

This surface is highlighted and the yellow
arrow shows the positive direction of the
draft rotation using the "Right Hand Rule"

*Arrows are displayed to
indicate the direction in which
the draft will be created.*

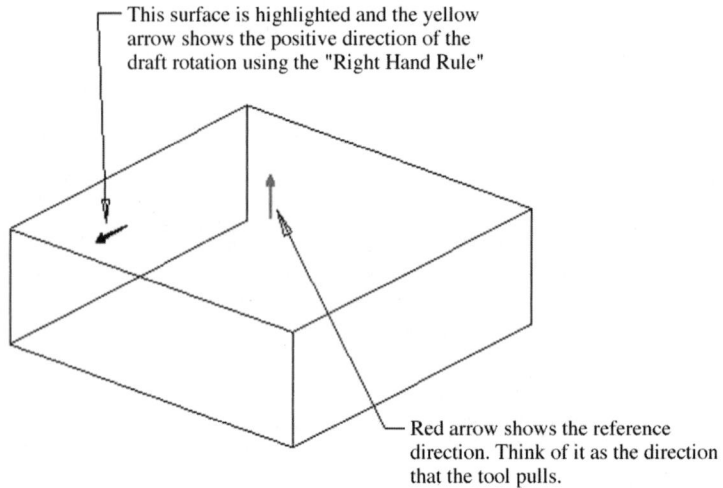

Red arrow shows the reference
direction. Think of it as the direction
that the tool pulls.

*Completed draft
feature.*

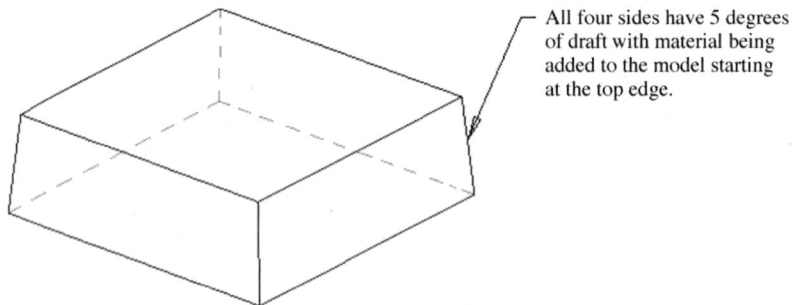

All four sides have 5 degrees
of draft with material being
added to the model starting
at the top edge.

Neutral Curve Drafts

The neutral curve draft works the same way as the neutral plane draft but
allows the draft surfaces to be pivoted about a curve or edge rather than the
intersection between the draft surface and the neutral plane. This gives you
the option of creating a draft surface that can pivot about a curve that is not
perpendicular to the reference direction. The other advantage of the neutral
curve draft is that the neutral curve, or where the draft surface pivots, can be
a nonplanar curve. With the neutral plane option, the pivot line or edge must
be planar. You will create a neutral curve draft in an exercise later in this
chapter. The following illustration shows an example of draft created with
the neutral curve draft type.

Draft created with Neutral Curve.

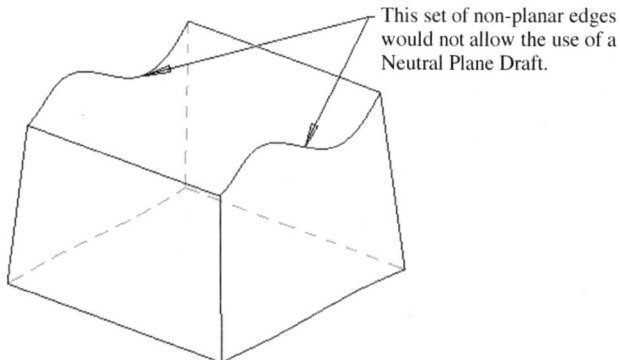

This set of non-planar edges would not allow the use of a Neutral Plane Draft.

Splits

A split draft is the line where two or more parts of the tool meet on a common surface. The term *split* is used because it is at this line the tool "splits" when opened. The split draft is important to the draft feature because the surface of the part with the split requires that it have two draft angles going in opposite directions. This is illustrated in the first of the following illustrations. The split draft can also be used to separate areas that may require different draft angles on the same portion of the tool. The second of the following illustrations depicts this second type of draft.

You will be creating the draft on this part in exercises 14-2 and 14-3.

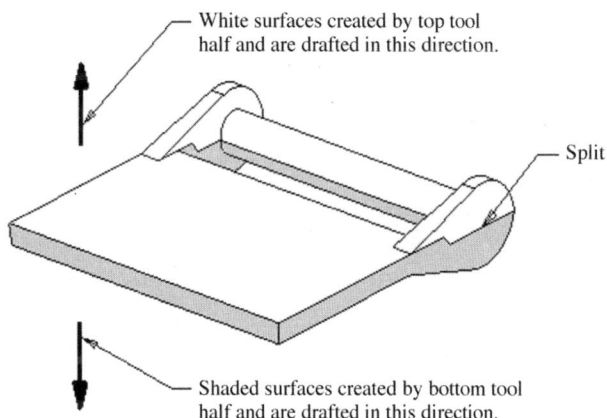

White surfaces created by top tool half and are drafted in this direction.

Split

Shaded surfaces created by bottom tool half and are drafted in this direction.

Split draft used to separate two areas with different draft angles but made by the same half of the tool.

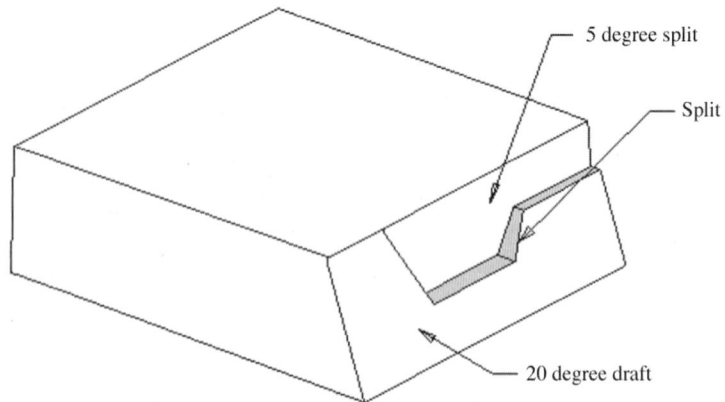

5 degree split

Split

20 degree draft

All surfaces created by the same half of the tool,
but some areas required different draft angles.

⇥ **NOTE:** *Pro/ENGINEER needs to know where to create the split. There are two ways Pro/ENGINEER can split a draft. The first is at a curve, and the second is at a surface.*

Neutral Plane Splits

There are two options for creating split neutral plane drafts: Split at Pln (split at an existing plane) and Split at Skt (split at a sketch at the time the draft feature is created). The Split at Pln option will create the draft split at an existing plane and will use that plane as the neutral plane.

Because the existing plane may not be planar to the draft surfaces, Pro/ ENGINEER prompts you for a reference plane from which to measure the draft angle, and asks for a planar neighbor. The term *planar neighbor* can be a little confusing. Think of it as an adjacent (shares an edge) and perpendicular model surface. With a Split at Pln draft, the draft angle applies to both sides of the split and will go in opposite directions. The following illustration shows a typical Split at Pln draft.

Elements required for creating a neutral plane split-at-plane draft.

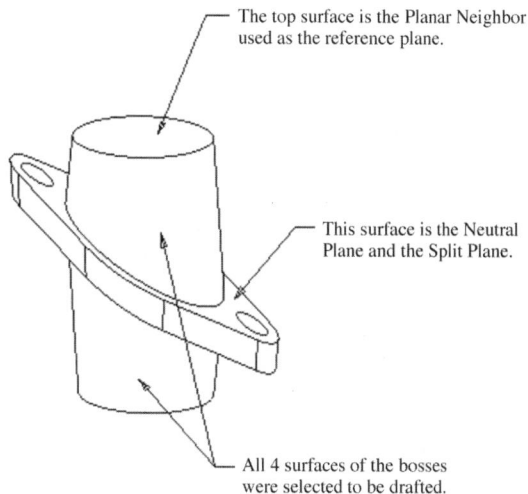

The top surface is the Planar Neighbor used as the reference plane.

This surface is the Neutral Plane and the Split Plane.

All 4 surfaces of the bosses were selected to be drafted.

The other split option for neutral plane drafts is the Split at Skt option. The split line is sketched as part of the draft feature and cannot be used as the neutral plane. The neutral plane is selected before the sketched split is created. You can sketch the split on the surface directly, or on a datum plane. If you sketch the split on a datum plane, it will be projected through the part and onto any surface selected. The following illustration shows the surfaces used for a sketched split draft.

Surfaces used in the sketched split draft.

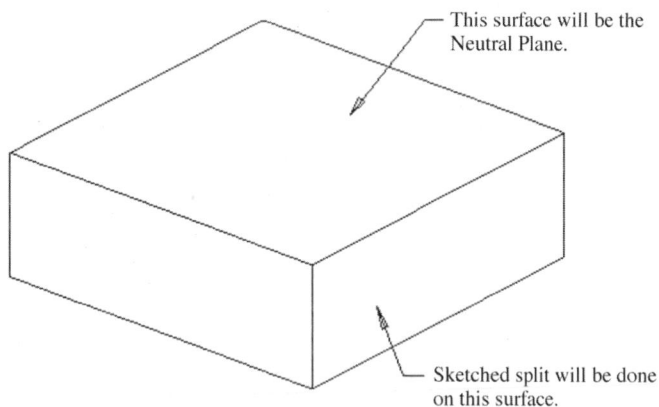

This surface will be the Neutral Plane.

Sketched split will be done on this surface.

The neutral plane is also the reference direction entity in a Split at Skt draft. Therefore, be careful about how you select the neutral plane when sketching splits. One additional difference between the two neutral plane splits is that each side of the split on the Split at Skt draft has a separate draft angle. The sign on the draft angle value determines whether the surface is rotated in to remove material or out to add material to the model, as shown in the following illustration.

Arrows indicating which side of the draft Pro/ENGINEER is working on and direction to rotate the surface.

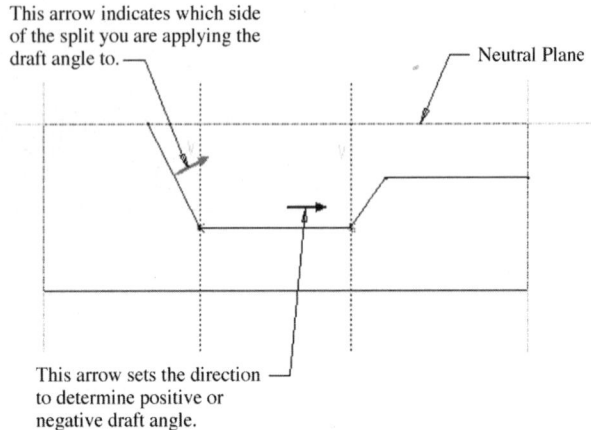

This arrow indicates which side of the split you are applying the draft angle to.

Neutral Plane

This arrow sets the direction to determine positive or negative draft angle.

The first of the following illustrations shows the completed draft feature. The second of the following illustrations is a top view of the completed draft. By viewing the model from the draft direction (red arrow when creating the draft feature), you can check to make sure the draft is correct. You must be able to see the entire surface, albeit foreshortened, for the part to be removable from the tool.

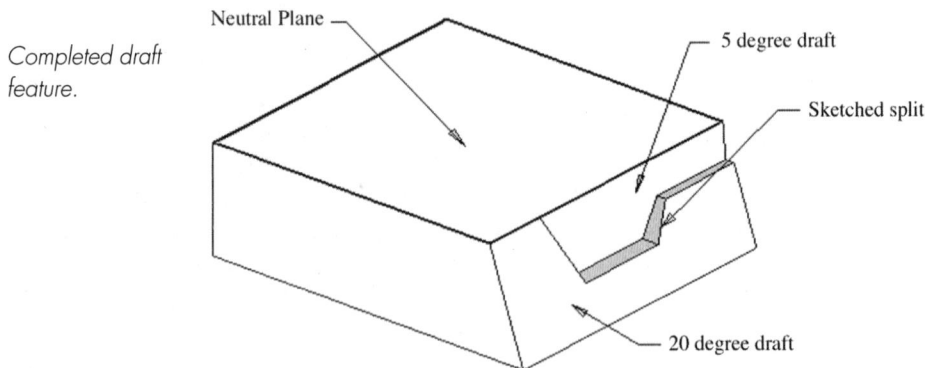

Completed draft feature.

Neutral Plane

5 degree draft

Sketched split

20 degree draft

Original block size and shape ⌐

*Top view shows that
all surfaces are visible
from the top, ensuring
draft is correct.*

— 5 degree draft surface

— Shaded area is the difference
between the draft angles at the
split.

— 20 degree draft surface

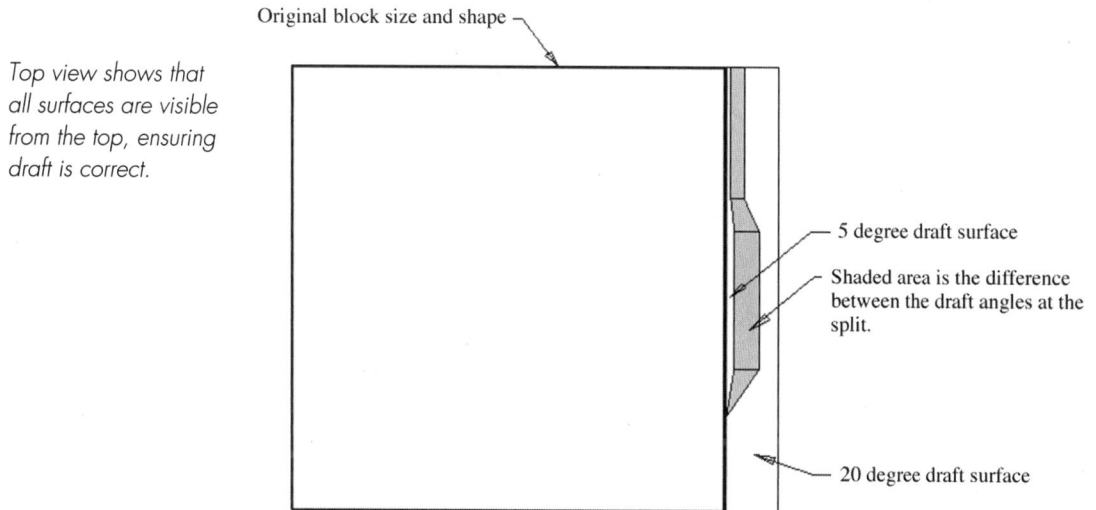

Neutral Curve Splits

The split functionality in neutral curve draft is similar to the neutral plane draft but is more flexible in the shape of the split and the variation of the split halves. Splits can be made at Split at Crv (Split at Curve) and Split at Srf (Split at Surface). Both the curve and the surface used to create the split need to be created before the draft. There is no option for creating the curve or the surface from a sketch inside the draft feature.

When you select a split from a neutral curve draft, you can create the draft on one or both sides of the split. If draft is being created on both sides of the split, you have the additional option of each side having an independent draft angle or each being dependent on the other. An independent angle means that each side of the split is treated as a separate draft. They have their own draft angle values and can even have different neutral curves. If the two sides are dependent on each other, there is only a single draft angle value and Pro/ENGINEER determines the remaining draft angle so that the two draft surfaces meet at the split.

Further Exercises

In this exercise section, you will be creating all of the draft on a single part. The part geometry shown in the following illustration is part of a plastic hinge pin. This will be the starting geometry for exercises 14-2 and 14-3. In exercise 14-4, you will use the variable section sweep feature to create the draft surfaces on an offset link part.

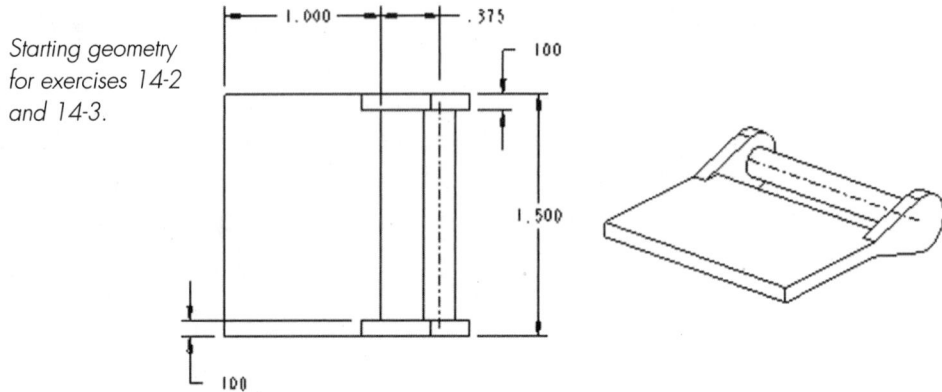

Starting geometry for exercises 14-2 and 14-3.

The draft requirements for this part are located in three definable areas. The first are the front and rear surfaces of the flat tab section. The other two areas of draft are the inside surfaces of the hinge pin ears and the hinge pin itself, and the outside surfaces of the hinge pin ears that also include the side of the tab section. The parting line for the part is at the top surface of the tab section, forcing the use of split parting line, and draft, to create the hinge pin. The following illustration shows the draft requirements for the hinge pin part.

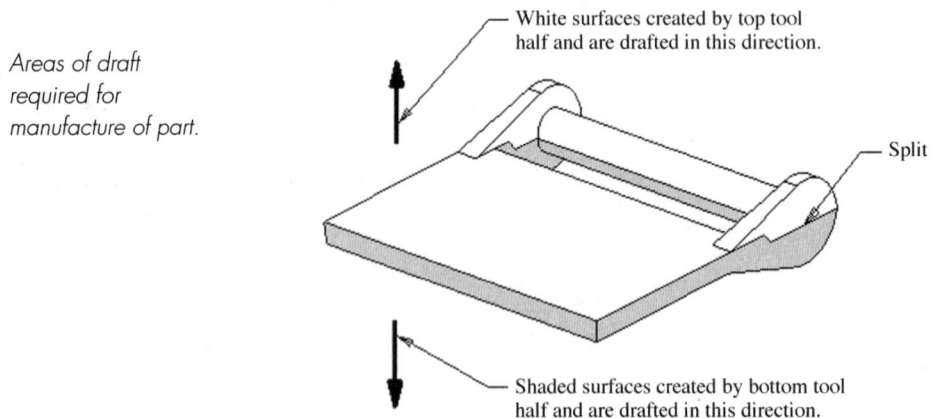

Areas of draft required for manufacture of part.

White surfaces created by top tool half and are drafted in this direction.

Split

Shaded surfaces created by bottom tool half and are drafted in this direction.

Exercise 14-2: Creating a Basic Neutral Plane Draft

The first draft you will create will be a neutral plane draft on the front and rear of the tab surface.

   1. Select **Feature** ➡ **Create** ➡ **Tweak** ➡ **Draft Neutral Pln** ➡ **Done** ➡ **No Split** ➡ **Constant** ➡ **Done**. Pro/ENGINEER will prompt you to select the surfaces to be drafted. Select the surfaces indicated in the following illustration. The next required item is the selection of the neutral plane. This will be the top surface of the tab.

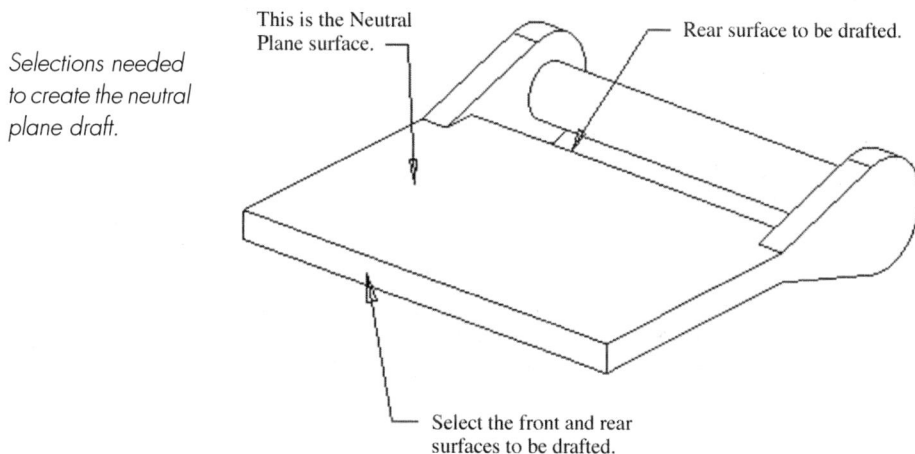

Selections needed to create the neutral plane draft.

This is the Neutral Plane surface.

Rear surface to be drafted.

Select the front and rear surfaces to be drafted.

   2. After the neutral plane has been selected, Pro/ENGINEER needs a reference plane that will be used to measure the draft angle dimension. This reference plane is always perpendicular to the surfaces being drafted, which often is the neutral plane. For this exercise, use the **Use Neut Pln** option and then select **Done**.

The following illustration indicates that two arrows will appear and one of the draft surfaces selected will be highlighted in yellow. The red arrow indicates the draft direction, and the yellow arrow shows the positive direction for the draft rotation. Use the "right hand" rule to determine the proper sign for the draft angle value. For the example shown in the following illustration, a positive angle will remove material and a negative angle will add material.

Arrows showing direction in which draft will be created.

Red arrow that shows draft direction. Think of it as the direction that the tool will pull.

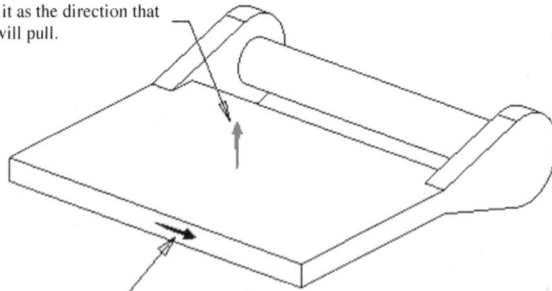

This yellow arrow sets the positive direction for the "Right Hand Rule" to determine the sign of the draft angle.

3. Enter a draft angle of *5* degrees that will *remove* material from the part. The following illustration shows the completed draft.

Side view of the part showing the completed neutral plane draft.

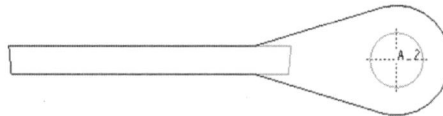

Exercise 14-3: Creating Neutral Curve Draft with a Split at Curve

In this exercise, you will use a neutral curve draft to draft the outside surfaces of the hinge pin ears and sides of the tab. This is because a split is required so that the tool can step down from the top surface of the tab to the center of the hinge pin, allowing each half of the tool to create half of the hinge pin. However, you need to first create the curve because this type of draft feature does not allow you to create the curve during the draft.

1. Create the curve that will be used to split the draft. Select **Feature** ➡ **Create** ➡ **Datum** ➡ **Curve** ➡ **Sketch** ➡ **Done**. Select the outside surface of the hinge pin ear as the sketching plane. Sketch the curve shown in the illustration at right.

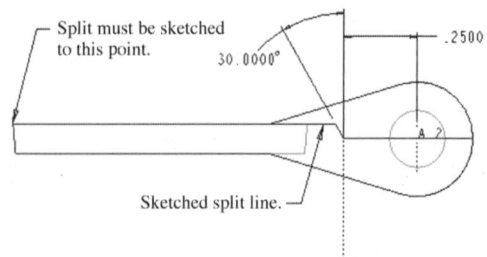

Split must be sketched to this point.

30.0000°

.2500

Sketched split line.

Sketch of the curve that will be used for the split.

2. With the curve created, you can now create the draft. Select **Feature** ➡ **Create** ➡ **Tweak** ➡ **Draft** ➡ **Neutral Crv** ➡ **Done** ➡ **Split at Crv** (highlight the **Both Sides**, **Independent**, and **Constant** options) ➡ **Done**. Select the outside surface of the hinge pin and select **Done Sel** ➡ **Done**. The next prompt is to pick the parting line curve. Use the **Curve Chain** option and pick the curve; then select the **Select All** option and **Done**. The last item to pick before entering the draft angles is the reference direction. This is going to be a plane perpendicular to the draft surfaces. Select the top surface of the tab, as shown the following illustration.

Selections made to create a neutral curve draft.

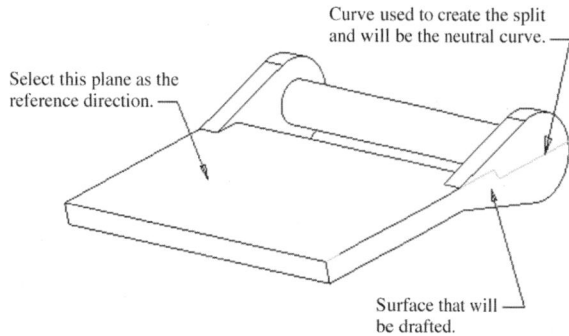

Curve used to create the split and will be the neutral curve.

Select this plane as the reference direction.

Surface that will be drafted.

3. The draft direction arrows are now displayed. Read the prompts in the message window carefully. They will help you determine which half of the split draft you are working with, as well as the proper draft angle value to enter. The following illustration shows the arrows displayed.

Arrows used to determine the proper values for the two drafts.

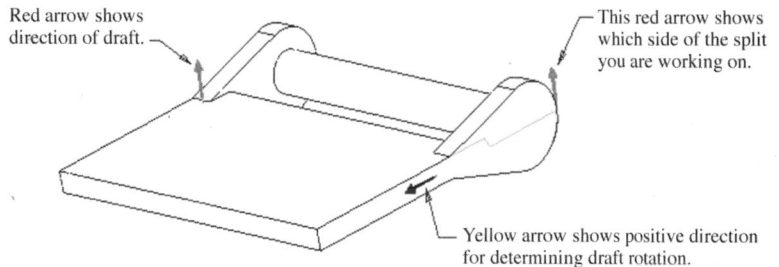

Red arrow shows direction of draft.

This red arrow shows which side of the split you are working on.

Yellow arrow shows positive direction for determining draft rotation.

4. Enter a 5-degree draft angle for the first side, and the same for the second side. Pro/ENGINEER prompted you for both sides because the Independent option was selected. Had the Dependent option been used, you would have been prompted for the first value only. The second side would have been auto-matically created at the same value as the first. Select the **OK** button to com-plete the draft feature, shown in the following illustration.

*Completed neutral curve draft
with a split parting line.*

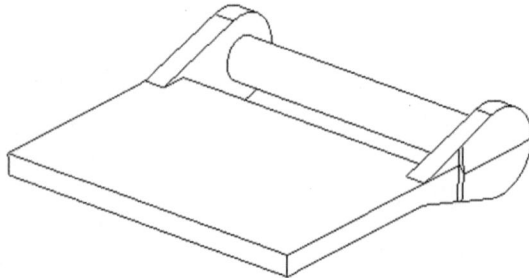

Use the same method to create the draft on the other three surfaces of the part, the other outside surface, and the two inside surfaces. Datum curves need to be created on these surfaces in order to create the drafts. The first curve to create is the curve on the outside surface of the opposite hinge pin ear. The best way to do this quickly, ensuring that the parting line is identical, is to project the datum curve used in the first draft onto this surface.

5. Select **Feature** ➥ **Create** ➥ **Datum** ➥ **Curve** ➥ **Projected** ➥ **Done**. Use the **Curve Chain** option and pick the curve used to create the split in the first draft; then select **Select All** ➥ **Done**. Pro/ENGINEER will prompt you to pick the surface to project the curve onto. Pick the outside surface of the other hinge pin ear; then select **Done**. From the PROJ TYPE menu, select the method to project the curve. Select **Norm to Surf**. The normal-to-surface option projects the selected curve normal to the surface the curve is being pro-jected onto. Select **OK** to complete the feature; then create the draft in the same way as in exercise 14-2.

You now need to create draft on the inside surfaces of the hinge pin ears. Creating the draft on these two surfaces is done in exactly the same way as the outside surfaces, with the exception of the method used to create the curve used to split the draft. A projected curve does not work in this case because the curve becomes discontinuous at the hinge pin, which would produce a "Geom Check" message. The workaround for this is to sketch the curve onto each surface. However, be sure to use the Use Edge method to ensure that the split line will be identical to the others and will update should the origi-nal curve be modified.

6. After the curves have been created, follow the instructions from exercise 14-2 to create the additional drafts. Use a 5-degree draft angle for all of the drafts. When all the drafts are complete, the part will look like that shown in the first of the following illustrations. The second of the following illustrations indicates that you can check that the draft surfaces are drafted in the proper direction by looking at the part in the direction from which the tool will pull.

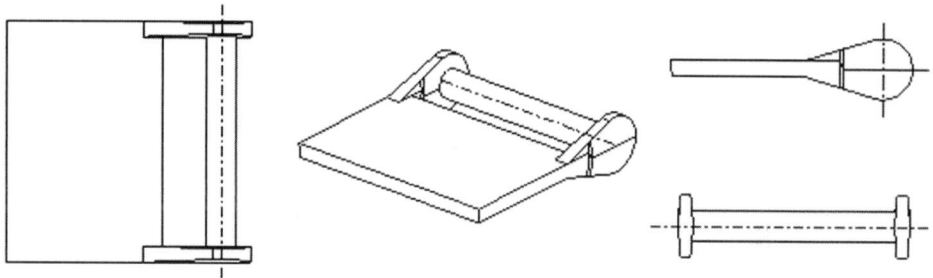

Various views of the completed part with draft.

Check for proper draft direction.

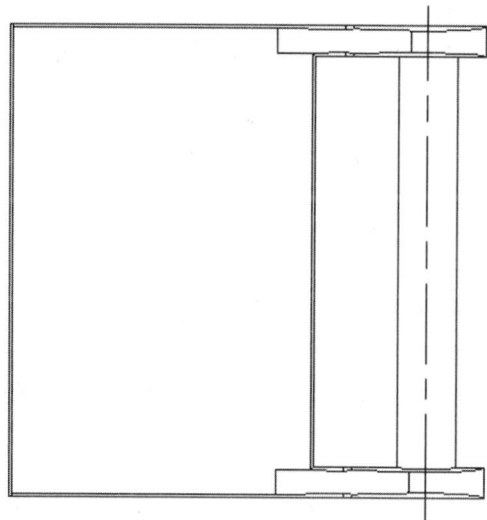

All tangency lines that are created by the draft features are visible in this bottom view. This ensures that the draft on this part is correct.

Exercise 14-4: Using a Variable Section Sweep to Create Draft Surfaces

Not all draft surfaces can be created using the draft feature alone. Nonplanar and curved surfaces cause the most problems when working with draft. A workaround for this situation is to create surfaces with the proper draft angle, using the surface feature and then replacing the undrafted part surface with the drafted one. In this exercise, you will create the draft surface using this procedure and the Variable Section Sweep feature.

The geometry you will be using for this exercise is an offset link, shown in the following illustration. The link contains geometry typical of a cast or forged part, with a center web of material and a short wall around the perimeter of the part above and below the center web for stiffness. The next illustration, on the following page, shows the datum curve to be used as the trajectory for the swept surface.

Starting geometry.

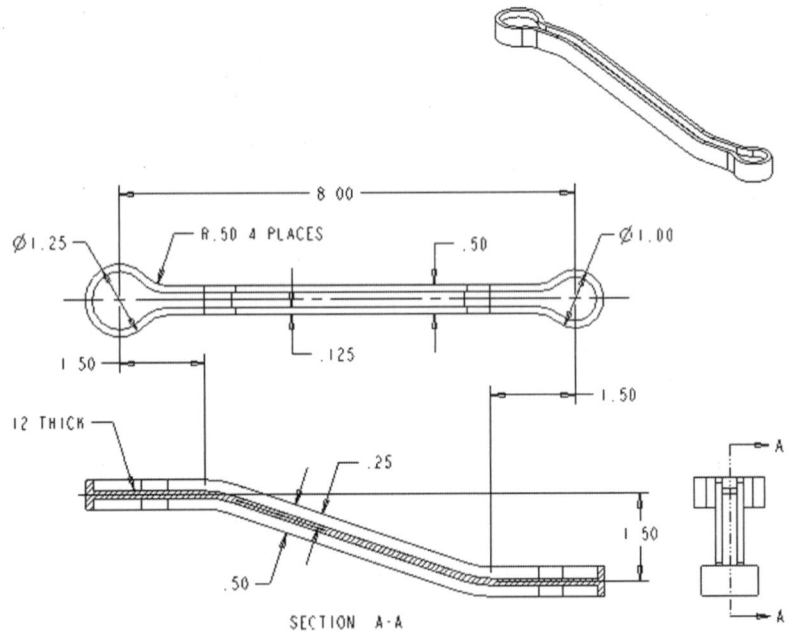

SECTION A-A

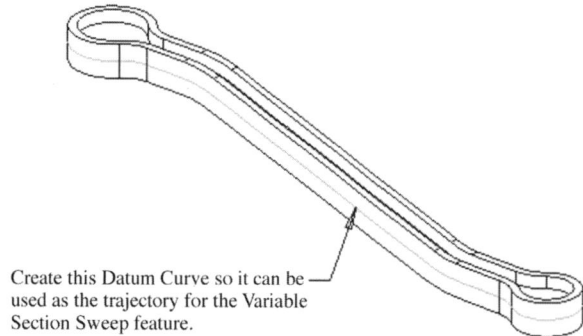

Datum curve for swept surface trajectory.

Create this Datum Curve so it can be used as the trajectory for the Variable Section Sweep feature.

With the starting geometry and datum curve created, you can begin to add the draft using a variable section sweep.

1. Select **Feature** ➥ **Create** ➥ **Surface** ➥ **New** (this menu choice will only be available if a surface has been previously created) ➥ **Advanced** ➥ **Done**.

As previously stated, you will use the Var Sec Swp option and then select Done. What makes this type of surface work so well for creating draft is the Pivot Dir (pivot direction) option. This option keeps the section used to create the surface normal to a selected plane as it follows the selected trajectory. This is exactly the way a draft surface needs to be constructed to keep a constant draft angle as it follows a step or split in the draft.

2. Select **Pivot Dir** ➥ **Done** and pick the default datum plane TOP, as shown in the following illustration. This is the plane the section will remain normal to while being swept around the trajectory. Therefore, it should be perpendicular to the "pull" of the tool. Select **Okay** when prompted for the "Direction of Operation."

Pivot direction plane needs to be normal to the direction the tool will "pull."

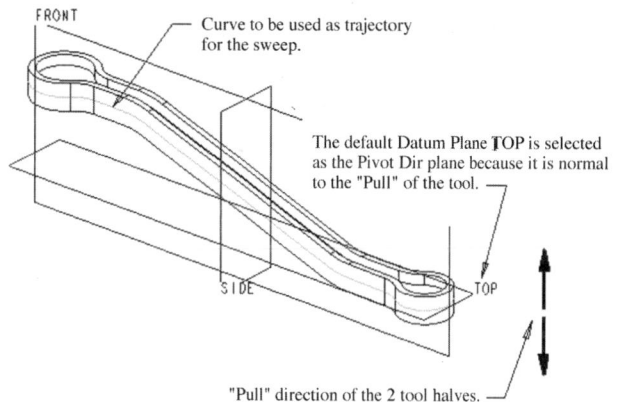

FRONT

Curve to be used as trajectory for the sweep.

The default Datum Plane TOP is selected as the Pivot Dir plane because it is normal to the "Pull" of the tool.

SIDE

TOP

"Pull" direction of the 2 tool halves.

3. Use **Select Traj** ➥ **Curve Chain** and pick the curve shown in the first of the following illustrations. The curve segment will highlight. Select **Select All** ➥ **Done**. Place the start point at the point shown in the illustration. This start point is not required in creating the surface but will help make the illustrations easier to follow. Select **Done** when prompted for another trajectory. Create the sketch shown in the second of the following illustrations. In this illustration, the 5-degree dimension is the draft angle. In addition, the lengths of the line segments need to be longer than the surfaces of the part. Select **Done** to finish the sketch and **OK** to complete the surface feature.

Arrow indicating sweep start point and direction of travel.

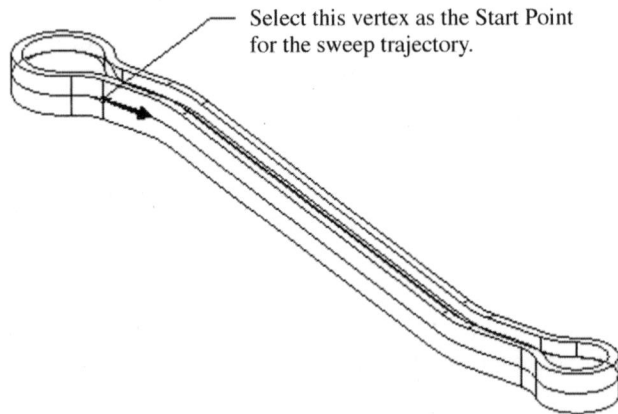

Select this vertex as the Start Point for the sweep trajectory.

Sketch of the swept surface feature.

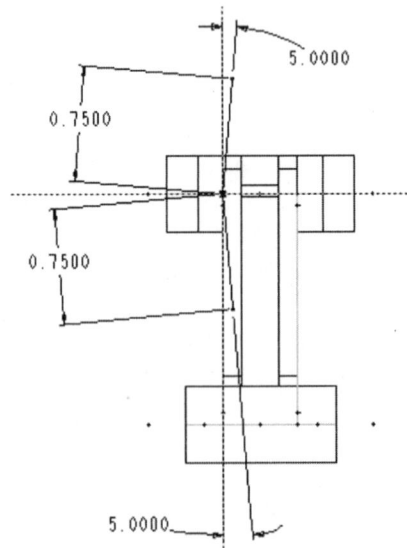

5.0000

0.7500

0.7500

5.0000

The last step in this exercise is to replace the undrafted surfaces with the swept surface that has the draft angle.

4. Select **Feature** ➥ **Create** ➥ **Tweak** ➥ **Replace**. The first prompt is to select the surface to be replaced. Pick the undrafted surface, as shown in the following illustration. Select the quilt the undrafted surface will be replaced with, pick the swept surface, and select **Done Sel** ➥ **OK**.

Completed variable section sweep surface ready to replace the undrafted part surfaces.

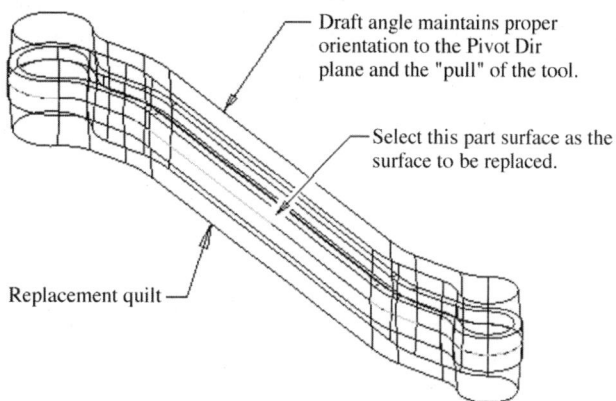

Draft angle maintains proper orientation to the Pivot Dir plane and the "pull" of the tool.

Select this part surface as the surface to be replaced.

Replacement quilt

The entire outside surface of the part in this exercise was replaced by the swept surface, and as indicated in the previous illustration the swept surface is no longer displayed. Look at the part from the top and notice that all of the tangent edges for the top half of the part are solid. This indicates that there are no undercuts in the part and that the draft has been created successfully. The illustration at right shows the completed draft.

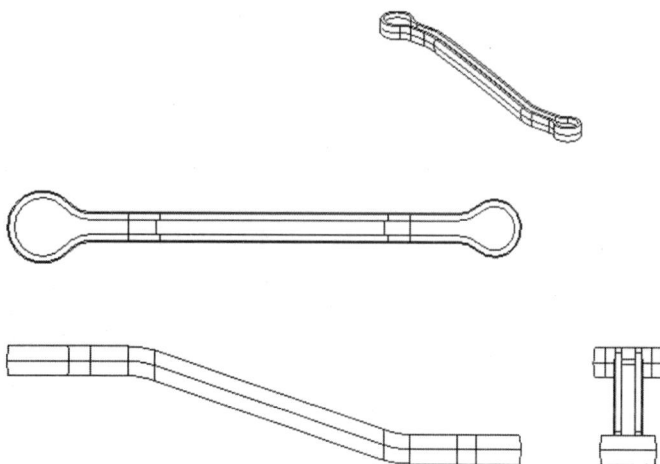

Completed draft on outside of link part.

Summary

This chapter reviewed the various types of draft features available in Pro/ENGINEER, from simple neutral plane drafts to the more capable neutral curve drafts. You also learned about split drafts and how they can be used to create surfaces formed by both halves of the tool and to separate areas for different draft angles made by the same half of the tool. In exercises 14-2 and 14-3 you put these basics to work by creating all of the draft features required for a hinge pin part. In exercise 14-4 you created draft surfaces without using the draft feature. Instead, you created a variable section sweep surface and replaced it with the original model surface.

Draft features are without doubt required modeling if a model is to be used by your tooling department to create the tool. Creating the features is, however, only part of the process. You also need to develop a working relationship with the toolmaker, whether they are a department in your company or an outside supplier. Keeping in close contact with them is the only way you can be sure you are creating the most efficient and accurate model that also meets the design intent.

Rounds

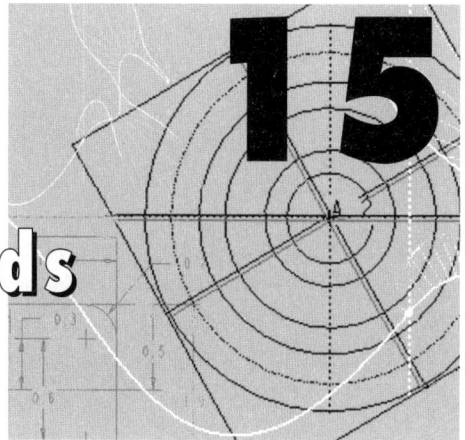

Few features in Pro/ENGINEER seem as easy to use and as difficult to understand as the Round feature. Rounds in most cases are considered detail features, which means that the round is not critical to the design of the part. However, as the model is being used more and more by manufacturing departments to write numerical control programs for the machining of parts, the round is becoming a more critical part of model geometry. Rounds combined with draft features are the sole reason most 100-feature parts end up being 300-feature parts by the time the design work is done.

This chapter explores the various types of rounds and explains some of the options you will see when working with rounds. In the latter part of the chapter you will learn some strategies to reduce the number, simplify the usage, and understand the behavior of rounds.

Round Feature Basics

To use the Round feature effectively you need to understand the fundamentals of rounds. This section of the chapter explains the various round types and selection methods that can help reduce the feature count of your models. You will also learn about the different options each round has and how they are used to create the desired round geometry. The better your understanding of the advantages and limitations of each round type, the better you will become at rounding your model to meet the design intent.

Round Types

Under the Round feature in Pro/ENGINEER, you will find options for two categories of rounds: Simple and Advanced. The main difference between them is the level of flexibility in their interaction with other rounds or features, but they both create the following four types of rounds, which are shown in the illustration following the list.

Constant	Rounds created along edges or between surfaces that have a constant radius.
Variable	A round placed in the same manner as constant rounds, but with a radius that can be varied at any number of points along the edge on which it is being created
Thru Curve	A round created between a surface and a curve that lies on a second surface
Full Round	A round or fillet that removes a surface when it is created

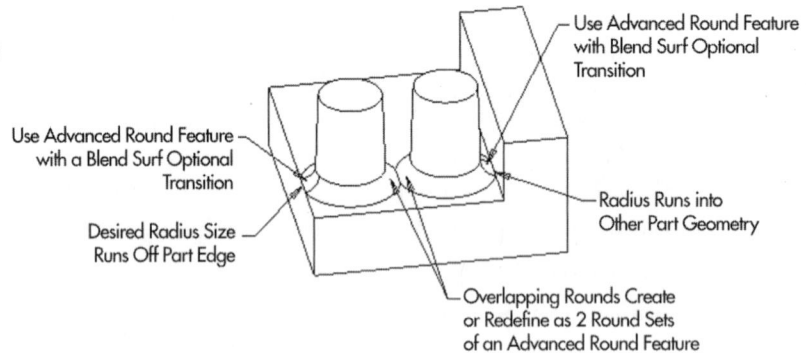

The four basic round types.

Selection Options

The various round types require different selection methods. The most common method of selection is Edge Chain. The Edge Chain selection method is used for the vast majority of rounds you will create in Pro/ENGINEER, consisting of the basic edge round and fillets.

When you select Edge Chain, you are given three additional options from which to choose. The One By One option allows you to select the edges for the round one at a time. The One By One method of selecting round references requires a lot of picks to create a round. The Tangent Chain option selects all tangent edges in both directions from the selected edge. This option is a true chain selection method. The Surf Chain option selects all edges of a selected surface and gives you the option of selecting which edges will be used for the round. That is, you can select all of the edges or any portion of the chain. The following illustrations show the various round selection options.

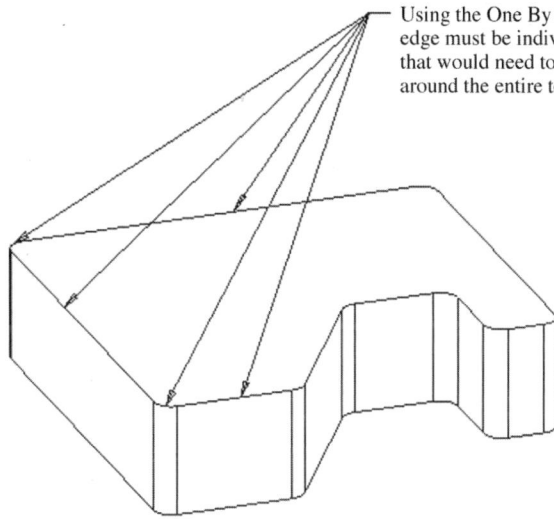

Using the One By One option of CHAIN selection each edge must be individually selected. There are 16 edges that would need to be selected to create a round that goes around the entire top surface.

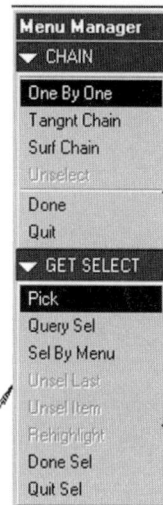

The CHAIN selection menu.

One By One method.

Tangent Chain option.

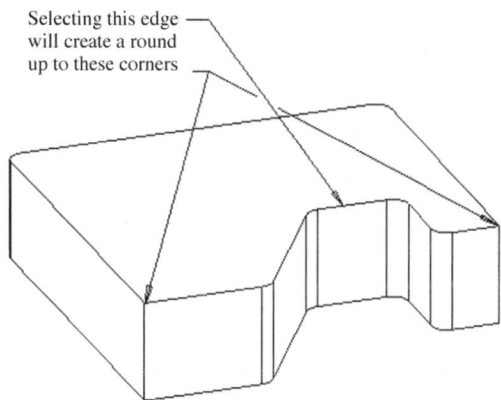

Selecting this edge will create a round up to these corners

The Tangent Chain option will select all edges until a non-tangent edge is encountered.

Selecting the top surface highlights
all of bounding edges of the surface.

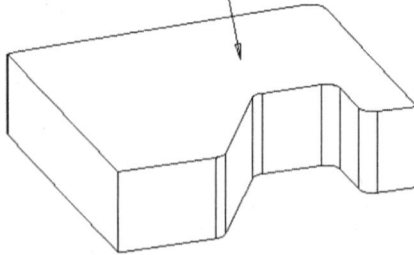

Menu Manager
▼ CHAIN

One By One
Tangnt Chain
Surf Chain
Unselect
Done
Quit
▼ CHAIN OPT
Select All
From-To

Surf Chain menu and the
edge selection options.

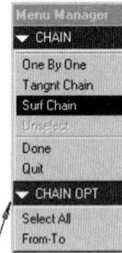

Surf Chain option selects all of the bounding edges of a
surface. You can then select which edges will be used to
create the round using the Select All or From-To options.

Surf Chain option.

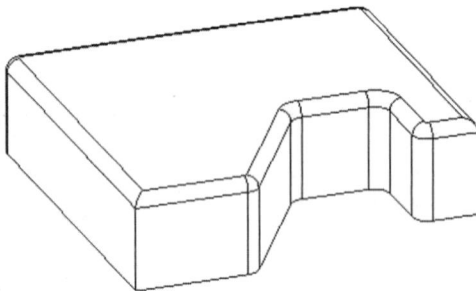

Round created using Select All option

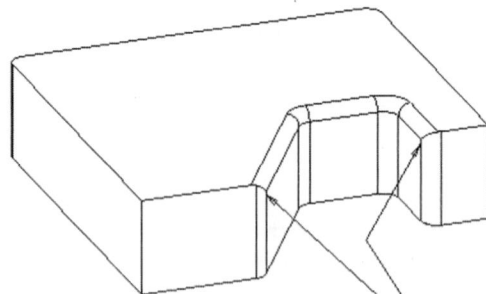

Using the From-To option and selecting
the 2 vertices of these edges, created
the round between the selected points.

Edge selection control using Surf Chain.

The second main type of selection method is Surf-Surf, which places a round
tangent to two selected surfaces. In the following section on dealing with multiple placements, you will use this method to create a round, and will learn
the technique for picking the surfaces.

Another method is Edge-Surf, which creates a round between a series of
edges and a surface. The last method, Edge Pair, is used to create a Full
round between two parallel edges. Examples of these two main types of selection methods are shown in the illustrations that follow.

Common application for an Edge-Surf round.

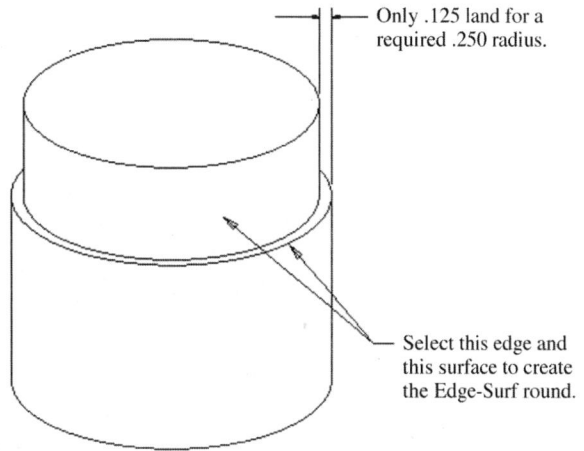

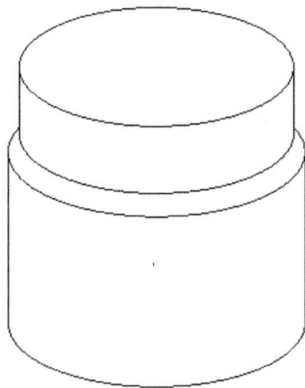

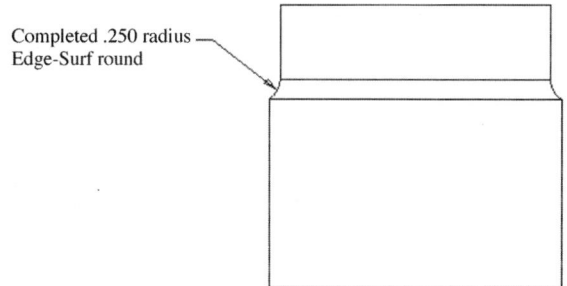

Only .125 land for a required .250 radius.

Select this edge and this surface to create the Edge-Surf round.

Completed .250 radius Edge-Surf round

Completed Edge-Surf round.

Placement of a Full round needed at the top of a rib.

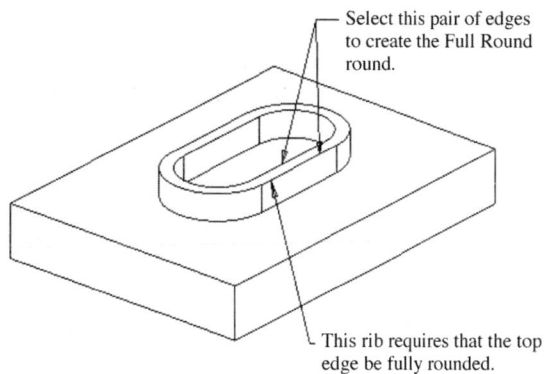

Select this pair of edges to create the Full Round round.

This rib requires that the top edge be fully rounded.

Round was created along
all tangent edges.

Cross-section through
full round at top of rib.

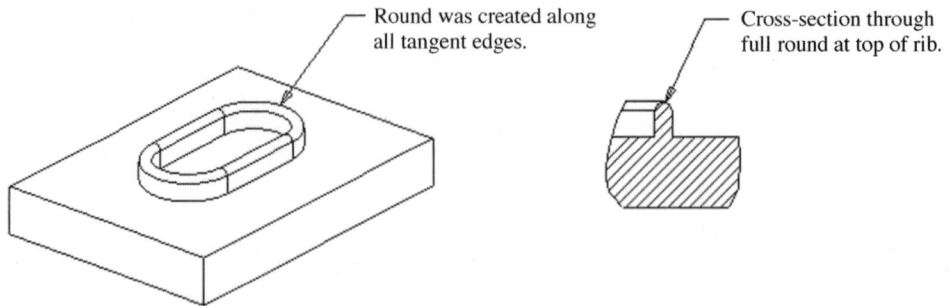

Full round created along all edges tangent to the selected pair. The distance between the pair of edges selected to create the round determines its radius.

Multiple Placement Options

Sometimes a round can be placed in more than one location, based on the entities selected to create the round. To deal with this type of problem, Pro/ ENGINEER highlights one of the options and displays a CHOOSE menu along the right side of the graphics window. The menu contains the options Next, Previous, and Accept. The Next option toggles through all possible placements. When the desired placement is displayed, you select the Accept option and the round is created. The following illustrations show the steps involved in this procedure.

*Using the Next and
Previous menu
picks, you can see
all the valid round
placements.*

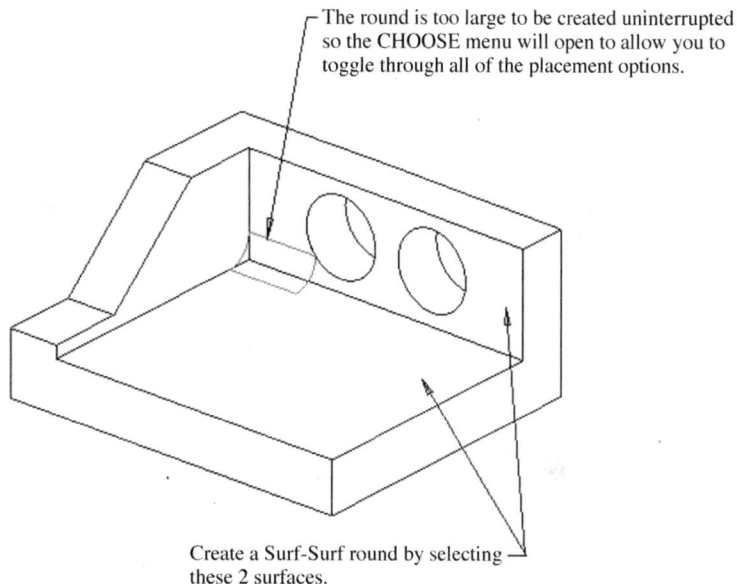

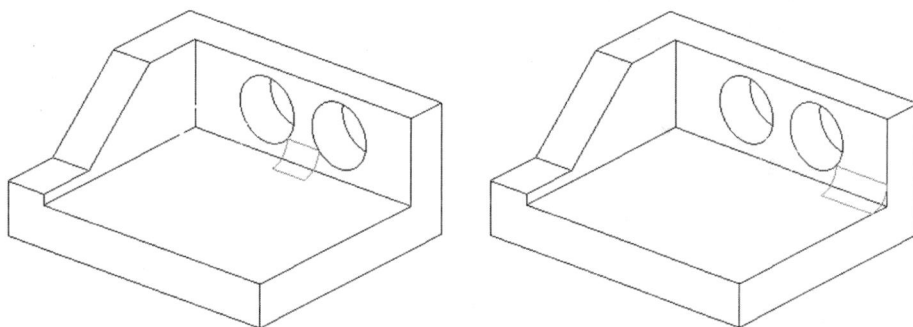

The round is too large to be created uninterrupted
so the CHOOSE menu will open to allow you to
toggle through all of the placement options.

Create a Surf-Surf round by selecting
these 2 surfaces.

Other valid round placements for this round.

When creating surface-to-surface rounds, you will at times run into the situation where the round will be interrupted due to geometry constraints and can be fit into more than one location. The Round Extent option can be used to create the round with the desired geometry. The Round Extent selection consists of two options: Term Surfs and Auto Blend. Term Surfs is short for "terminating surfaces." Auto Blend creates blended surfaces between areas where the round can be created. These two options can be used separately or together on the same round, as in the exercises that follow.

Exercise 15-1: Using the Auto Blend Option

The following steps show you how to create a simple round between the base surface and the surface with the two holes in it. Begin with the block geometry used in the previous round extent example. The block is shown in the first illustration at right.

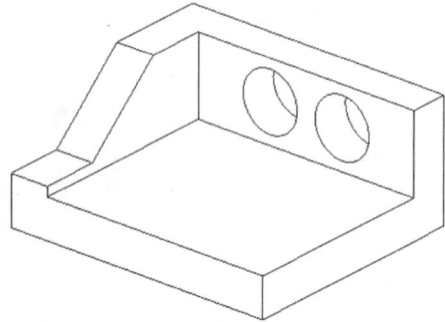

Block geometry used in the round extent examples.

1. Select **Feature** ➡ **Create** ➡ **Round** ➡ **Simple** ➡ **Surf-Surf** and select the two surfaces. The CHOOSE menu will open.

2. To become familiar with how this menu works, toggle through the optional placements by selecting **Next**.

3. Select **Accept** for any of the placements and scroll down to the **Round Extent** element from the ATTRIBUTES menu, shown in the second illustration at right, and select it; then select the **Define** button.

List of round elements for a simple round.

4. Check the **Auto Blend** option in the ROUND EXTENT menu and complete the round by selecting the **OK** button. The completed Surf-Surf round is shown in the following illustration.

Completed Surf-Surf round with the Round Extent element set to Auto Blend.

Radius value is reduced to fit the round under the holes.

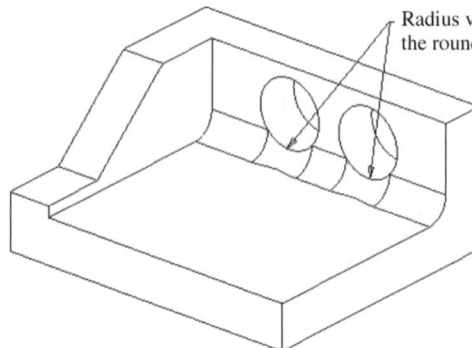

As you can see in the previous illustration, the blended surfaces around the holes go from the full radius value to a radius that will fit under the hole and then back to the full radius.

Exercise 15-2: Using Term Surfs and Auto Blend Together

This exercise demonstrates how the two round extent options can be used together to get the desired round geometry. Starting with the same block from the previous exercise, along with the round created along the bottom left edge, you will add a new round to the edge under the two holes. With the first round already in place, the new round will want to follow the radius in the corner and round the vertical edge, as shown in the following illustration. However, the design intent calls for you to keep the new round confined to the bottom edge.

Using both Round Extent options to get the round to match design intent.

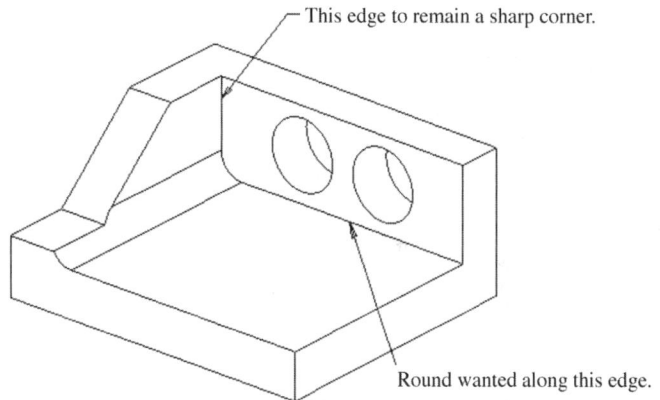

This edge to remain a sharp corner.

Round wanted along this edge.

1. To create this round you would make the same selections as in exercise 15-1. However, when the ROUND EXTENT menu is open, select both the **Term Surfs** and **Auto Blend** options.

2. After the two options have been selected, Pro/ENGINEER prompts you to pick the terminating surfaces. Select the first round, as shown in the following illustration.

Round wants to continue up the vertical edge.

Select this round as the terminating surface.

Round needs to be terminated at first round and blended to be placed under the holes.

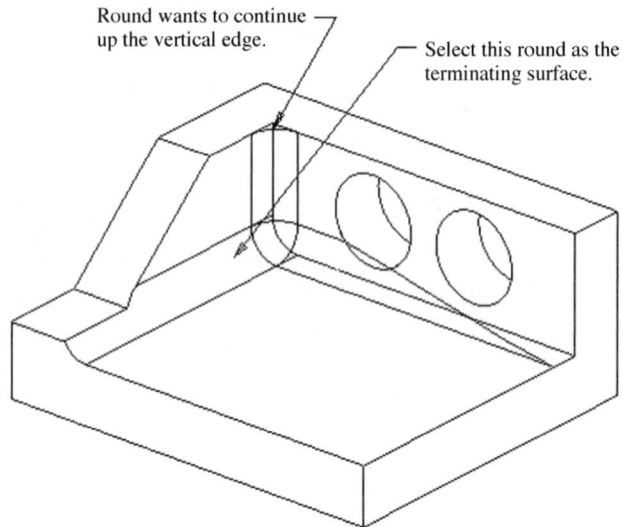

3. Complete the round feature. The finished round should look like that shown in the illustration at right.

By selecting the round as a terminating surface, the round is prevented from continuing up the vertical edge.

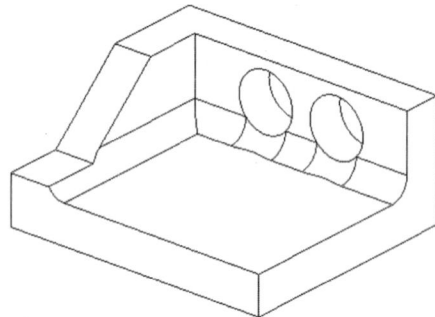

Completed round to match design intent.

Advanced Rounds

The overriding difference between simple and advanced rounds is the additional control of round intersection geometry and transitions gained by using advanced rounds. Transitions are the areas where a normal round cannot be continued. Advanced rounds also give you the option of changing the shape of a round. The round shape can be created with a rolling ball or with a sweep normal to a spine. The cross section of the round shape can also be set to allow for a conic rather than circular shape. The other obvious difference is that advanced rounds are created in a series of "round sets" rather than as single features, as with simple rounds.

In order to have the control of how round corners will be created, Pro/ENGI-NEER needs to know what the round geometry is going to be for each of the edges leading into a corner. This is where round sets come in. A round set is the geometry that will be created for each edge round that leads to the corner. Each round set has its own set of attributes. Round attributes include the round type (constant or variable), the selection method (such as Edge Chain or Surf-Surf), the radius value, and the round extent setting.

You can create as many round sets as you want within a single advanced round feature, and then use the Transitions element to define how the intersection of the round sets will look. In sections to follow, you will work through exercises involving transitions, including corner transitions and round sets and their use. In exercise 15-3, you will create an advanced round.

Exercise 15-3: Creating an Advanced Round

The steps that follow take you through the process of creating an advanced round.

1. Select **Feature ➡ Create ➡ Round ➡ General Rnd ➡ Done ➡ Advanced ➡ Done**. Select **Add** and then pick the edges for the first round set, as shown in the following illustration. The illustration shows a standard advanced round created from three round sets and using the default round options. Select **Done**. Pro/ENGINEEER will now prompt you to enter a radius value. Enter *.125*. The ROUND SET 1 dialog box shows the values for the current round elements. Select **OK.** The first round will be displayed on the model.

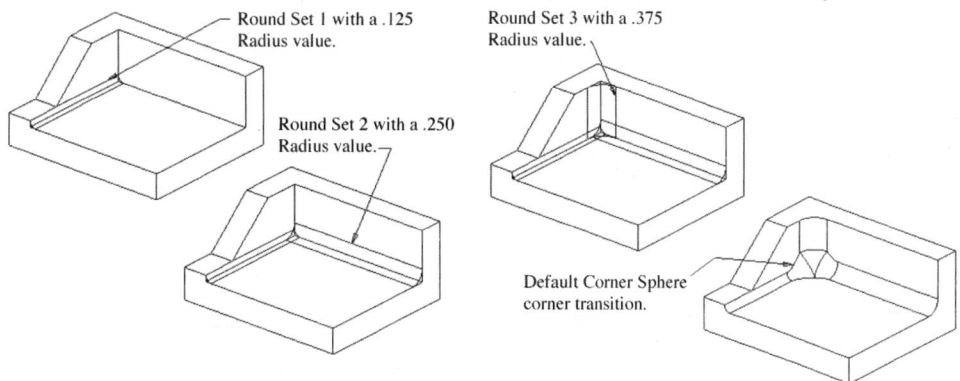

Round Set 1 with a .125 Radius value.

Round Set 3 with a .375 Radius value.

Round Set 2 with a .250 Radius value.

Default Corner Sphere corner transition.

Standard advanced round.

2. To create the second round set, select **Add ➡ Done**, pick the edge shown in the previous illustration, and select **Done** again. Enter the radius value *.250* and select **OK**. The second round will now be added to the display.

3. Select **Add** to start the process for creating the third round set, and then select **Done**. Enter the radius value *.375* and select **OK**. All three round sets are now displayed, but no corner transition is shown.

4. To complete the advanced round, select **Done Sets**. The corner transition will now also be displayed. Select the **OK** button to create the round geometry.

5. Modify the radius values for each of the round sets so that each round has the same radius value. Notice how this affects the shape of the transition in the corner.

Exercise 15-4: Using the Round Shape Element

The Round Shape element consists of three attributes, and each of these attributes has its own set of options.

- The round trajectory, or how the round is swept about the references.
  - Rolling Ball (default)
  - Normal to Spine

- Round cross section
  - Circular (default)
  - Conic

- Corner transition geometry type
  - Corner Sphere (default)
  - Corner Sweep
  - Patch

Each of these attributes has an effect on how the round will be constructed. This exercise looks at modifying the round trajectory and round cross section attributes to demonstrate how each changes the shape of the round. To modify the trajectory method, perform the following steps.

1. Select the **Round Shape** element from the Round dialog box and select the **Define** button. Select the **Norm to Spine** option; then select **Done** and **OK** to finish the round.

The round geometry at this point does not look any different from that created with a rolling ball, but it does have some advantages in more severe geometry, such as that discussed in the following Tip.

✓ **TIP:** *Try using the Normal to Spine option when working with tightly curved references that the rolling ball option may have trouble following. A common error message for this type of geometry is "Fillet surfaces are overlapping."*

   2. Using the same default round geometry from the previous exercise, change the round's cross section to a conic. Select the **Round Shape** element from the Round dialog box, and select the **Define** button. To change the cross section type, select **Conic ➥ Done ➥ OK.** Now modify the radius value of the round. You will notice that the round has a conic parameter dimension in addition to a radius value. This conic parameter value can range from .05 (an ellipse) to .95 (a hyperbola). A value of .50 represents a circle.

The following illustrations show various shapes of rounds based on the conic parameter values indicated. The first illustration shows a default round shape. In the second illustration, the round has been redefined as a conic and the conic parameter dimensional value has been modified to .2. This gives the round an elliptical shape. The radius value as well as the conic parameter dimension of this round can be modified. The radius value sets the size of the round, whereas the conic parameter dimension determines the shape of the round. The third illustration shows a detail of the conic indicating its elliptical shape and showing that it is tangent to two surfaces. The fourth illustration shows the much tighter hyperbolic look of a conic round with a parameter value of .8.

Default round shape.

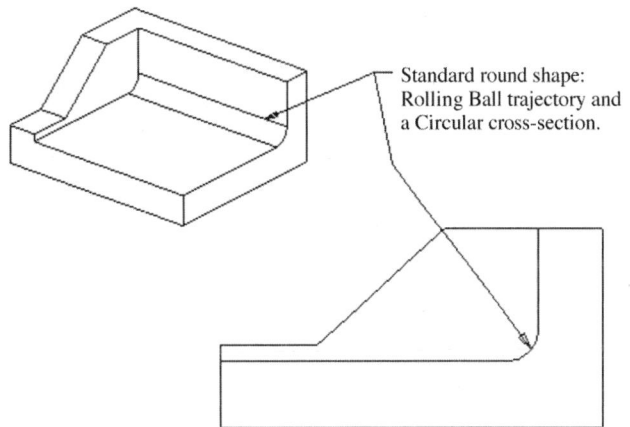

Standard round shape:
Rolling Ball trajectory and
a Circular cross-section.

R .25

.2D CONIC PARAM

Round shape redefined as a conic.

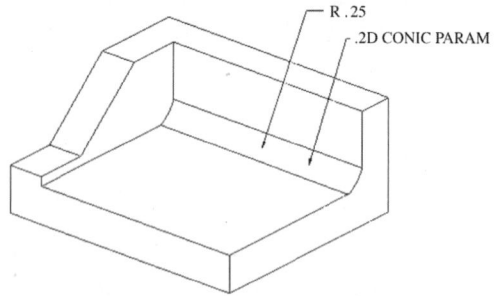

Elliptical round tangent to two surfaces.

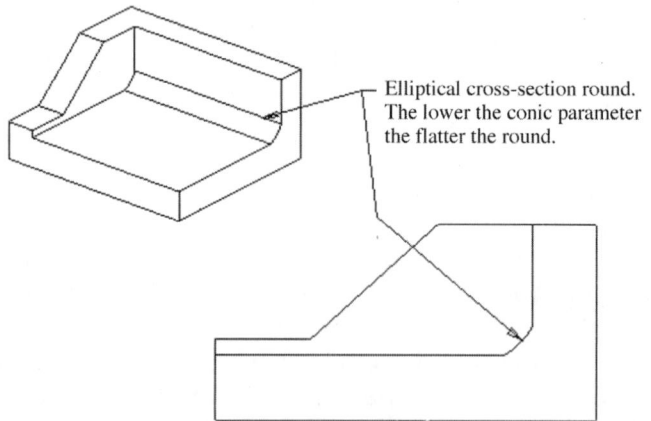

Elliptical cross-section round.
The lower the conic parameter
the flatter the round.

Round has a hyperbola shape.
The larger the Conic Parameter
value the sharper the corner.

Conic round with parameter value of .8.

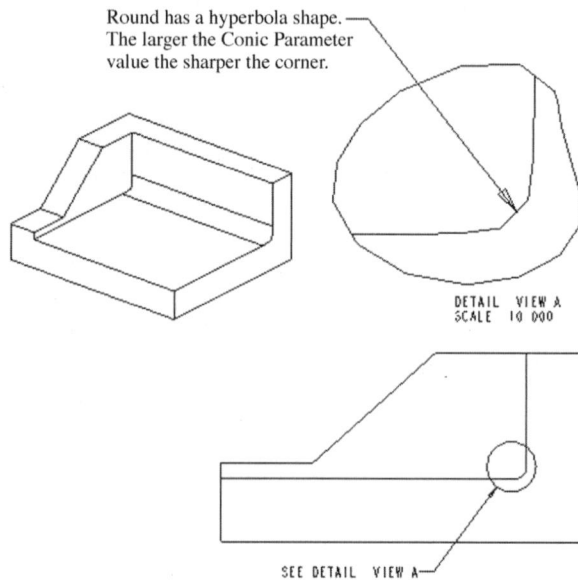

DETAIL VIEW A
SCALE 10 000

SEE DETAIL VIEW A

You would use conic cross section rounds when blending stylized surfaces. They tend to have a smoother-looking transition between the surface and the round than the default circular round. Hyperbolic rounds will spread out the inflection points of a round so that the part has the look of a tight radius but appears over a larger area. This is especially noticeable when the model is shaded. The opposite is true of the elliptical shape. It has the effect of shortening a large, blunt radius round. Experiment with various conic parameter values to see the effect they have on the part geometry, and how you can make them work to your advantage.

Transitions

The Transitions element includes the following seven transition options.

Stop	Allows you to select the surface or point where a round will terminate.
Blend Srfs	The "blend surfaces" option creates blended geometry between two round sets. This option works similar to the same option in a simple round.
Intersct Srfs	The "intersect surfaces" option terminates two round sets at their intersection.
Continue	Creates a round that looks as if it had been created first and the features that caused the interruption created later.
Corner Sphere	Tthe default Corner Transition option. It creates a spherical corner with the radius of the sphere equal to the radius of the largest round.
Corner Sweep	Also a Corner Transition option. This option sweeps the two smaller radii rounds about the largest round.
Patch	Another Corner Transition option. This option creates a corner round between round sets, blending up to four round sets.

The previous list represents the terminology and round functionality in Pro/ENGINEER in brief. In the sections that follow, you will go though some exercises that demonstrate how these round elements work and that explore some of the options available within these elements. In these exercises, you will be adding the Pro/ENGINEER transition types to various types of model geometry as indicated in the following illustration.

Advanced round geometry for
application of various corner
transition types.

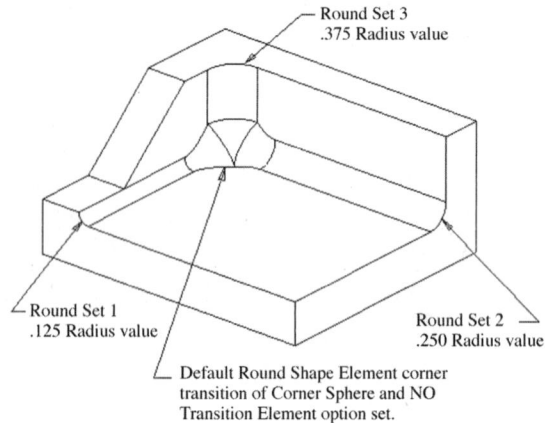

Round Set 3
.375 Radius value

Round Set 1
.125 Radius value

Round Set 2
.250 Radius value

Default Round Shape Element corner
transition of Corner Sphere and NO
Transition Element option set.

Exercise 15-5: Adding a Transition to an Existing Round

In the exercises that follow, you will be adding the transition types to various types of
model geometry using the Redefine option. The basic steps for redefining a round to
add a transition follow.

1. Select **Redefine** and pick the round; then select the **Transitions** element and
the **Define** button.

2. Select the required transition type and follow the instructions for which edges
to select.

3. After the transition edges have been picked, select **Done** ➡ **Done Trans** ➡ **OK**
to complete the round.

Stop and Blend Surfs Transitions

These two types of transitions work the same way as the Round Extent options
Auto Blend and Term Surfs, previously discussed.

Exercise 15-6: Adding an Intersct Srfs Transition

In this exercise you will redefine a round to add an Intersct Srfs ("intersection of surfaces") transition.

1. Select **Redefine** and pick the round; then select the **Transitions** element and the **Define** button.

2. Because you want to all three round sets to intersect, pick all three green edges in the corner. You only need to select the green edges of the round surfaces you want to intersect.

3. Select **Done** ➥ **Done Trans** ➥ **OK** to complete the redefine of the round. There is no blending of the edges in the corner. Sharp edges are created where each round set intersects the next, shown in the illustration that follows.

No blending is done between the rounds sets where they intersect.

Corner using the Stop transition option from the Transitions element.

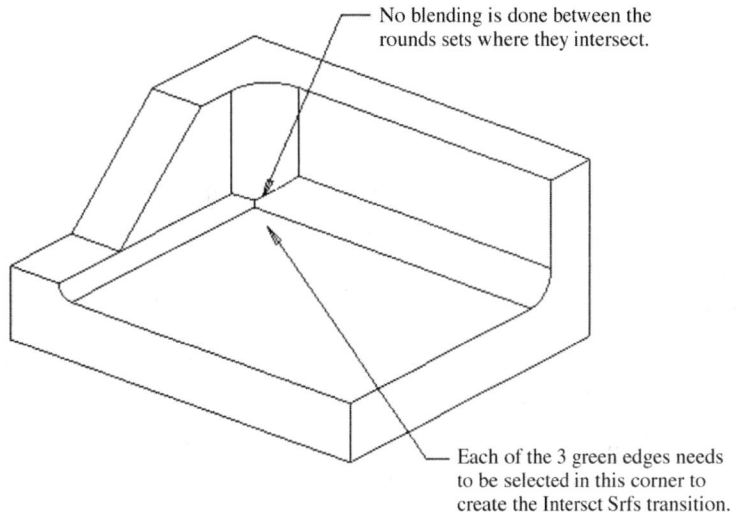

Each of the 3 green edges needs to be selected in this corner to create the Intersct Srfs transition.

Continue Transition

Unlike Blend Surfs or Auto Blend, the Continue transition maintains the radius value when making the transition. A Continue transition will make the model look as if the round were created before the interfering geometry. Two Continue transitions are shown in the illustration that follows. In both cases, the transition was not created explicitly, but rather as a standard transition by Pro/ENGINEER.

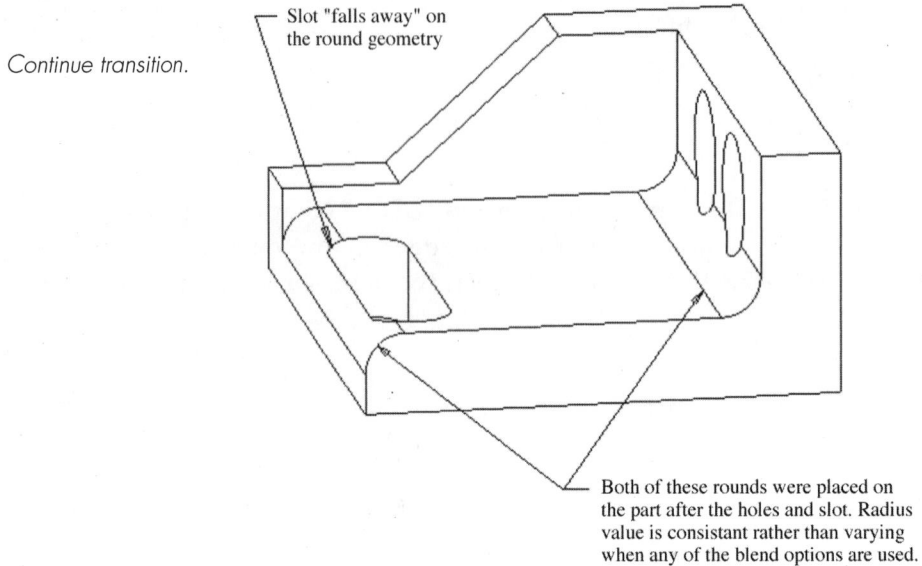

Continue transition.

Slot "falls away" on the round geometry

Both of these rounds were placed on the part after the holes and slot. Radius value is consistant rather than varying when any of the blend options are used.

Corner Transitions

There are three basic categories of corner transition: Corner Sphere, Corner Sweep, and Patch (including Modified Patch). A round can be created with any of the three types by selecting the round shape element and selecting the desired transition type from the ATTRIBUTES menu. The list that follows describes the corner transition types.

Corner Sphere	The default corner transition. It creates a spherical corner, with the radius of the sphere equal to the radius of the largest round.
Corner Sweep	Sweeps the two smaller radii rounds about the largest round.
Patch	Creates a corner round between round sets. This option blends up to four round sets.

Corner Sphere

The Corner Sphere option has been used in several exercises in this chapter. The following illustration shows an example of a Corner Sphere transition in which three round sets have different values. You can vary the round radii to see the effect it has on the shape of the transition. You can modify the look of a corner sphere transition by modifying the spherical radius or the length of the transition areas.

*Corner Sphere
transition.*

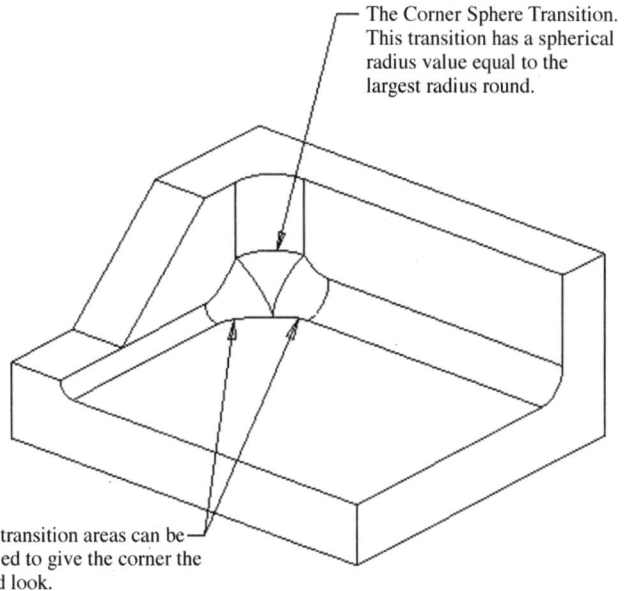

The Corner Sphere Transition.
This transition has a spherical
radius value equal to the
largest radius round.

These transition areas can be
modified to give the corner the
desired look.

Corner Sweep

The Corner Sweep transition sweeps two smaller round sets around a third. This gives the corner the appearance that the two smaller rounds were created after the larger one. This all works out well when two of the round sets are smaller than the third and equal to each other. However, a different situation exists when there are different combinations of radius values. If all three radius values are different, as in the illustration that follows, the transition simply blends the two smallest rounds as it sweeps around the third. A more involved problem exists when two of the rounds have an equal radius value and are larger than the remaining round. In this special case, Pro/ENGINEER will create a corner sphere transition.

Corner Sweep transition.

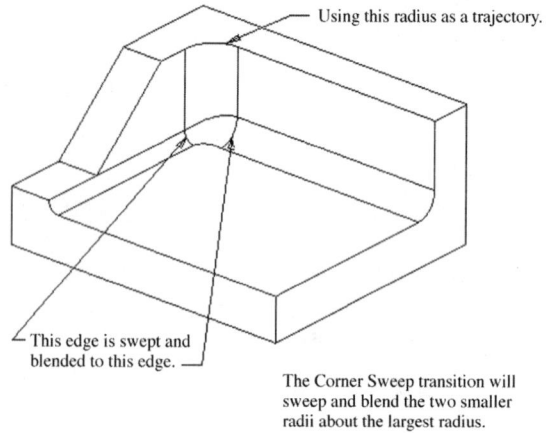

Using this radius as a trajectory.

This edge is swept and blended to this edge.

The Corner Sweep transition will sweep and blend the two smaller radii about the largest radius.

Patch

The Patch transition can be thought of as a surface created using the three edges of the rounds as the boundaries. These boundary edges are defined by the three vertices created by the intersections of all tangency lines of the rounds. The following illustration shows how a patch transition works.

Items used to create a Patch Corner transition.

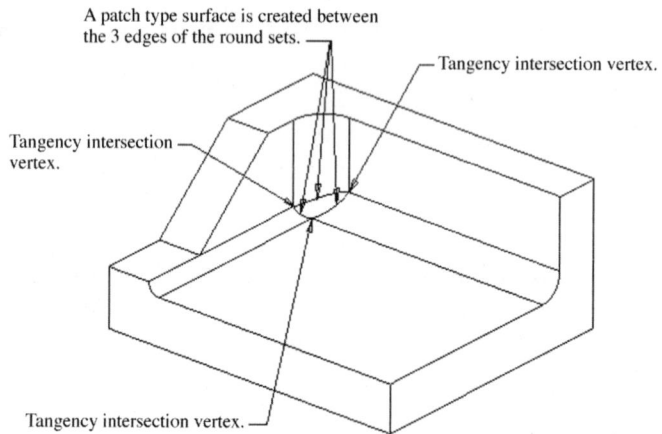

A patch type surface is created between the 3 edges of the round sets.

Tangency intersection vertex.

Tangency intersection vertex.

Tangency intersection vertex.

Exercise 15-7: Creating a Modified Patch

The Modified Patch transition (shown in the following illustration) is another option you have to create the type of corner geometry you need to meet design intent. To modify a patch transition, you add a fillet between two of the round tangency lines. The steps for creating a modified patch follow.

1. Create an advanced round with a patch transition. Before you complete the round, select **Modify** from the PATCH menu and then **Done**.

2. Pro/ENGINEER will prompt you to pick the surface the fillet will be created on. Pick the surface and then select **Done Sel**.

3. The selected surface will highlight and you will be asked to enter a radius value for the fillet. Enter the radius value and select **Done Trans**; then complete the round.

*Modified Patch
transition.*

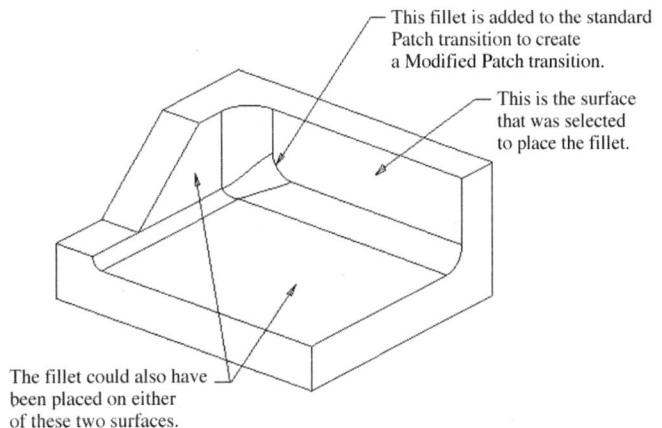

This fillet is added to the standard
Patch transition to create
a Modified Patch transition.

This is the surface
that was selected
to place the fillet.

The fillet could also have
been placed on either
of these two surfaces.

Now that you know the basics of round creation, the section that follows presents guidelines for the use of round features in models.

Guidelines for Using Rounds

In this section you will learn some facts about rounds, design guidelines associated with rounds, and some tips and techniques to use when working with round features. The following are general guidelines regarding rounds.

- Rounds are never standalone features. They will always have parents, and their children can cause difficult regeneration problems.

- Whenever possible, add rounds to your model later in the design process.

- Whenever possible, add draft to features before adding rounds.

- If rounds fail during creation, try adding them at a smaller value than you need and modify their size until they are the size you need.

- Rounds created across multiple features are typically less stable during model modifications than those that are not.

- Avoid adding fillets and rounds to complex features in Sketch mode. They generally cause dimensioning problems and difficulty in regenerating the sketch.

- Create rounds on vertical feature segments first, making it possible to loop the face edges.

- If you need to use edges in Sketch mode, avoid using round edges. Rounds need to be as childless as possible to avoid unwanted relationships downstream.

- When creating surfaces to later be made into protrusions, try to add draft and rounds prior to the protrusion step. This will make patterning and modifications easier.

Design Sequence and Feature Reduction

When adding rounds to a part, try to reduce feature count by selecting multiple edges or surfaces for each feature. In the following exercise you will learn some selection techniques for lowering feature counts, quickening entity selection, and simplifying dimensional modifications.

Exercise 15-8: Creating a Base Part Model

In this exercise you will be creating the part geometry shown in the following illustration.

Basic part geometry.

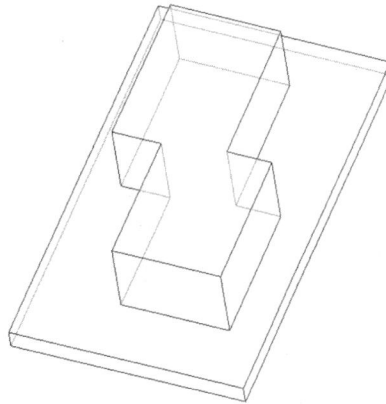

1. Draft the four surfaces along the base part, shown in the illustration at right. You could select each surface individually or use draft tools that allow you to group the surface selection. To select all four faces in the fewest picks, use the **Loop Surfs** option. This allows you to pick only the bottom surface of the model and get all four side surfaces automatically, as shown in the following illustration.

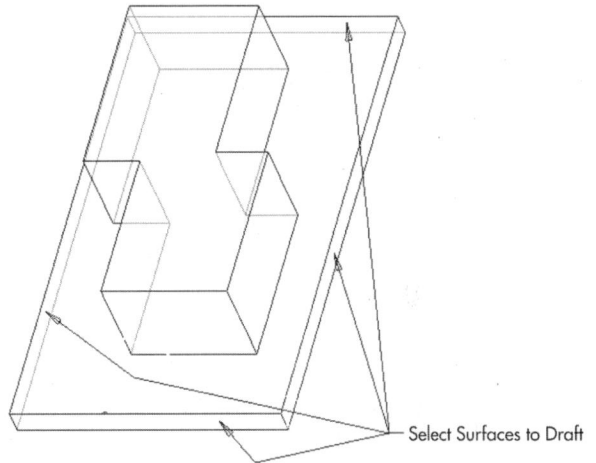

Select Surfaces to Draft

There needs to be draft added to the sides of the base.

Using the Loop Surfs selection option.

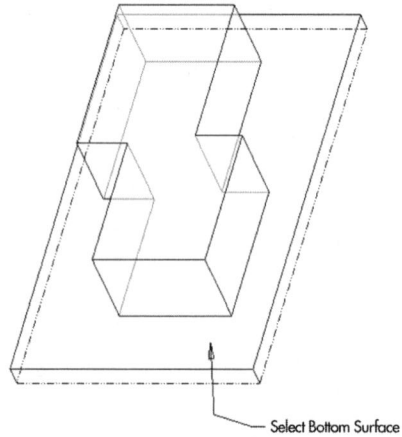

Select Bottom Surface

2. For the next set of surfaces, use the same selection method. However, this time you will be prompted for an edge that bounds the necessary surfaces. Pick this edge, shown in the second illustration at right.

3. Add the remaining round features and complete the part with a shell. When selecting edges for rounds, first pick all vertical edges (shown in the following illustration) that will have the same value. There is really no way to reduce the initial picks (12 selections) while creating the vertical rounds. However, once the vertical rounds are complete, you have lessened the picks needed for the horizontal rounds.

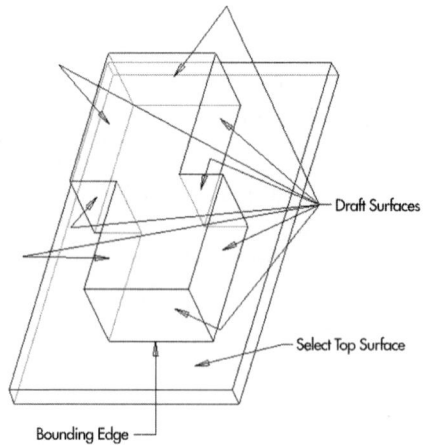

Draft Surfaces

Select Top Surface

Bounding Edge

Selecting the indicated edge as the bounding edge.

●

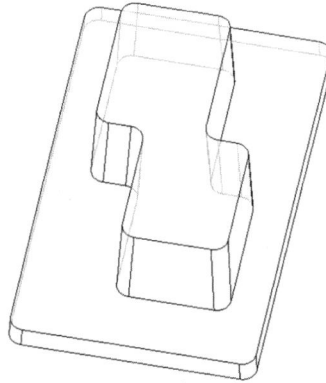

Selecting the vertical edges to round first will ease the selection of the horizontal rounds later.

Part After Vertical Rounds are Added

4. Use the **Edge Chain** or **Loop** (Loop selection is available under the *config* setting ALLOW_OLD_STYLE_ROUND) selection methods to complete the round features. Select all three of the edge chains shown in the following illustration. Selecting the three edge chains will complete the horizontal rounds in a single feature.

Selecting the three edge chains.

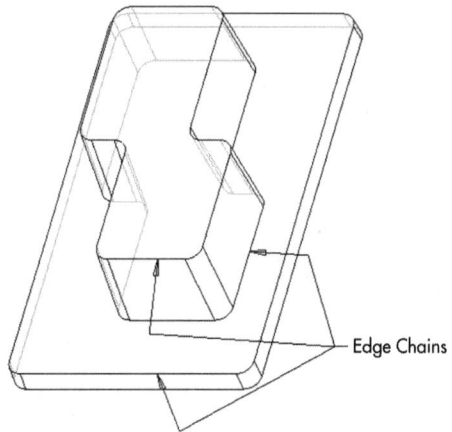

Edge Chains

5. Complete the part by adding the shell feature, shown in the following illustration.

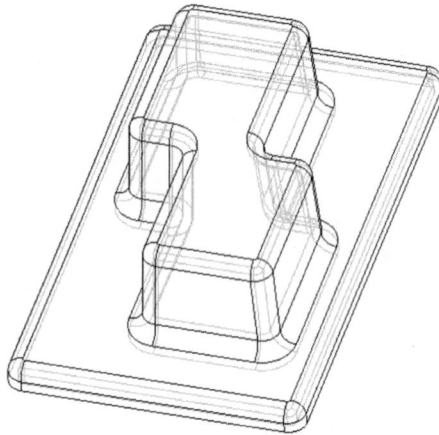

Completed part with shell feature.

↝ **NOTE:** *The config.pro option ALLOW_OLD_STYLE_ROUND should be set to YES. There are still some rounds that cannot be created using simple or advanced rounds, but that can be created using the old-style rounds. Try to create the round with various options of simple and advanced rounds. If these do not work, try the old-style round.*

In the exercise, there are only ten features, three of which are the default datum planes. If the rounds and drafts were created as individual features after the shell, there could have been up to 74 features. This illustrates quite clearly the need to group these "pick-and-place" types of features.

Rounds Before Draft

Generally, rounds should be added to a part after draft has been added. However, there are exceptions to this rule. In the exercise that follows, you will see some of the problems that can occur when rounds are added in Sketch mode or prior to draft.

Overlapping geometry is most often the cause of an error message when you draft the surfaces a round is a child of. (See the first of the following illustrations.) What happens is that as the surfaces taper due to the draft, and the value of the round becomes too small, the round folds over itself, as shown in the second of the following illustrations. This problem, however, never occurs in a model in which draft is added prior to the rounds.

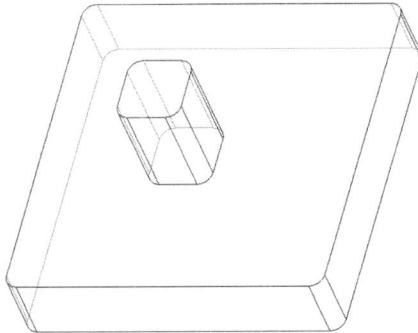

Rounds added to protrusion before any draft features.

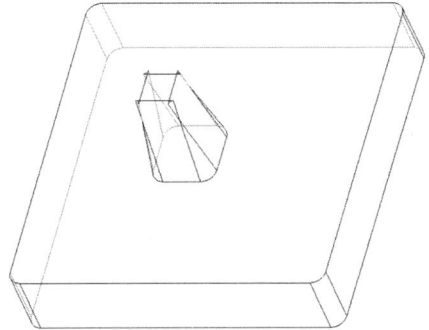

Overlapping geometry problem.

In this case, if the rounds had been made in Sketch mode, you would have had to either make the rounds larger by modifying the sketched value or redefine the sketch, removing the rounds. If this feature were one of several simple bosses built early in a large model, and the draft added late in the model, the redefinition could have resulted in multiple feature failures and subsequent reroutings of child features.

Exercise 15-9: Creating Rounds and Draft with Surface Protrusions

Sometimes it is necessary in the creation of complex parts to merge surfaces and then protrude on or cut away from existing solid geometry. However, as previously mentioned, you should avoid this technique if possible. In the following exercise, you will work with a part generated with curves that define its mating surface, and deal with some of the challenges this situation produces.

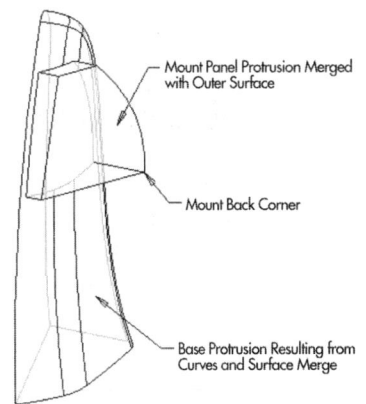

Mount Panel Protrusion Merged with Outer Surface

Mount Back Corner

Base Protrusion Resulting from Curves and Surface Merge

This exercise is based on the mirror mount shown in the illustration at right. This part will be mounted in the fly window area of a truck. The back face and front edges need to match the window, as well as the A-pillar along the windshield. This data was imported as IGES curves. The remaining issues concern the dashboard and side window vent areas. These two areas will affect the mount back cor-

Mirror mount.

ner and mount panel flat face. To get clearance over the dashboard, you will be drafting the mount panel's side walls, and adding a large radius to the mount back corner.

First deal with the draft issues. The parting line will be at the merge between the
mount panel and the base protrusion.
This draft will be a neutral curve-driven
draft pulled in the direction of datum A.
All outer side surfaces will be included,
and the draft value will be constant, as
indicated in the first illustration at right.

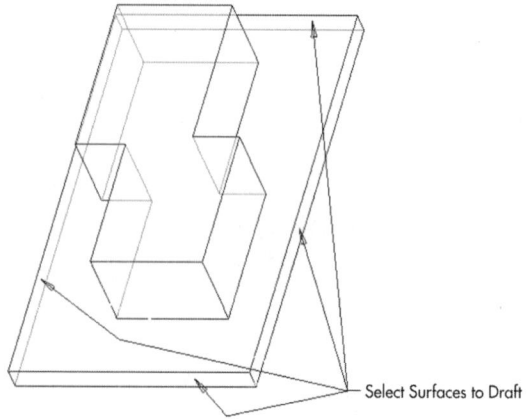

Creating the draft surfaces before the rounds can be added.

1. Select **Neutral Crv** from the
   draft menu; then pick the side
   surfaces and neutral curve
   edge. Because rounds have not
   yet been added to the mount
   panel, select **One By One**. You
   will notice that the four-sided
   surface has eight edge entities
   that need to be selected—one
   edge entity for each intersecting surface portion.

2. Look for small edge segments
   that may cause overlapping
   geometry once draft is applied.
   If your draft does result in over-
   laps, you have a couple of
   options for correcting the situation. First, you could redefine
   either surface to move the
   resulting intersections further
   apart, increasing the segment

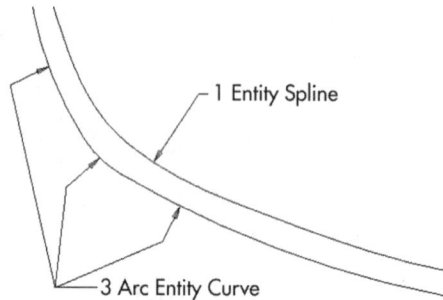

Arc set closely matches the single entity spline.

sizes. Second, you could redefine the parent curve to have fewer segments,
using splines to achieve this. The spline may not exactly match the arc sets but
may still meet the needed part aesthetics. The second illustration at right
shows how closely the arc sets can match the spline. In this exercise, the part's
segments are large enough to avoid this problem.

3. Start adding the rounds. The back corner will need a larger round than the other three edges; therefore, you need to make two separate features. Start with the vertical edges and move on to the base rounds of the protrusion, as indicated in the illustration at right.

4. The three .25-inch radii go in with no problem, even though the upper segmented portions become very small. However, there is a problem with the mount back corner. As you increase the radii on that corner, it fails because of the segment along the bottom. Correct this problem using a surface-to-surface round (shown in the first of the following illustrations)

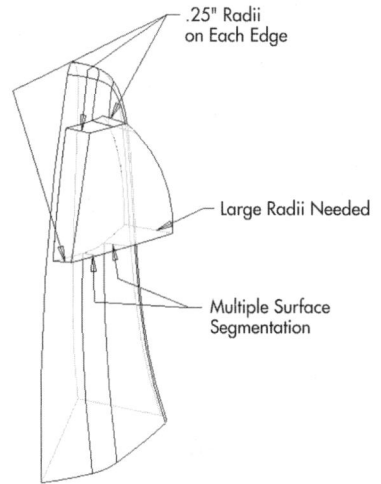

Rounds to be added.

or by redefining the mount panel extruded surface and including radii in the sketch. If you opt for redefining the sketch, you will need to redefine the draft to include the round as a draft surface. The fixed "problem round" is shown in the second of the following illustrations.

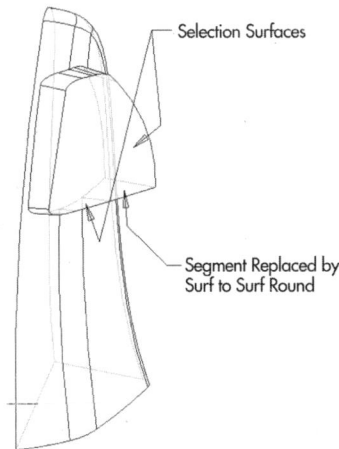

The round in the back corner requires a Surf-Surf round.

Using the Surf-Surf option completes the troublesome round.

↝ **NOTE:** *There is a small graphical difference between the General Rnd and the Allow_Old_Style_Round surface-to-surface round types. A small segmentation occurs with the old style, where the general type blends the surfaces more consistently. The difference is shown in the following illustrations. Potential problems are related to STL and IGES file creation.*

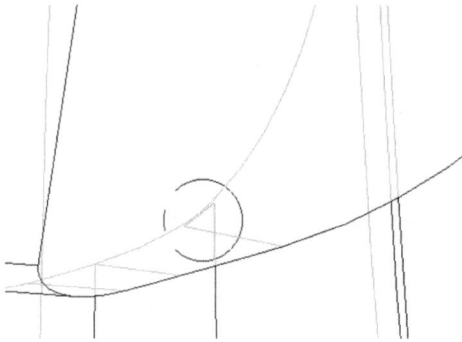

Small segmentation problem that can occur with Old Style rounds.

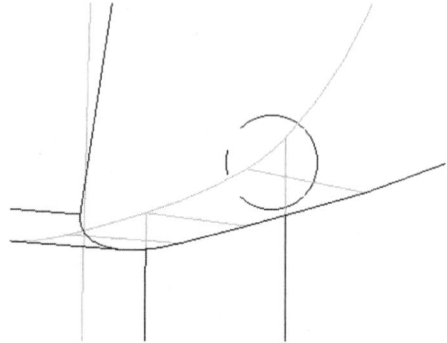

The General round does a better job of blending into the model geometry.

5. When you have completed the rounds and draft, shell the part to maintain constant wall thickness, and make final cuts to the bottom and back mating surfaces to finish the design. The finished mirror mount is shown in the illustration at right.

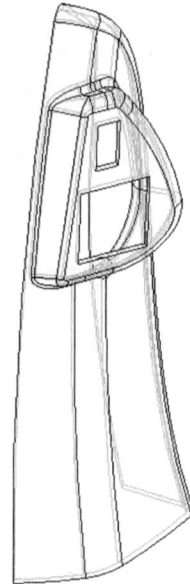

Completed mirror mount.

Summary

This chapter has presented round feature basics, including the types of rounds and the various selection methods associated with them. The chapter also covered the creation of round sets for advanced rounds, as well as the transitions available for both round geometry and corners.

As with most features within Pro/ENGINEER, the order in which rounds are created is very important. The chapter included guidelines for creating rounds that help reduce problem geometry and feature count in the model. The chapter also included real-world geometry exercises in which you worked with both draft and round features.

↝ **NOTE:** *The following PTC Web site has many excellent examples of solutions to common geometry situations that cause rounding problems: www.ptc.com/cs/rounds_helper. To use the Web site, you must register with PTC. To register, all you need is a Pro/ENGINEER serial number and an e-mail address. The process is simple and is free of charge. The Web site contains a great deal of useful information on all aspects of Pro/ENGINEER, and includes a customer support area.*

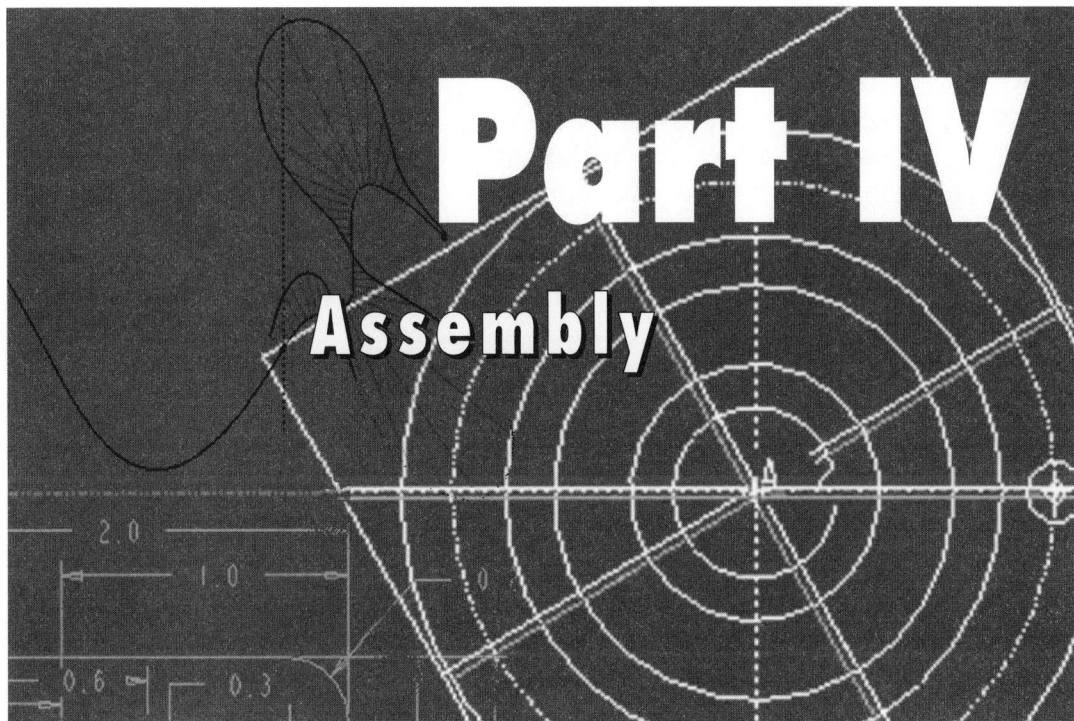

Part IV

Assembly

The ability to model parts and components and manipulate them is what you should have been striving for to this point. However, what brings the modeling process together is the ability to assemble components into subasdsemblies, and subassemblies into assemblies. Part IV deals with various aspects of assembly.

Chapter 16 expolres Assembly mode in Pro/ENGINEER, beginning with a discussion of the difference between the "bottom up" and the "top down" modeling approaches, which determine how a product is assembled. With the approach to assembly understood, the chapter moves on to the basics of assembly using Pro/ENGINEER. The chapter also covers increasing assembly speed, design verification, and other concepts related to assembly basics.

In Chapter 17, you are introduced to the concept of "skeleton models and motion." This refers to the fact that Pro/ENGINEER provides tools under the Model Analysis feature that allow you to examine the relationships among components as they will be placed and as they are intended to move in final assembly. Exercise 17-1 takes you through the use of a skeleton model.

Assembly works only as well as the planning and modeling that precede it. The master model technique, the subject of Chapter 18, is Pro/ENGINEER's solution to relatively painless finalizing of matters of style and appearance when changes need to be made late in the design process and/or throughout a complex model before or after assembly. In exercise 18-1, you create a master model.

The last chapter of Part IV, Chapter 19, presents additional assembly techniques. These include techniques related to working with casting and machining models, including using duplicate models, using family tables, and the merging technique. In exercise 19-1, you are shown how to use the Merge feature. In exxercise 19-2, you create the skin of an assembly model using the various techniques discussed.

16

Assembly Mode

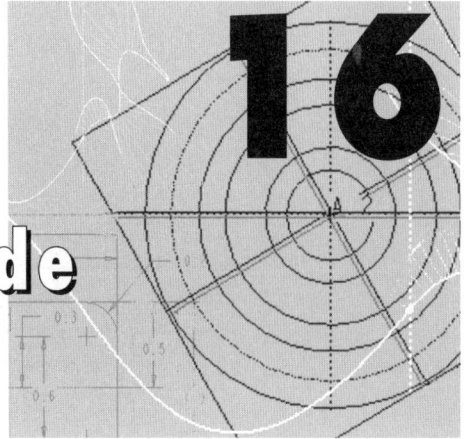

Throughout this book you have been reading and learning about design intent as it applies to planning, sketching, and part modeling. Now you are ready to create assemblies, which involves the same concerns you have encountered thus far: capturing design intent, creating modifiable assembly models, and communicating your intent to others.

Assembly mode within Pro/ENGINEER is minimally a means of putting multiple parts together to form subassemblies and complete product assemblies. Assemblies can help others visualize the completed product, with little interpretation necessary. However, Assembly mode represents much more than this advantage.

"Bottom Up" and "Top Down" Modeling

Currently there are two design methodologies in use in the creation of assemblies: "bottom up" and "top down." The "bottom up" method starts with an overall concept of the product (e.g., a car) and then moves to work on individual components and subassemblies. With this method, you begin modeling tires, wheel hubs, struts, springs, and so on. Each part model is then added to the next until you end up with a subassembly, wheel linkages, frames, and body panels. After you assemble all subassemblies, you have a complete product (in this case, a car). The concept of "bottom up" design is simplified in the illustration that follows.

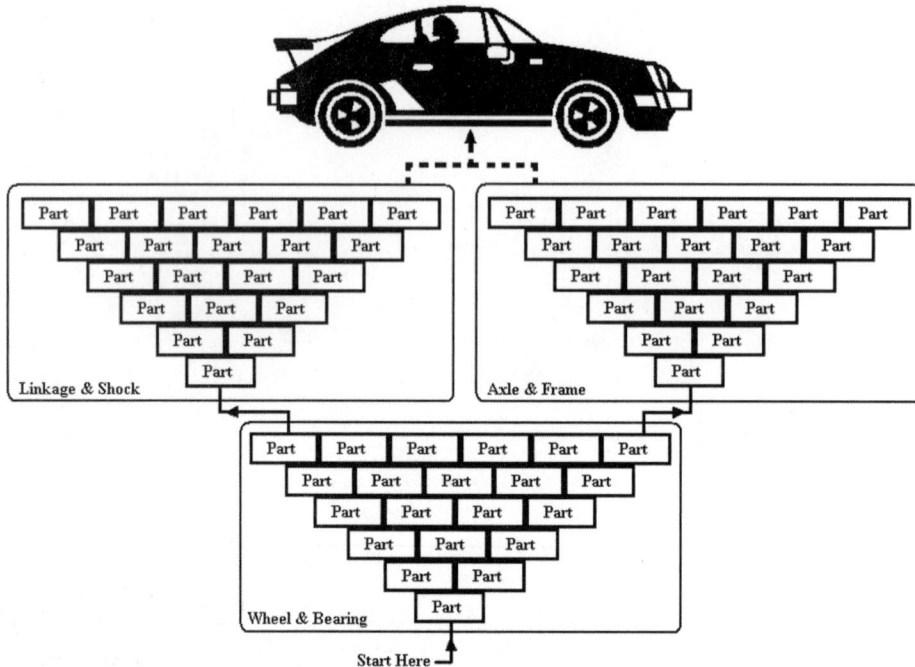

"Bottom up" method.

The previous illustration indicates that with the "bottom up" method you are stacking parts on top of each other to form each subassembly until you complete the total assembly. Because the total assembly is dependent on how each part is assembled to the previous part, the result can be a very precarious overall assembly. This type of assembly can work, but it is absolutely critical that the parts and subassemblies are built with change in mind. The use of a start assembly model consisting of the default datum planes and coordinate system is a must. In many cases in this type of assembly method, the use of coordinate systems for location of part models is very helpful.

The "top down" method, on the other hand, focuses more on planning, overall assembly management, and capture of design intent. As with the "bottom up" method, in the "top down" method you begin with an overall concept. The car again serves as an example. The first thing you need to do is plan. The following are all part of this planning process.

- Plan how you will capture design intent.

- Plan major product components.

- Plan how the major components will be located relative to one another.

- Communicate the design intent and overall assembly structure to your project groups and obtain buy-in.

- Plan specific subassembly and model relationships to major components.

- Create part models capturing the design intent.

With this approach, you are looking at the big picture (i.e., the total car). Then you are breaking down the whole into its major components (i.e., frame, drive train, body, and so on). Once you have divided the car modeling task into manageable pieces and have determined how components will go together, you can begin creating the specific parts and subassemblies. The concept of the "top down" approach is simplified in the following illustration.

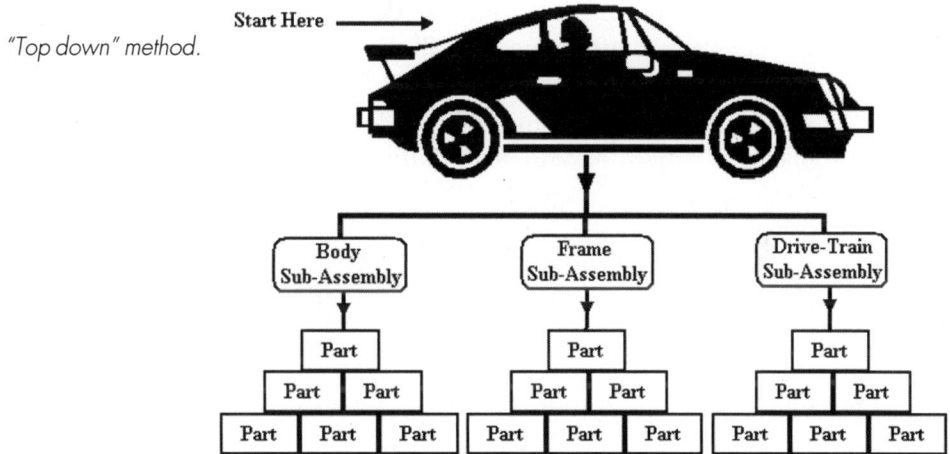

"Top down" method.

The "top down" approach creates an overall stable platform for your product assembly. With a little up-front planning, it will also help you achieve a more streamlined, user-friendly model that captures and communicates design intent.

Basics

The indispensable rule of assembly creation is to plan what you are doing. Regardless of the methodology you subscribe to in your overall assembly creation and maintenance, you need to know how you are going to put your part models together in the assemblies.

There are very few products in existence today that consist of a single part, as do the boomerang and the club. This being the case, you should become familiar with how to locate parts with respect to one another. Whether you are working at the top-level assembly or one of the subassemblies, you should begin with a start assembly model consisting of default datum planes and a coordinate system.

Your start model should also have some saved names and layers, to make orientation and layering easier. The first component assembled should then be located to this assembly start model. Placing and constraining components has become quite easy in Pro/ENGINEER with the introduction of the Component Placement dialog box, shown in the following illustration.

The Component Placement dialog box contains all placement options and information necessary to constrain components, monitor placement status, preview placement, and modify the position of components. The area that seems to give most people problems in Assembly mode is the orientation of the components relative to the red or yellow side of selected datum planes. That is, if you select the wrong side of a datum plane, the component flips in an undesired way.

The main thing to remember when mating datum planes is that you can use the Preview button to verify constraints. This allows you to modify your selection before finalizing the component's placement. While mating planes, it is sometimes difficult to remember which side you have selected. Here again, the Component Placement dialog box is of use. The Component and Assembly reference areas show the selected datum plane and side you have picked.

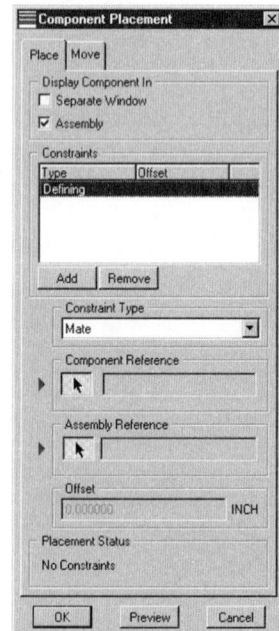

Component Placement dialog box.

When a component has been fully constrained, its position has been locked relative to three degrees of freedom: X, Y, and Z. Locating or locking a component can be done with multiple combinations of constraints (at minimum, one constraint, using coordinate systems; Two with Align or Insert in combination with a mate; or three mating surfaces). Within the Constraint Type section you are given options for the following placement methods: Mate, Mate Offset, Align, Align Offset, Insert, Orient, Coordinate System, Tangent, Edge on Surface, Point on Surface, and Default. The following table describes each constraint type.

Constraint Types

Constraint	Description
Mate	"Mate" means touching or facing each other. Mates can be set up between any planar surfaces, model faces, or datum planes. When a mate is specified to a datum plane, you must include whether you want the red or yellow side of the datum to be used in the placement. If you select the yellow side of a datum plane on one component to mate to the yellow side of a datum plane on another component, the datum plane yellow sides will face each other, orienting the model accordingly. The first of the illustrations that follow this table shows mated yellow sides.
Mate Offset	Mate Offset works the same as the Mate option, with the exception that you can enter an offset dimension between selected surfaces to complete the placement.
Align	Align makes planar surfaces coplanar and facing the same direction. If you select the yellow side of two datum planes to align, they would face the same direction, as shown in the second of the illustrations that follow this table. The Align option will also let you select axes and revolved features and make them coaxial.
Align Offset	The Align Offset option works the same as the Align option, with the exception that you can enter an offset dimension between selected surfaces to complete the placement.
Insert	Insert allows you to select revolved features or axes, which the option makes coaxial. The Insert option will lock a component in two degrees of freedom, and will need an additional Mate or Align to fully constrain the model.
Orient	Orient is similar to the Align option but does not make selected surfaces coplanar. Selected yellow sides of a datum plane will be parallel and face the same direction.
Coord Sys	Coord Sys fixes two coordinate systems, locking all three degrees of freedom in a single pick. The option aligns all coordinates in the X, Y, and Z directions. Vehicle manufacturers often use this type of assembly, wherein all components are located and created with respect to the rear axle coordinate system.

Constraint	Description
Tangent	The Tangent option is used to assemble revolved features by their tangency. This option will cause selected surfaces to be touching and facing each other. Keep in mind that additional constraints will need to be assigned to obtain proper component orientation.
Pnt On Srf	The Point on Surface option allows you mate a planar surface, datum, or model face to a datum point from a part or assembly. The term *mate* is fundamentally correctly used in this context because you are attaching the planar face to the datum point.
Edge On Srf	The Edge on Surface option allows you to assemble components by attaching a linear edge from one component to a planar surface on another.
Default	Default assembly's components using the default coordinate systems of both components.

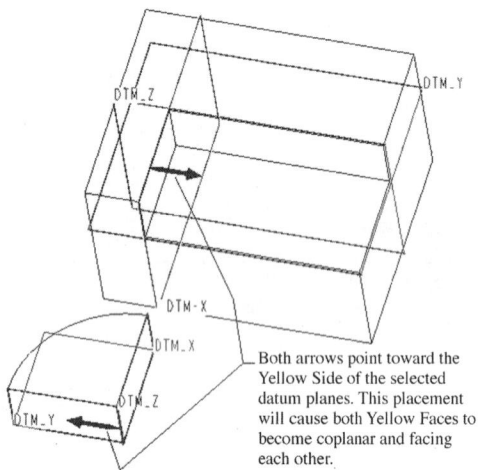

Both arrows point toward the Yellow Side of the selected datum planes. This placement will cause both Yellow Faces to become coplanar and facing each other.

Yellow sides mated.

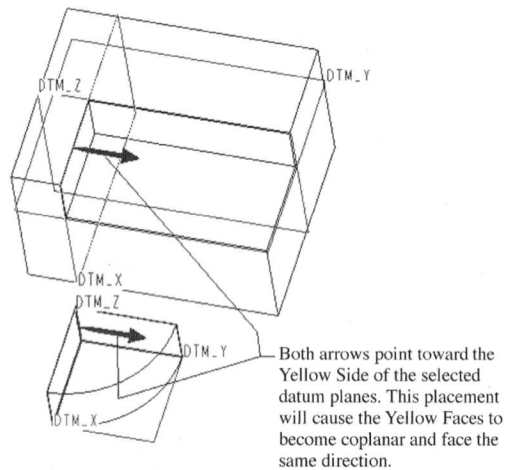

Both arrows point toward the Yellow Side of the selected datum planes. This placement will cause the Yellow Faces to become coplanar and face the same direction.

Yellow sides aligned.

Anytime you start assembling components, you are given information about the components' Placement Status. A component does not need to be fully constrained to be placed. At any point during the constraining process, you can click on the OK button and the component will be placed as a "package." The Package option is useful for experimenting with a component's location and configuration. A packaged component is not located to any other component, however, and therefore cannot be referenced for geometry creation or placement in the assembly. Once the position of a packaged component is finalized, the component should be locked down.

Speeding Up Your Assembly

One of the biggest problems facing designers and engineers is system speed. Just as in the individual models you have built, the more features and complexity you add, the slower the system will seem to run. In Assembly mode, this becomes apparent when you initially retrieve assemblies containing a large number of components and then try to make model changes or simply rotate your view. Regeneration time increases dramatically.

There are a few things that can help alleviate the problem of slow speed. The first is to get a faster computer with a better graphics card, but unfortunately that is not always possible and really does not address the main problem. The problem is that you are displaying all of the product's graphical geometry but in most cases working on a small portion of the entire assembly.

The solution here is really quite simple. Use layers to blank unnecessary components, and create simplified representations to remove excess geometry. Both of these techniques focus on displaying only the necessary geometry and components for the work you need to accomplish. The use of simplified representations does have one advantage over layering in that layered components that are blanked are still retrieved into the work session and therefore affect the machine's speed.

> ➝ **NOTE:** *For a detailed look at simplified representations, see Chapter 6 of the* Pro/ENGINEER Assembly Modeling User's Guide.

Design Verification

Now that you have put all of your components together, what can you do with the assembly? If the only thing you are doing with your solid models and assemblies is getting a 3D visual image, you are far ahead of the 2D world. However, within Pro/ENGINEER's Assembly mode is a rather impressive set of tools and capabilities that allows you to do much more.

The products you design will typically need to adhere to form, fit, and functionality constraints. The product has to look good to meet marketing needs, must go together easily on the shop floor, and must work the way it was designed to. This is where the benefits of Pro/ENGINEER come into play. You need to be able to answer some questions before you let the product go out the door. Does the product fit the styling constraints? Do all components go together properly? Is it manufacturable? Does it function properly? Have you planned enough space for motion and internal routings? Some of the answers will come as a result of prototyping and testing, but prior to that there are a few things you can determine.

With the use of the Model Analysis tools under the INFO pull-down menu, you can measure an assembly's mass properties and volume to determine its overall weight, center of gravity, and inertia. In addition, you have the ability to determine the clearance and interference between component pairs, as well as globally throughout the entire assembly. In some cases, a design calls for interference fits, but where they are not called for it is very handy to locate them before expensive tooling changes are needed. When determining interference between components, the results are displayed and expressed as a volume. That volume is highlighted for you, and you are given the opportunity to create a solid model of the interference zone.

This ability to create components from an interference volume can come in very handy in certain cases. A real-life example of this involved the design of an integrated hood and grill for a large tractor. Styling and a maximum size for the hood and grill was determined, but with farm tractors there are various tire sizes, turn radii, tire pitch angles, and axle pivot angles. Combined, these factors caused different sized tires to intersect the hood and grill of the tractor. However, in this case, the ability to make a solid model of the interference volume allowed for a plan that incorporated a localized styling dent in the grill to accommodate most tire sizes and to specify turn stops for others.

In many cases in the product design, the assembly has not been fleshed out enough to have any specifics for major components. You may only have a relative mounting location and a basic size. How do you plan around large voids? One of the best solutions is to create an envelope of space, which is a model of about the right size and shape and contains the information that is known. If that model is placed into your assembly, it serves two purposes. First, it represents an area where you cannot route pipes, wires, or hoses. Second, it holds a space in the assembly so that you do not forget to add the component, which also helps you keep a more accurate bill of materials. As you begin to flesh out the area around the envelope, you inevitably determine more detail for the components in the envelope, thus completing its design.

In answering the questions related to motion and component interaction, you must plan your assembly to display your intended motions. To display motion in an assembly is relatively simple, but you need to plan and design for it to work properly. Build a schematic to display the motion your models require; then model it. The model you are creating for the display of motion is a "skeleton," to which you can mount components, as well as modify to display movement. The following illustration shows an example of a skeleton model. It consists of two datum points, which are created on two curves. Each datum point has an axis created through it and the two axes have a datum plane created through them.

Sample skeleton model.

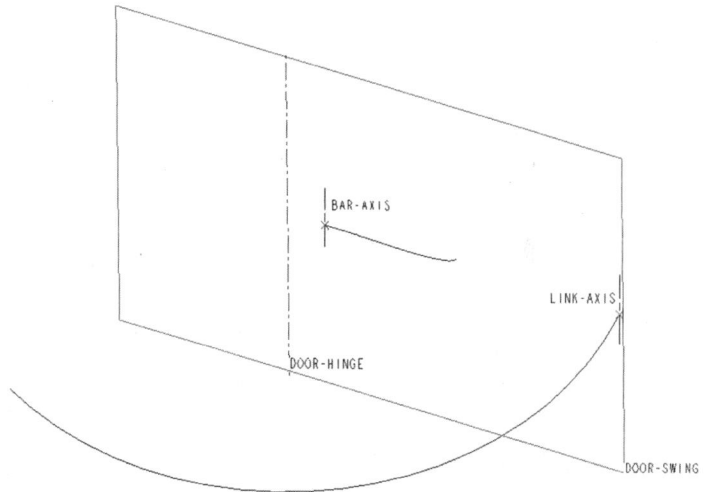

The features in the previous illustration will be used to control the position of a wind latch in a door assembly as the door is opened. The large arc represents the swing path of the latch bar as the door is opened around the DOOR-HINGE axis. In a skeleton such as this, you build the key assembly mount locations into the skeleton model and then assemble the skeleton as the first component of your assembly. You would next assemble the specific components to that skeleton. Then, by modifying the skeleton model you change the associated components' positions. In Chapter 17 you will be creating this skeleton and a program to simulate the motion of a door opening, as well as a wind latch rotating to lock the door in its fully open position.

Additional Concepts

As you have seen, Assembly mode is used for more than just putting a product together. Assembly mode is also used for design verification, motion study, and mass property analysis. In addition, you can use the assembly functionality as a vehicle to drive modeling references and associativity. By using the Merge and Cut Out commands under the ADV COMP UTL menu, you are using assembly functionality to create a geometrical link between parts.

The Merge command is used to add geometry to, and Cut Out to subtract geometry from, another model. These two commands can either Copy or Reference their geometry to another model. If Copy is used, the geometry and dimensions will be copied to the new model and changes to the original model do not affect the new model's features. With use of the Reference option, changes to the new model's geometry are made by modifying the original model geometry.

Casting and machining models are good examples of the advantageous use of the Merge command. The vast majority of modeling time and effort goes into the creation of the casting model. However, the casting is only useful when the excess material is machined away. The most difficult question of the casting-to-machining part modeling process has been what to do with all of the casting geometry. Because the machined model is the one used in the assembly models, it takes a lot of space to maintain two models with duplicate features and feature counts.

The best process developed so far for working with castings is the "casting-to-machining" process. In this process you create the casting model, then create a blank part as the machining model. You next use Assembly mode to copy the casting model's geometry into the blank machining part as a reference, using the Merge command. The final step is to add your machining cuts to the merged machining model. The following are the main benefits of this process.

- A lower, overall feature count, because the merged casting geometry counts as only one feature.

- The machining model references the casting so that changes to the casting are driven into the machining model (i.e., there is only one model to modify).

- In Drawing mode it is easier to locate show dimensions for the machining cuts.

  ↬ **NOTE:** *The casting-to-machining process is examined in more detail in Chapter 19.*

Another technique that uses the Merge functionality in Assembly mode is the "master model" technique. By the use of a master model, you are able to capture a product's overall appearance and relay it to all associated visually sensitive components. To build your product using this method, you first create a single block model that contains all aesthetically important features, as in the model shown in the following illustration.

Master model for a mouse.

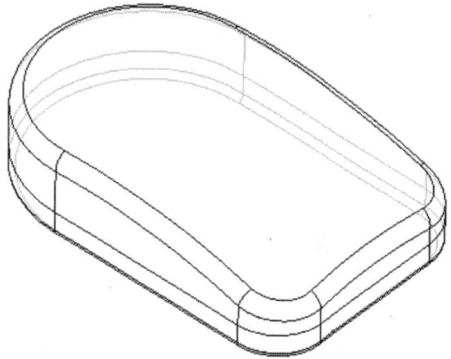

The model in the previous illustration is a mouse you will be creating in Chapter 18. The model contains only the necessary geometry to capture the styling intent for the mouse. This model needs additional reference geometry to be used to map out the mouse's components. In this case, the geometry will be a set of curves outlining the three buttons and the parting line, as shown in the following illustration.

Component mapping.

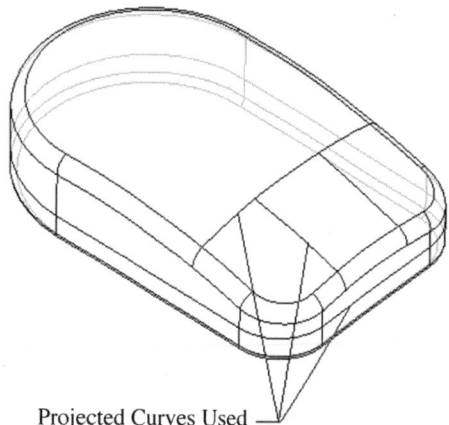

Projected Curves Used
Mapping Components

Once the master model's components are planned out, you can reference copy the master model's geometry to all mapped components. The last step in this technique is to retrieve each component, which will contain all geometry in the master model and cut away all excess geometry, leaving only the individual component. The real benefit to this type of modeling technique comes when changes are made to the product's styling. Because all components reference the master model, any necessary changes can be made to the master alone and driven to every component during regeneration. This saves a lot of model modification time and helps avoid the possibility of forgetting to update any models.

Summary

In this chapter you have read about some of the basic tenets of assembly design. It is important that you as designer and engineer have a good understanding of the product you are working on and plan your design intent through the assembly to the subassemblies and parts. By doing this, you leave a clear path for someone else to follow through your designs.

Some of the earliest assemblies you built using Pro/ENGINEER probably consisted of a series of parts stuck together. Such assemblies can produce a correct bill of materials, look correct, and show all necessary clearances, but may not represent how the parts will actually be assembled. This is where planning comes in. Break up the total assembly into its major subassemblies, make sure those subassemblies contain the correct mounting information and assembly references, and then create the individual component parts with their proper mounting information. By doing this, you drive your design from the top down, and incorporate design intent, form, fit, and functionality in every piece of geometry you create. This makes you more efficient and decreases product design time. It also makes modifications much easier.

Planning is the most important step in the design process, and time-saving techniques for specific design issues are vital for effective modeling. Techniques such as casting to machining, master modeling, and motion modeling help ease difficult design issues, and help you verify your design in a virtual environment prior to cutting any tooling orders.

17

Skeletons and Motion

Part of your job in project design is to make your products easy to assemble and to attempt to locate and correct errors before you get into production. Pro/ENGINEER contains a couple of tools that will assist you in achieving these goals. The first is a series of Model Analysis tools that deal with clearance and interference. By using the tools under the Model Analysis feature you are able to verify your component interrelationships in their assembled positions. The second tool that assists you with the verification process is the skeleton model.

Skeleton Models

The Model Analysis tools are of great benefit in the verification process; however, they are not the complete answer. In most cases, your products contain some moving parts, which means that you will need to verify clearance in more than one location. Skeleton models aid you in planning and capturing your design intent. By using skeletons in your part and assembly design, you can plan your assembly constraints and demonstrate how your product will function.

Skeletons in Pro/ENGINEER can consist of datum points, curves, surfaces, axes, or planes. The key, however, is to keep the skeleton simple. You are modeling motion paths and assembly constraints only. The skeleton may not look like much more than an axis and two datum planes, but if you use them correctly they can really capture and display your design intent. In the following exercise you will be creating some part models to use with a skeleton part in assembly. You will then be modifying the assembly to show the parts in motion.

Exercise 17-1: Modeling Using a Skeleton

Your goal in this exercise is to model an electrical cabinet, including door, frame, and wind latch linkage. The model will show the door opening position up to 120 degrees (fully open), demonstrating how the wind latch linkage will move as the door is opened. The illustration at right shows the part models you will be creating. In this view the model is assembled, with the door opened about half way.

Door hinge and link pin models have been omitted from this exercise. In addition, you will be modeling only a portion of the door frame and door to reduce excess geometry on the screen.

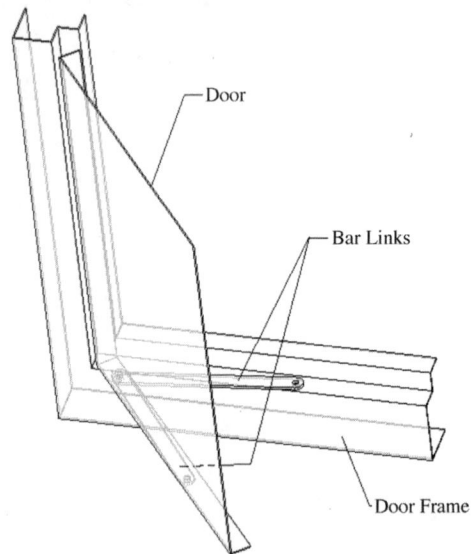

Partially open door on cabinet assembly.

When planning for motion in your assemblies, you first need to figure out which parts are stationary and which are motion parts. You then need to determine how much motion is needed, as well as how the motion parts will move. Taking the previous illustration as an example, first assess and categorize the necessary motion. Your assessment might consist of the following points.

- The door will need to be shown in positions from closed (0 degrees) to fully open (120 degrees). A datum axis, and a datum plane created through that axis at an angle, are required to accomplish this.

- The linkage consists of two links, one of which needs to be anchored to the frame, and the other to the door. The two links are joined at an axis, the location of which is moving as the door is swung open. This motion will be much more complex to show than the door opening, requiring a datum axis at the intersection of two arcs of equal radii.

- The radius value will be the same as the link hole, center-to-center distance, and is determined by dividing the chord length of an arc generated by the door link anchor axis as the door is opened from 0 to 120 degrees. The anchor points for the link on the door and frame are located 9 inches from the door hinge axis, as shown in the following illustration.

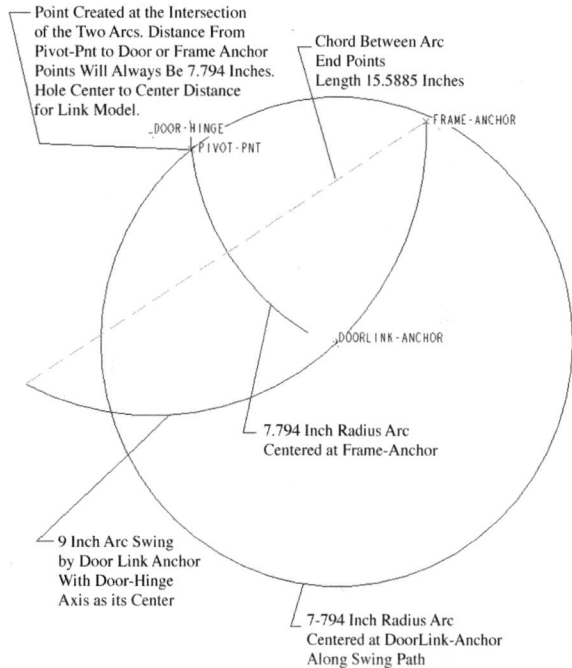

Link axes and pivot path.

Point Created at the Intersection of the Two Arcs. Distance From Pivot-Pnt to Door or Frame Anchor Points Will Always Be 7.794 Inches. Hole Center to Center Distance for Link Model.

Chord Between Arc End Points Length 15.5885 Inches

_DOOR-HINGE
PIVOT-PNT

FRAME-ANCHOR

DOORLINK-ANCHOR

7.794 Inch Radius Arc Centered at Frame-Anchor

9 Inch Arc Swing by Door Link Anchor With Door-Hinge Axis as its Center

7-794 Inch Radius Arc Centered at DoorLink-Anchor Along Swing Path

Now that you have determined the motion paths and key feature locations, you need to model them as a skeleton part.

1. Create a new part and name it *Skeleton*. Start the Skeleton part with the default datum planes and a default coordinate system. Select **Datum ➥ Axis ➥ Two Planes**. This will create an axis at the intersection of two datum planes. Pick DTM1 and DTM2 for the axis creation. This will be used to locate the door hinge and to create the datum plane, which will be used to open the cabinet door. Change the name of the new axis to *Door-Hinge*.

➥ **NOTE:** *Naming important features in your skeleton is a must. It helps you locate important features and maps them out for others who might work on your designs.*

2. Create a datum plane through the Door-Hinge axis at an angle to DTM2. Enter *30* as your angle value. Change the name of your new plane to *Door-*

Swing. By modifying the angle value for this plane, you will be opening and closing the door component. This plane also represents the outer surface of the door itself.

In the previous two steps you created all geometry necessary to show the first motion, opening and closing the door. In the following steps you will be defining the geometry needed to control the link mounting and motion controls.

3. Create a datum curve representing the path the Door-Anchor will follow. The curve will be a 120-degree arc centered on the Door-Hinge. However, because this will be used as part of your component assembly, you will need to plan for material thickness and external feature locations. Select DTM2 as your sketch plane and DTM3 as the bottom reference; then sketch the section shown in the following illustration. Make sure you do not use the Door-Swing datum plane as a reference in this feature's creation, because you do not want the door-opening angle to affect this path's length. Name this feature *DoorLink-Path.*

DoorLink-Path.

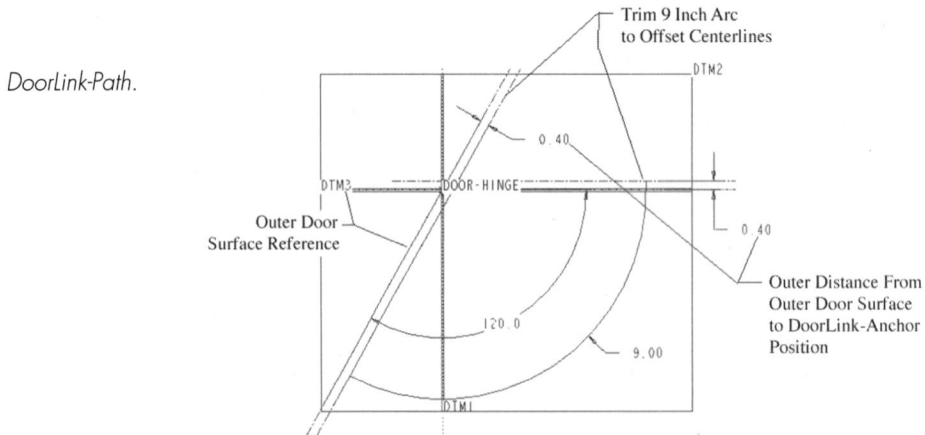

4. Create a **Datum Point** on the DoorLink-Path curve. Use the **Length Ratio** option and enter *.5* for the length ratio. This should place a datum point at the middle of the curve feature. Name the new point *Door-Anchor.*

5. Create another **Datum Point** on the DoorLink-Path curve. Again, use the **Length Ratio** option; however, this time enter *1.0* for the length ratio. This should place a datum point at the 0-degree end of the curve feature. Name the new point *Frame-Anchor.* Your skeleton model should now look similar to that shown in the following illustration.

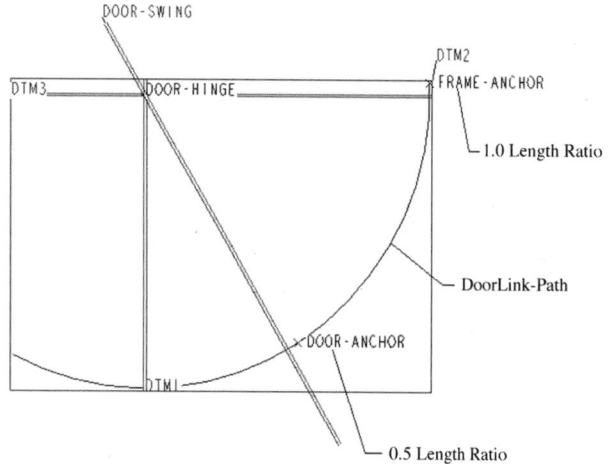

Skeleton with link anchor points.

6. Determine the position of the pivot point for the two links for your assembly. Use the **INFO** pull-down menu and select the **Measure** option. The Measure dialog box will pop up. From the **Type** section, select **Distance**, and then pick the Frame-Anchor point and the vertex at the 120-degree end of the Door-Link-Path curve, as shown in the first illustration at right.

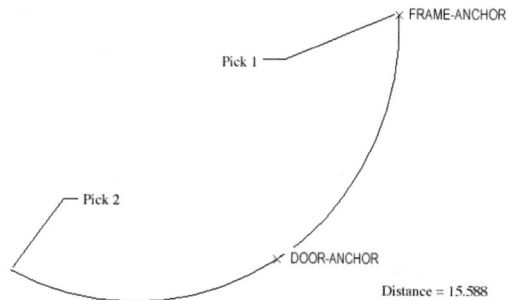

Distance measurement screen picks.

7. The distance between the Frame-Anchor point and the vertex at the end of the DoorLink-Path curve should be 15.588 inches. Half of this length will be the radius value of the two arcs you will need to locate the linkage pivot point. Create the first arc as a datum curve centered on the Frame-Anchor point. Create the section shown in the second illustration at right. Use DTM2 as the sketch plane and DTM3 as the bottom horizontal reference. Name this feature *Arc1*.

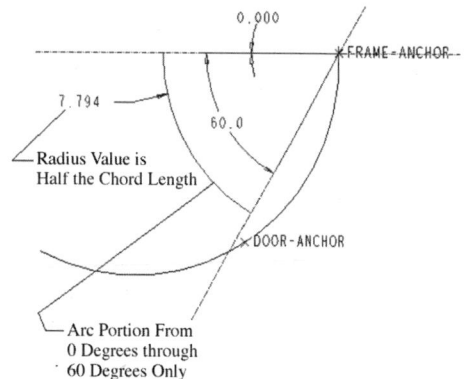

Linkage pivot Arc 1.

8. Create the second arc to use for locating the linkage pivot point. This will again be a datum curve, but its center will be on the Door-Anchor point and it will be a complete circle with a radius of 7.794. Use DTM2 as the sketch plane and DTM3 as the bottom horizontal reference. Name this feature *Arc2*. The section is shown in the first illustration at right.

9. Create a **Datum Point** at the intersection of the two curves Arc1 and Arc2. Name the point *Pivot-Point*. After creation of the Pivot-Point, your skeleton model should look like that shown in the second illustration at right.

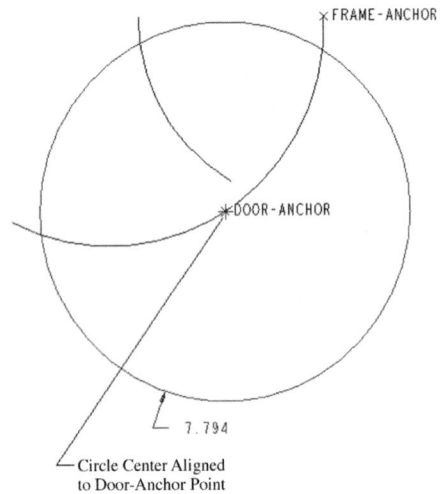

Linkage pivot Arc2.

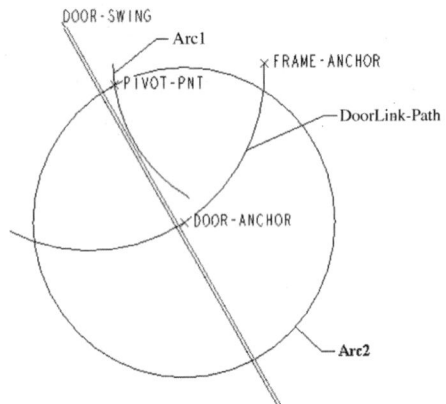

Skeleton model with Pivot-Point.

You need to perform three more operations to complete your skeleton part. First, you need to add intelligence to your design. You have named all of your features and have left a very nice map, but you still need to create relationships between some of the features so that the motion of one affects the geometry it is supposed to. Second, you need to test the skeleton to make sure everything is working the way you intended it to. Third, you need to add the necessary features to aid in the assembly of the modeled components. In this case, the features you will be adding are datum axes to be used to locate the links in assembly.

Because of the way you have created your datum curves and points, you have already set up a pretty smart model. For example, if you move the Door-Anchor point, Arc2 will move with it, shifting the position of the Pivot-Point. This is exactly what you want, but you still need to tie the motion of the Door-Swing datum plane to the location of the Door-Anchor point. To do this, you need to write a Relation between the Door-Swing angle dimension and the Door-Anchor ratio value. The relation will force the Door-

Anchor ratio value to vary from 1.0 to 0 as the angle changes from 0 to 120 degrees. This is another good time to do some naming of important information.

10. Use the **Modify** option and pick the Door-Swing plane and Door-Anchor point to view their dimensions. Select **Dim Cosmetics** ➥ **Symbol** and rename the datum plane angle dimension to *OpenAngle* and the point ratio dimension to *PathPoint*. This should clear up any confusion during future modifications.

11. From the PART menu, select the **Relations** option, and add the following relation.

```
PathPoint=1 - (OpenAngle/120)
```

The reason you are subtracting the OpenAngle ratio from 1 is that the Path-Point ratio value for the fully open door needs to be 0. Therefore, when the OpenAngle is 120, this will be the case.

12. Test your design. Create a datum curve to represent the links you will be adding in assembly. Do this by sketching two lines aligned to the Frame-Anchor, Pivot-Point, and Door-Anchor points, as shown in the following illustration. In sketching this datum curve, it will be helpful if the OpenAngle is larger than 10 degrees.

Link representation curves.

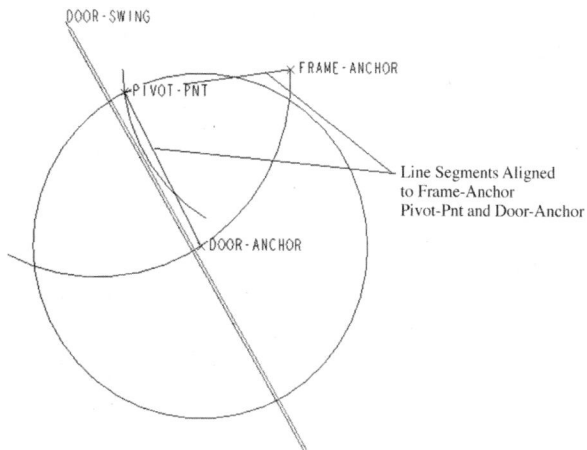

DOOR-SWING

PIVOT-PNT

FRAME-ANCHOR

Line Segments Aligned
to Frame-Anchor
Pivot-Pnt and Door-Anchor

DOOR-ANCHOR

13. Modify the OpenAngle value between 0 and 120 to examine how this affects your skeleton model and its curve and point locations. If everything is working properly, the two straight-line segments representing the latch linkage and the

Pivot-Pnt should move relative to the door's OpenAngle. You now have a pretty smart skeleton model and are ready to add the datum axes you will need for your component assemblies.

14. Create three axes using the **Pnt Norm Pln** option. Make each axis normal to DTM2 and through each point: Frame-Anchor, Pivot-Pnt, and Door-Anchor. This will mean three separate axis creations. Name each axis (as shown in the illustration at right) of the completed skeleton part. The names are *FrameLink-Anchor, Link-Pivot,* and *DoorLink-Anchor.*

15. When the axes have been named, **Save** the part and open a new part window.

16. Now you are ready to create the component parts for your assembly. Create the Link component using the following illustration. Make sure you have axes in the hole centers for use in the assembly.

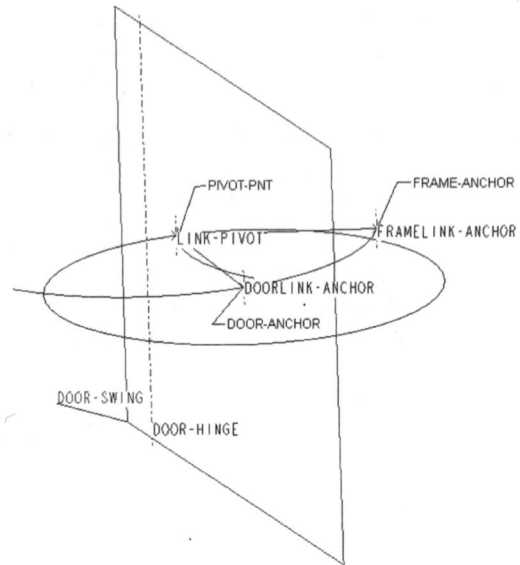

Completed skeleton part.

Two views of the link component.

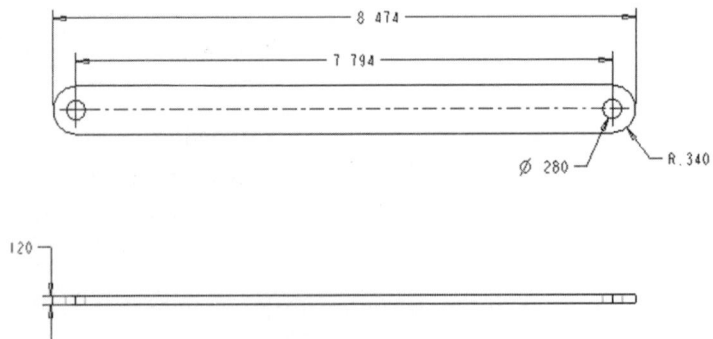

Link Component

17. Save the Link component model and open a new part window.

18. Create the door component. Make sure to include an axis representing the hinge location. This axis will be in Assembly mode to **Align** the door and skeleton model. Create the hinge axis using the **Two Planes** option, and pick DTM1 and DTM3. Name the axis *Hinge*.

The following illustration shows the first protrusion, which is 0.98 inches thick. Notice the offset values from the default datum planes. These provide gaps for the two link arms below the door and a theoretical hinge location along the left side.

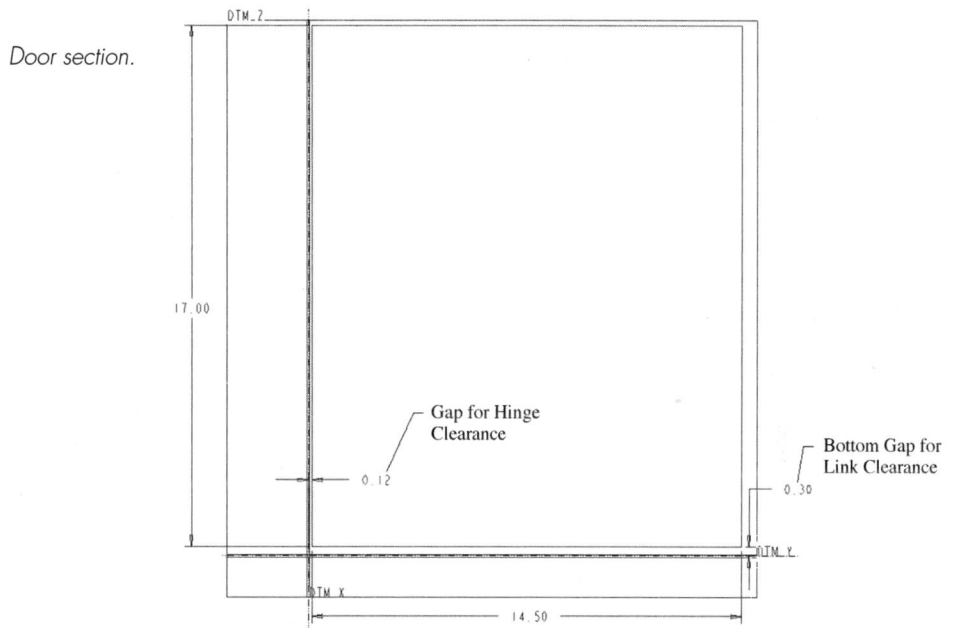

Door section.

19. Shell the door protrusion to a *0.06*-inch wall thickness. The door will also need a hole located 9 inches from the axis and offset from the front surface by 0.40 inches. Locate the hole as shown in the following illustration.

Door anchor hole position.

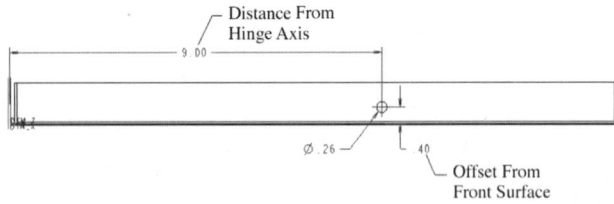

20. **Save** the door component model and open a new part window.

21. Create the Frame component model. Starting with the default datum planes and default coordinate system, create the first protrusion, as shown in the following illustration. Enter a depth of *2.0*.

Frame protrusion section.

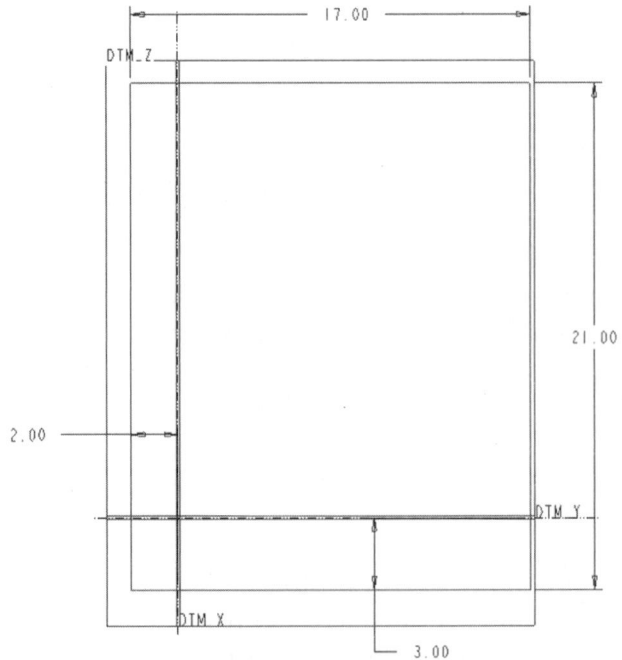

22. Remove the area the door will fit in. The cut depth is *1.0* inches. Refer to the following illustration for section details.

Door opening cut section.

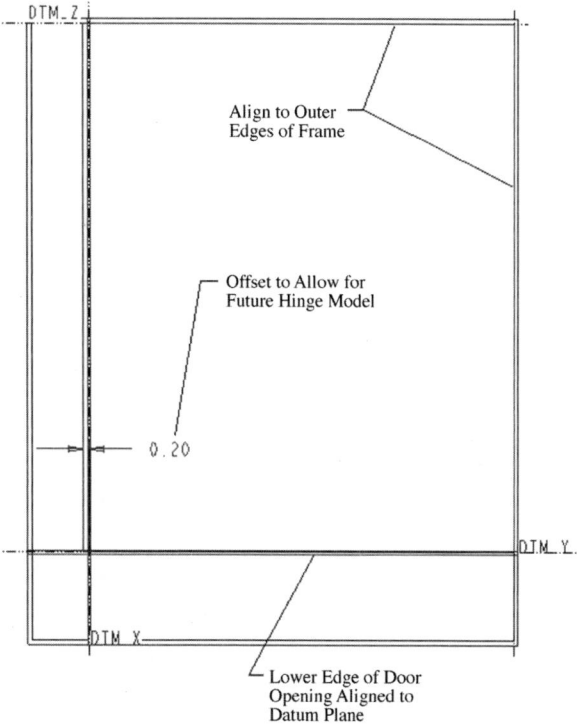

23. Use the **Thru All** option to add a cut creating the upper door lip. Refer to the following illustration for section details.

Door lip cut section.

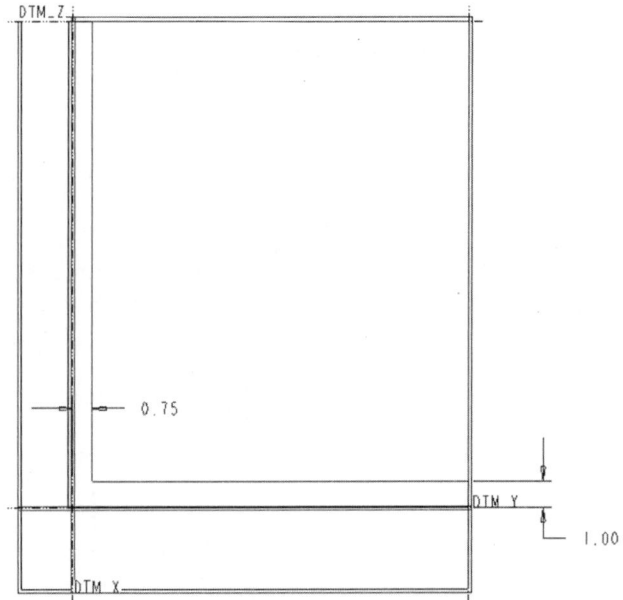

24. Shell the model to a *0.06*-inch wall thickness and add the hole for the Frame-Anchor axis, as shown in the following illustration.

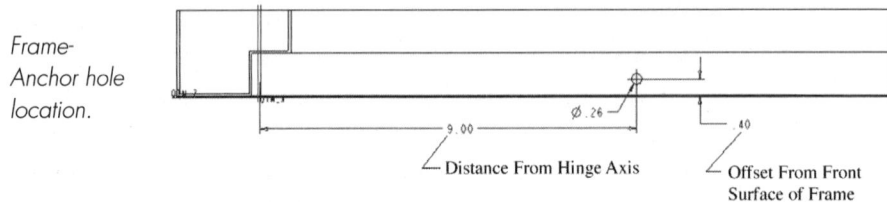

Frame-Anchor hole location.

Once you created the Frame-Anchor hole, the frame component is complete.

25. **Save** the frame model.

You have now created the three component models that will be assembled to your skeleton model. The component models are shown in the following illustration.

Three component models.

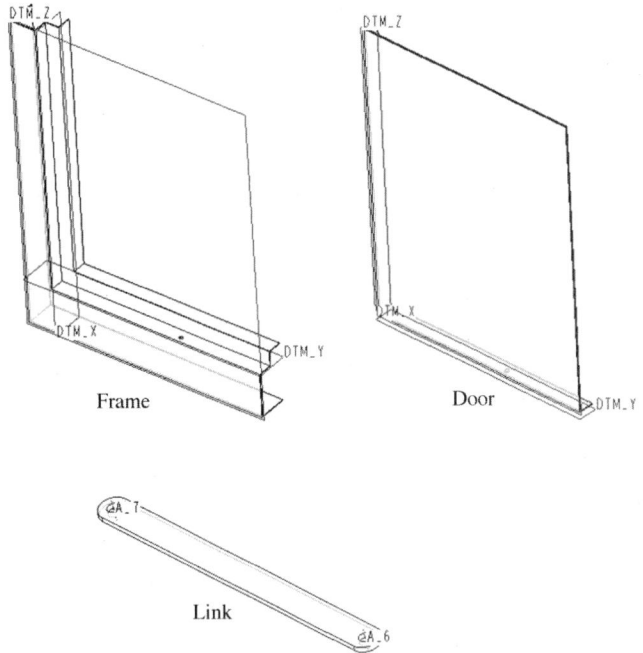

Frame

Door

Link

You are now ready to assemble the component models and verify that the assembly works.

26. Create a new assembly containing the default datum planes and coordinate system. Name this new assembly *Door-Sub.* Select **Component** from the ASSEMBLY menu, then **Assemble**. From the Open dialog box, pick your skeleton part to assemble. From the Component Placement dialog box, Constraint Type section, select **Default,** then click on the **OK** button at the bottom of the dialog box. This will assemble the skeleton model to the assembly, and the default coordinate system of the assembly to the default coordinate system of the skeleton.

•• NOTE: *The name you choose for the assembly model in the real world should represent your company's naming convention, for use in the product BOM.*

27. Assemble the Frame component to the Skeleton model. Because the frame will not move in the assembly, use the **Default** option from the Constraint Type section, as you did for the Skeleton model. Click on the **OK** button at the bottom of the dialog box to complete the placement.

The next three components will move in the assembly, so they will need to be placed more specifically.

28. From the ASSEMBLY menu, select the **Modify** option. From the ASSEM MOD menu, select **Mod Part**. Pick the skeleton model to modify; then pick the Door-Swing plane and change the open angle to *90* degrees. From the MODIFY PART menu, select **Regenerate**. Select **Done** to complete the part modification process.

The reason to open the door wider is to make the DoorLink-Anchor and Link-Pivot axes more easily seen and easier to pick. The assembly should now look like that shown in the following illustration. The model is a little cluttered, but still workable.

Assembly ready for moving parts.

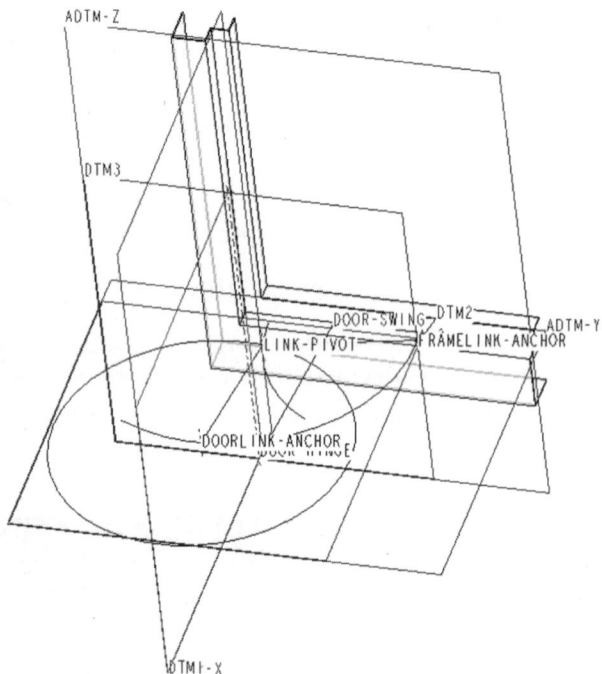

29. Assemble the Door component. This part will need to be assembled to the Door-Hinge axis, the Door-Swing plane, and ADTM-Y (ADTM2). Select the **Align** option from the Constraint Type section of the Component Placement dialog box. Pick the Hinge axis from the Door part and the Door-Hinge axis from the Skeleton part. Select **Mate** for a Constraint Type, and pick the outer surface of the Door part (this could also be DTM3, yellow side) and then the

Door-Swing plane (red side) from the Skeleton part. Use **Mate** again and select the datum along the bottom of the Door Part (DTM2, red side) and ADTM-Y (ADTM2, yellow side). The component should be fully constrained. Click on the **OK** button in the dialog box to complete component placement. The following illustration shows what the assembly should now look like.

Assembly with door attached.

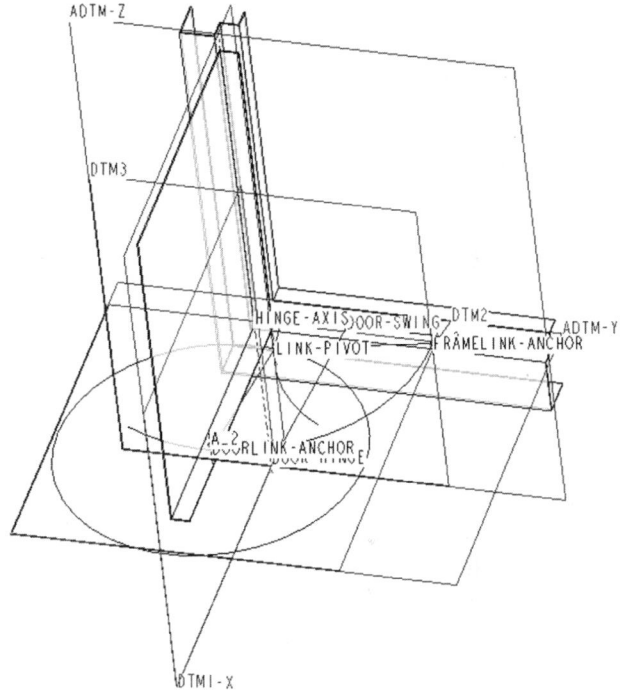

At this point, the screen is too cluttered. Therefore, it would be a good practice to start layering off components not necessary for additional part placements.

30. Create five layers in the Top Level of the assembly. Name them *Skeleton, Frame, Door, Link1,* and *Link2.* Place the Frame, Door, and Skeleton on their named layers and set the display of the Frame and Door to **Blank.** Once you have repainted the display, it should look as it does in the following illustration.

Door and Frame components blanked.

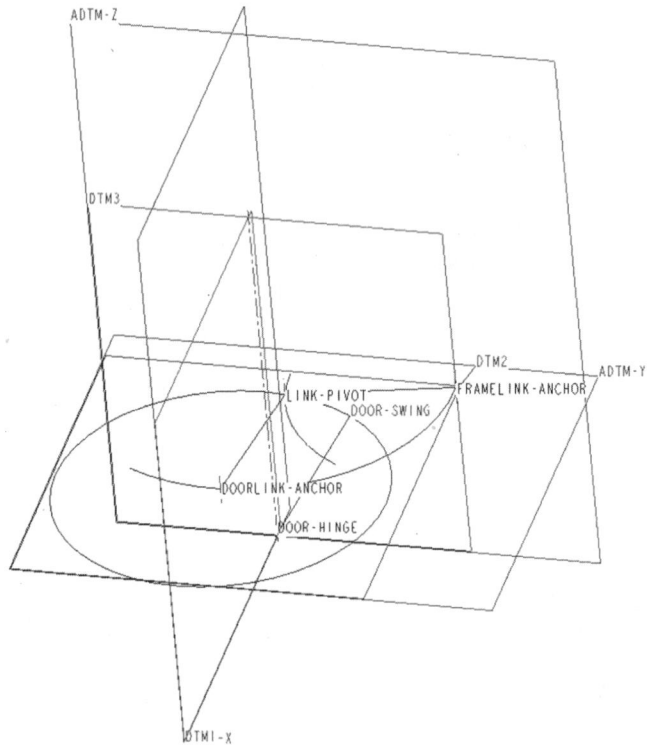

Now you can more easily see and place the Link components.

31. Using the **Align** option, assemble the Link component to the Skeleton part. Align one of the hole axes to the FrameLink-Anchor and the other to the Link-Pivot axis. Use the **Mate Offset** option to offset one of the flat surfaces of the Link part from DTM2, yellow side, of the Skeleton part. Enter *0.03* for the offset value. Click on the **OK** button in the dialog box to complete component placement.

32. Assemble the second Link component to the Skeleton part. You will be placing the Link part twice in this assembly. Again, use the **Align** option. Align one of the hole axes to the DoorLink-Anchor, and the other to the Link-Pivot axis. Use the **Mate Offset** option again and select the bottom surface of the new Link part and the top surface of the first Link part. Enter *0.03* for the offset value. Click on the **OK** button in the dialog box to complete component placement. With the two links assembled to the Skeleton part, the assembly should look as it does in the following illustration.

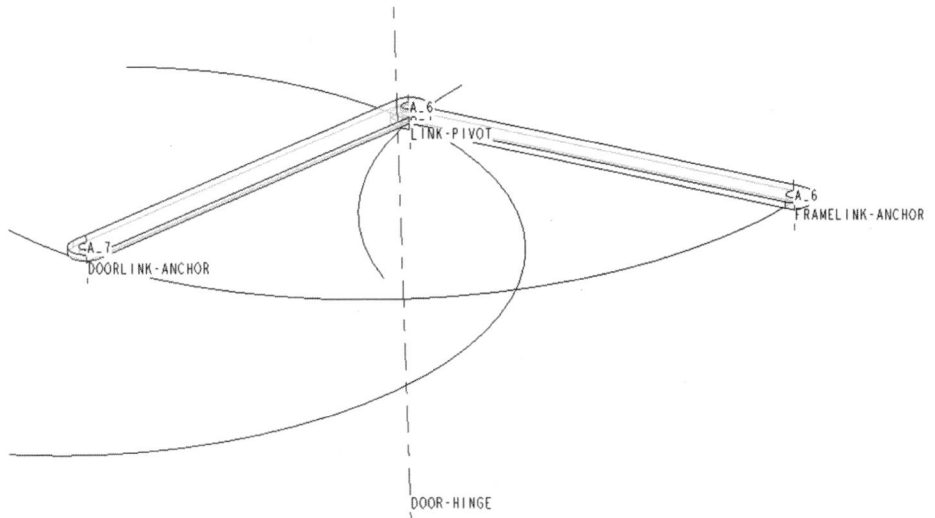

Two assembled link components.

33. Place the two Link components on the assembly layers Link1 and Link2 (one on each layer). **Unblank** the Frame and Door components and repaint the screen. The entire assembly should be visible and should look like that shown in the illustration at right.

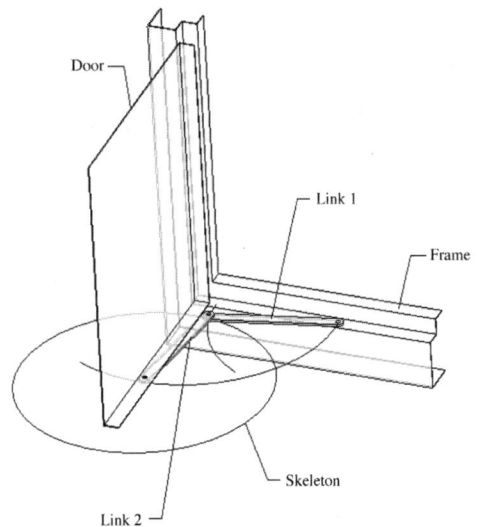

Completed assembly.

34. To show the Door at various degrees of open, select the **Modify** option. From the ASSEM MOD menu, select **Mod Part**. Pick the skeleton model to modify; then pick the Door-Swing plane and change the open angle dimensional value from 0 to *120* degrees. From the MODIFY PART menu, select **Regenerate**. Select **Done** to complete the part modification process. The following illustration shows views of the door closed and fully open.

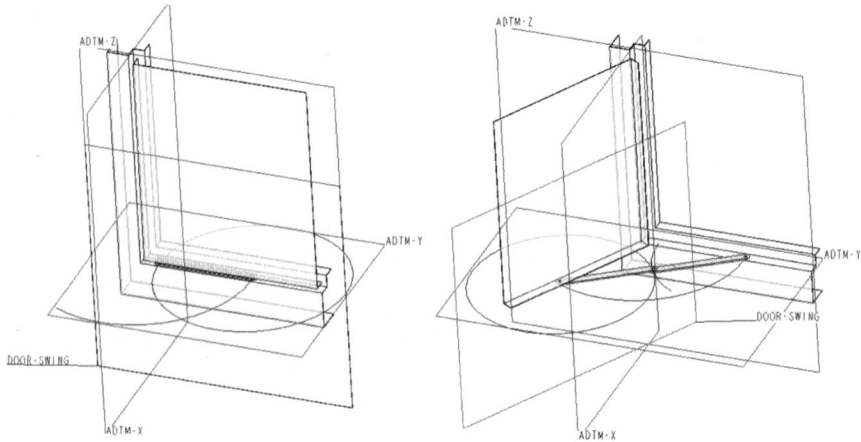

Door closed and fully open.

Summary

As you have seen through the exercise in this chapter, assemblies can be used for more than just putting parts together. Through planning and a little practice, you can both capture the design intent of your product and bring it to life. Through the use of skeleton models, you can define and test the movements of components. In addition, because the geometry in the skeleton is extremely lightweight, regeneration time is very short.

In using a skeleton, you do not have to track down several model variables to modify for motion simulation. All variables are contained in one place. In the exercise, you planned and constructed a skeleton part to be used in an assembly, but you could have just as easily created the skeleton as a set of assembly features to obtain the same results. The major advantage to creating the skeleton as an assembly feature is that you could then use Pro/PROGRAM to create a program file for the assembly, which you could then use to prompt for the open angle and regenerate the assembly automatically.

With the use of skeleton models, relations, and programs, your product design takes on another level of complexity. Therefore, relaying your design intent becomes more difficult, and all the more important, because not just anybody can start modifying your design without knowing the path you took during the design. This is why naming features and dimensions is so important. In this exercise, you renamed several important features and dimensions, and the names were fairly long. The reason for the length and detail of these names should be quite obvious. By using names that do not have to be interpreted, you clearly convey your design intent.

18

The Master Model Technique

In your everyday design life you are confronted with numerous categories of problems and design constraints, but styling is typically the area that presents the most challenging workarounds. When a new product is created or an old product reworked, matters of appearance and style are usually the most difficult to finalize, but also the only things that will end up set in stone.

Master Model Technique

Pro/ENGINEER handles the problem of finalizing matters of style and appearance by incorporating a pretty handy procedure that simplifies style modifications across the entire product without any major headaches. The procedure is called the "master model technique." The following are the basic steps of this procedure.

1. Build a single model that captures the styling and overall design intent.

2. Create a blank start model as a component.

3. Use Assembly mode to copy the styling information to that component.

4. Copy the new component with its merged geometry to create the new component model you will need.

5. Cut unnecessary geometry out of each component.

6. Assemble the components as the finished product.

7. Change the master to propagate changes through all assembly components.

In the following exercise, you will be using the master model technique to create a mouse for your new computer product. The steps you will be taking are those previously listed. The main goal here is to become familiar with how the master model technique can help you control and maintain your styling and design intent throughout the various key components in your product's design. The last part of this exercise drives home the benefits of this process, when unexpected changes occur.

Exercise 18–1: Creating a Master Model for a Mouse

The illustration at right shows a master model for a three-button mouse with an upper and lower cover. It is currently one solid block with datum curves projected on it to map the buttons, split location, and mounting screw holes.

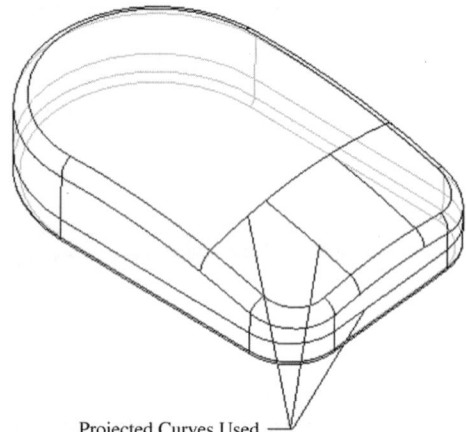

In this exercise, you will be creating a master model, three button models, and top and bottom cover part models. You first need to create the master model part.

Projected Curves Used
Mapping Components

Mouse master model.

    1. Using the default datum planes and default coordinate system as your start part, create the solid model using the following illustration as a guide. Name your new part *Master-Mouse.*

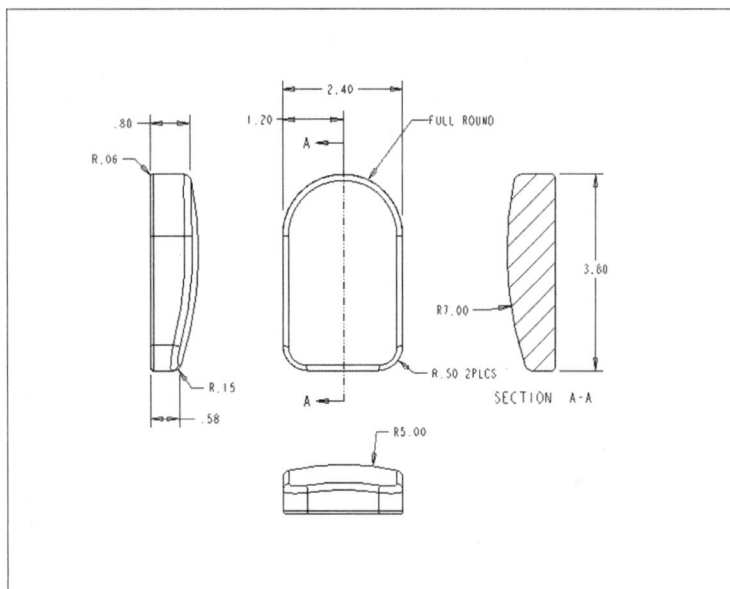

Mouse drawing.

Once you have completed the initial part model, it is time to start mapping out the components for the product. The model you have created contains and controls the product appearance and will be used to drive any changes to its design through the component part you create from it.

2. Map out the three buttons for the mouse. To do this, create a **Projected Datum Curve** on the top surface of the mouse model. When selecting surfaces to project the curve onto, select the top curved surface. The radius around the top is optional.

↪ **NOTE:** *If you are unfamiliar with projected datum curves, see exercise 12-2 for a step-by-step procedure.*

The curve you will be projecting represents the nominal button perimeter (i.e., a line-on-line condition with the top cover and adjacent buttons). The section for the projected curve is shown in the following illustration.

Note in the illustration at right that the projected curve section exceeds the surface boundary. This is to avoid potential problems with projecting the curve onto the tangent edges of the top surface of the mouse.

3. Create a datum plane **Offset** from the bottom flat surface by *.25* inches. The plane represents the parting line for the top and bottom cover. The following illustration shows the model as it exists so far. Rename the datum plane *Parting-Line* so that your design intent and purpose for the plane are unmistakable.

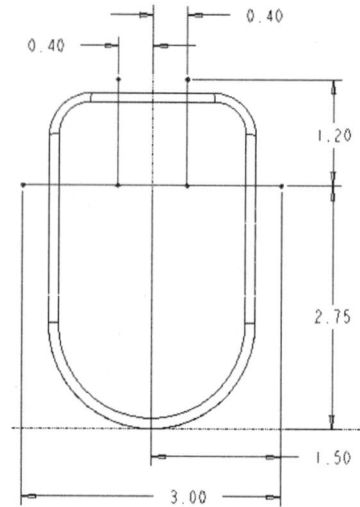

Projected curve section.

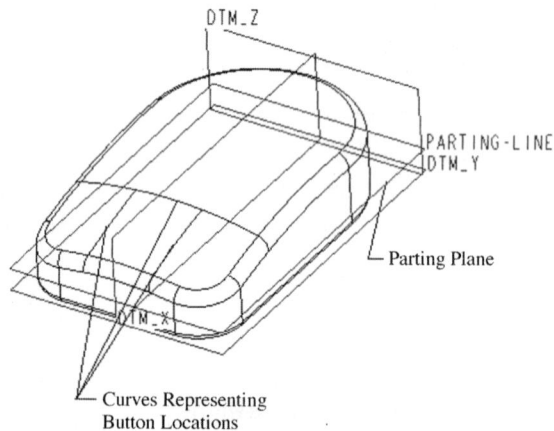

Master model after parting plane.

Now that your master model is basically mapped out, you are ready to start creating the mouse component parts.

4. Create a new part model consisting of the default datum planes and default coordinate system. Give your new blank part the name *Top-Cover*. Create a new assembly. Select **File** from the pull-down menus along the top of your screen, then **New ➡ Assembly**, and accept the default assembly name. For this proce-

dure there is no need to use an assembly start model because the assembly you are creating is temporary and will not be saved. Its only purpose is to act as a vehicle for you to copy the master model geometry to each component part.

5. From the ASSEMBLY menu, select **Component ➡ Assemble**, and select the blank part model named *Top-Cover*. The part will now be placed into your assembly. Select **Component ➡ Assemble** again and pick the Master-Mouse part to assemble to the Top-Cover part. To place the Master-Mouse part, select **Coord Sys** from the **Constraint Type** area; then pick the Master-Mouse default coordinate system as the **Component Reference**, and the Top-Cover default coordinate system as the **Assembly Reference**. The components are now fully constrained. Select the **OK** button from the Component Placement dialog box to complete the placement.

6. Copy the master model information into the Top-Cover part. From the COMPONENT menu, select the **Adv Utils** option. Here you will be using the Merge command. The order in which you select the models to be merged is important. The first model you pick is the one you will be adding information to, and the second model is the one the information is coming from. This is true for both the Merge and Cut Out commands under the ADV COMP UTL menu. Select the **Merge** command and pick the Top-Cover model as the component to perform the merge process on. Select **Done**. You will now be prompted "Select reference parts for MERGE process." Pick the Master-Mouse part, and then **Done Sel**. You will be presented with the OPTIONS menu, shown in the illustration at right.

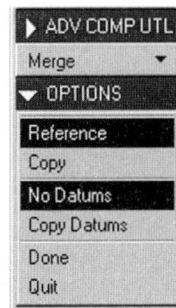

Menu for merge options.

The following list describes the options available under the OPTIONS menu.

| Reference | The Reference option keeps associativity between the new merged model and the original parent model, so that as you make changes to the original those changes are propagated through the child models. |
| Copy | The Copy option copies all feature information to the merged model. This option does not propagate changes from the original. |

| No Datums | No Datums prevents duplicate datum planes from being created in the merged model. |
| Copy Datums | Copy Datums copies all parent datum planes to the merged model. This can cause a proliferation of datum planes in your merged model, with some planes existing on top of each other. Because of this, use the Copy Datums option only when necessary. |

7. Select **Reference** ➡ **No Datums**. Select **Done**. You will be prompted "Detach reference part MASTER-MOUSE from the assembly?" For this exercise, you can select Yes or No; both will give the same results. Quit the assembly window and *do not* save the current assembly model; you will not need it again.

8. Open the Top-Cover model again. The model should look like that shown in the illustration at right, and it should contain all of the feature geometry from the Master-Mouse model except for the Parting-Line datum plane. This is because you selected the No Datums option.

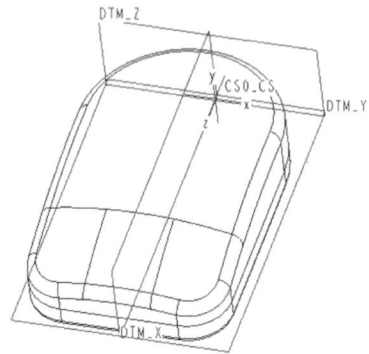

Top-Cover model.

The Top-Cover model consists of five features: three datum planes, one coordinate system, and the merge. Modifications can be made to this model in two ways. The first is to modify the Master-Mouse part and then regenerate it and the Top-Cover models to incorporate the change. The second way to modify the Top-Cover model is to add features to it. However, such changes will affect this model only.

Another characteristic you will notice about this model will show up when you try to layer off the surface datum curves. Upon selecting the curves, the entire model will highlight. Do not let this throw you. You are selecting the datum curves to add to the layer, but because the model's geometry is one feature, it will highlight when picked. This also holds true when surfaces and datum points are part of the merged geometry.

9. Now that the Top-Cover merge is complete, create the additional four parts you will need for your mouse product. The easiest method for creating the remaining models is to use the **Save As** option from the **File** pull-down menu. Save the Top-Cover with the following names: *Bottom-Cover, Left-Button, Right-Button,* and *Middle-Button.*

You should now have five identical models and one master model. All that remains is to cut away the unnecessary geometry from each of the component parts.

10. Complete the Top-Cover model. Use the datum curves on its surfaces as a blueprint for two cut features. The first cut will remove all geometry below the parting line. Create a cut for the section shown in the first illustration at right.

11. Make the cut that removes the bottom portion of the model with an offset from the parting line, to allow for a styling edge as well as clearance in the real-world assembly.

12. Remove the button portion of the Top-Cover with another cut by offsetting the button curve by *.03* inches, as shown in the second illustration at right.

13. **Shell** the model to a *.06-* inch wall thickness. When you have completed this step, the model should look like that shown in the third illustration at right.

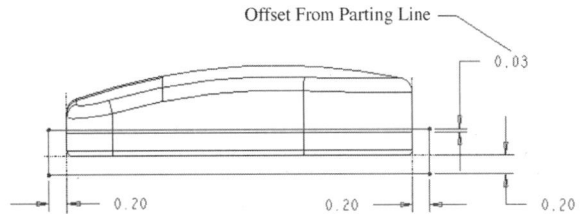

Offset From Parting Line

0.03

0.20   0.20   0.20

Parting line cut section.

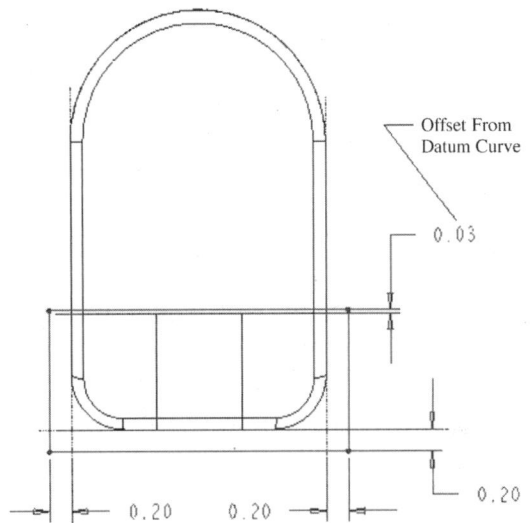

Offset From Datum Curve

0.03

0.20   0.20   0.20

Button removal section.

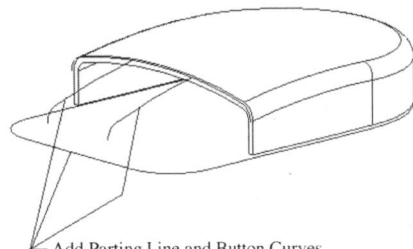

Add Parting Line and Button Curves to a Layer and Blank Them to Remove Clutter From the Display.

Shelled top cover.

14. Complete the Top-Cover model by adding a tapered lip to the bottom surface of the part. The first illustration at right shows the details of the lip section.

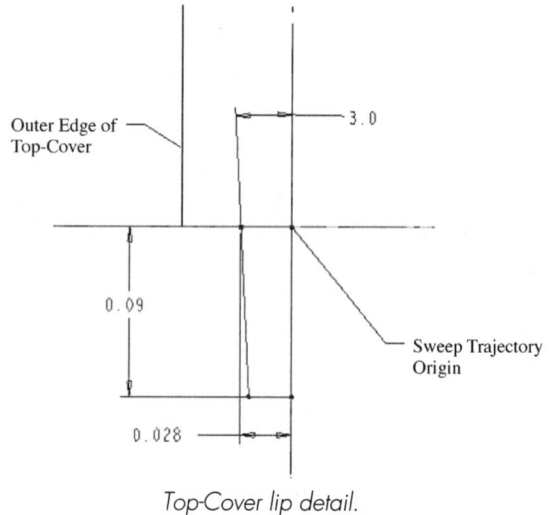

Top-Cover lip detail.

15. When finished adding the lip, **Save** the Top-Cover model, then open the model named *Left-Button*. As with the Top-Cover model, you will be cutting away all unnecessary geometry. First cut away all geometry except the button area. The cut should be a *.03*-inch offset above the parting line along the base curve for the buttons. Remove the geometry outside the section, as shown in the second illustration at right.

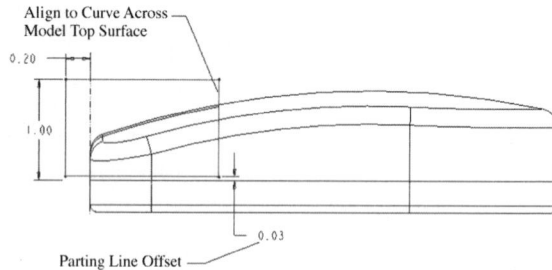

Left-Button geometry removal.

16. Remove the excess button geometry with another cut, leaving the left button geometry only. The cut should be offset from the middle button curve by *.01*. Refer to the following illustration for more information.

*Removal of excess
button geometry.*

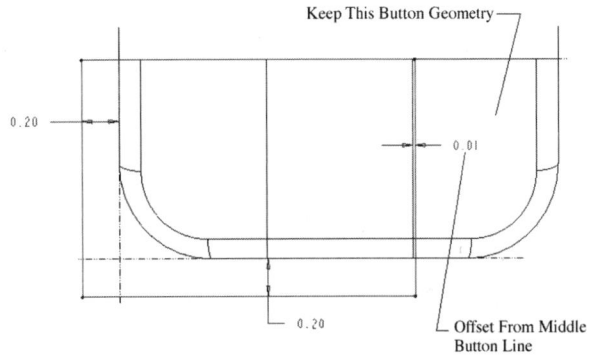

Keep This Button Geometry

0.20

0.01

0.20

Offset From Middle
Button Line

17. Shell the button to a *.06*-inch wall thickness, remove the surfaces shown in the following illustration, and save the resulting model.

Left-Button shelling.

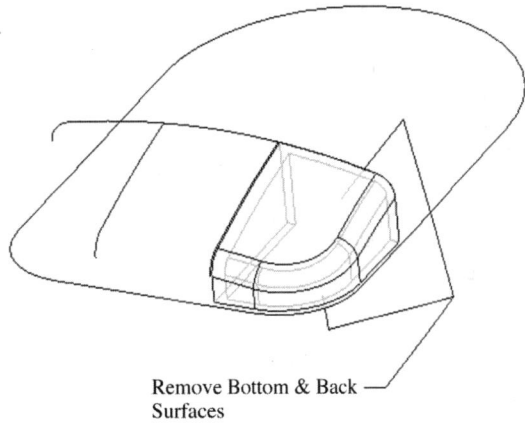

Remove Bottom & Back
Surfaces

18. Complete the Middle-Button and Right-Button by cutting away the unnecessary geometry, just as you did for the Left-Button. The only difference is that you will be removing different portions. Shell each button to a *.06*-inch wall thickness and save the resulting models. The following illustration shows the three completed button models.

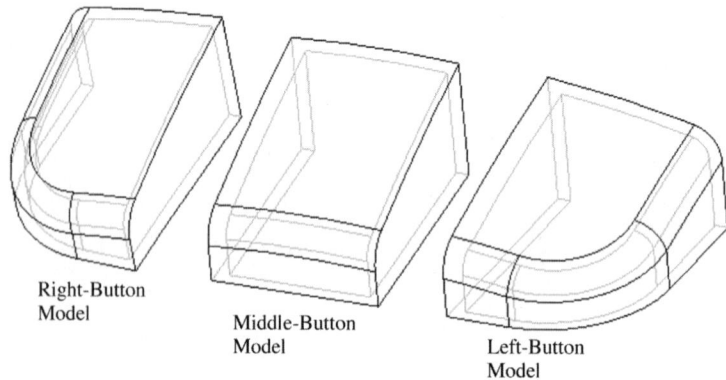

Completed mouse button models.

Right-Button Model

Middle-Button Model

Left-Button Model

19. Complete the last model, the Bottom-Cover. Cut off all geometry above the parting line; then shell the model to a *.06*-inch wall thickness. Finish the model by cutting a lip into the outer edge, as shown in the following illustration.

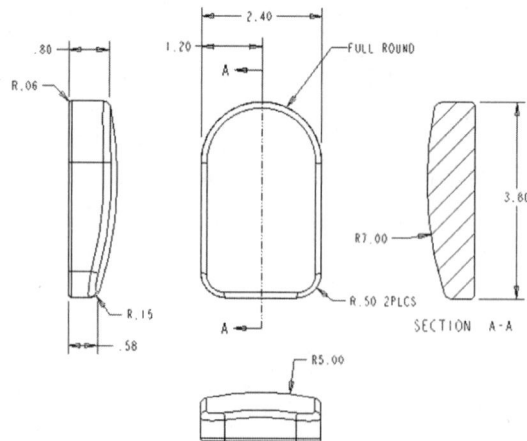

Bottom-Cover part.

20. To see the new mouse product assembled, create a new assembly named *Mouse*. Use a start assembly model or create an assembly containing default datum planes and coordinate system; then assemble each model to the assembly using the part and assembly default coordinate systems. The finished assembly should contain the following models: Top-Cover, Bottom-Cover, Left-Button, Middle-Button, and Right-Button. There is no need to assemble the Master-Mouse part in this assembly. The completed assembly should look like that shown in the following illustration. **Save** the Mouse assembly.

Completed assembly.

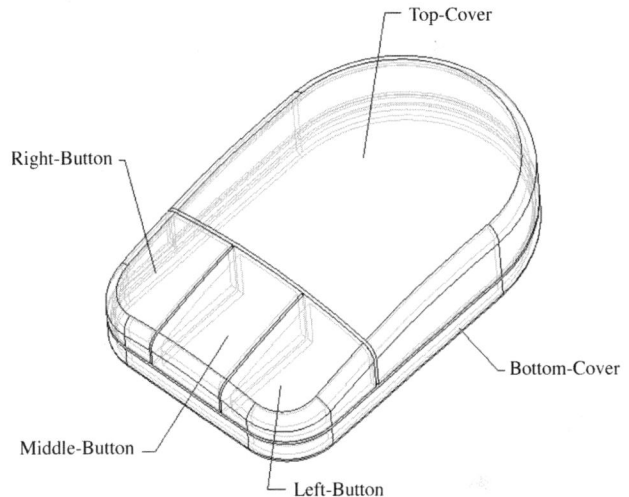

Top-Cover

Right-Button

Bottom-Cover

Middle-Button

Left-Button

Dealing with a Late Styling Change

Just when you thought you were done with the new product and ready to move on to the next project, you are confronted with a styling change. The project manager was muttering something about ergonomically correct hand position as he told you about the change, but luckily it will only affect four of the five models directly. This is when the master model technique really earns its keep. The following steps pick up with the mouse created in the previous series of steps to show you how the master model technique allows you to incorporate and adjust for design changes that can occur well into the modeling cycle. In the following steps, you will make changes to the mouse and most of its component parts.

21. Close the Mouse assembly window; then open the Master-Mouse part model.

22. The new mouse contour raises the three buttons while lowering the hand palm area on the mouse. From the PART menu, select **Feature ➡ Redefine**; then pick the cut you used to shape the top surface of the model and redefine this feature's trajectory. Select **Trajectory** and **Define** from the Cut dialog box. From the SECTIONS menu, select **Modify ➡ Done**. Select **Sketch** to change the trajectory's section. The following illustration shows the new cut trajectory section. Do not delete the current trajectory. Use the **Replace** option under **GEOM TOOLS** to substitute the new trajectory. By using Replace instead of

Delete you cause all reference features to follow the new geometry. If you use Delete you could have regeneration failures, which will force you to Reroute the failed geometry to the new references.

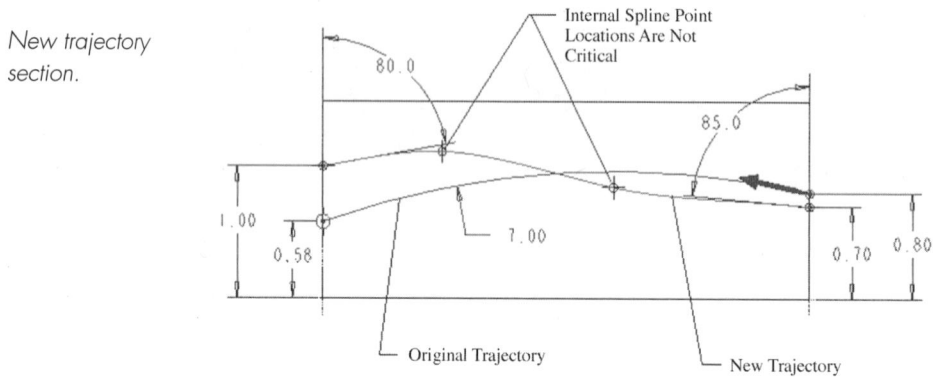

New trajectory section.

Internal Spline Point Locations Are Not Critical

80.0

85.0

1.00

0.58

7.00

0.70   0.80

Original Trajectory

New Trajectory

Use the Replace Option from the GEOM TOOLS Menu to Replace the Original Trajectory with the New Trajectory

After the Redefine is complete, the new Master-Mouse part model should look similar to that shown in the illustration at right.

23. Save the new model; then close the Master-Mouse part window and re-open the Mouse assembly. Because all of the mouse component parts are currently in session, the change you made to the top surface will not affect them until you **Regenerate** the assembly. You can either select

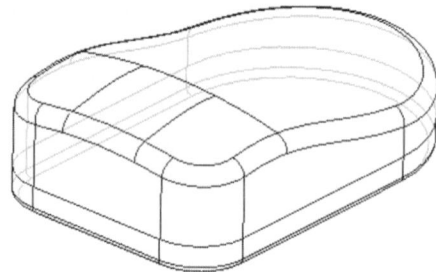

New Master-Mouse model.

Custom and pick each model to regenerate or select **Automatic** to pick them all. Once you have regenerated the assembly, all parts will update and the assembly should look like that shown in the following illustration.

Regenerated mouse assembly.

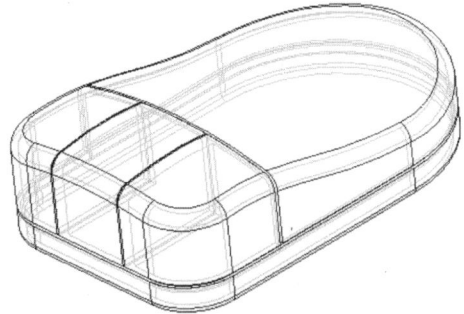

Summary

In this chapter you have worked through the master model technique. Thus far, the only drawback to this assembly/part associativity method is that for the technique to work properly, you must keep the master model so that the system can update the part geometry. This means that you will always need to keep that one extra model in your project folder to maintain the link. Other than that extra requirement, this method can save a lot of time and effort in model modifications because one change to the master is propagated through all associative parts.

Some care needs to be taken in the dimensional scheme you set up on each part, relative to the merged model's outer boundaries. In exercise 18-1, you dimensioned the cuts for the Left-Button from the parting line to the top of the cut. If the dimension had not been large enough to accommodate the change to the top surface that occurred later, you would have had an additional sliver of material above the cut. If you had dimensioned to the original top surface tangently to the top of the button cut, that cut would have failed after the spline replaced the original cut trajectory. This is because the system would not be able to solve for tangency to a spline.

Unfortunately, when designing a product you are not privy to all of the changes that can and will occur. Keeping this fact in mind and using the master model technique will help you avoid many of the problems that could occur.

19

Additional Assembly Techniques

This chapter focuses on two special assembly techniques that solve specific problems. The first technique, covered in the first exercise, addresses a situation most companies must deal with: how to work with models and drawings of parts that go through various stages before the part is complete and ready for production. Examples of this situation would be the cast part to be machined and a plastic part that is over-molded in a secondary operation. There are several methods of handling these situations, and in this chapter you will explore three of them in detail, discovering which method works best for certain situations.

The second exercise involves another specialized assembly technique. You will learn how to create the exact envelope of an assembly, which was a technique developed to solve problems for the supplier of an automotive manufacturer. This technique can also be used to create an inside or outside envelope that requires less model information than working with a full assembly. The technique therefore helps speed up manipulating model geometry.

Casting and Machining Models

The following sections present three solutions to modeling problems associated with cast-and-machined parts. The first solution involves creating two models, the second involves using family tables, and the third involves using the Merge option.

Creating Two Separate Models

Cast-and-machined parts are often given different part numbers for the various steps in producing the part due to the need to stock materials for, or track costs of, the part at each stage of manufacture. The obvious solution to this problem is to create two separate models. One part number is assigned to the casting part, and a second to the machined part. However, this solution separates the models as well, and if a change is made to the casting, the same change must be made to the machined part. This solution also creates twice the work, and leaves room for error when making the change. That is, the change may be done differently between the models, or missed on the machined part altogether.

For these reasons, you might think there is little use for this method, but it can, and is, being used in special situations. One such situation is where an unfinished assembly is purchased, then painted in house under a different part number, which is then used on the final product assembly. In this scenario, to create the purchased assembly, the internal components are modeled as parts in Pro/ENGINEER to the desired level of detail, and the assembly is created.

The assembly is created using the part number of the purchased component. To produce a model of the finished component, a second assembly is created, using the finished component's part number, with the purchased assembly assembled to the final assembly as its only component. Because the paint is not modeled, the finished part is the same as the unfinished part. A drawing can then be made of the finished component containing the required paint specifications and any other information the company might require. The beauty of this method is twofold. First, any changes made to the unfinished purchased assembly will be reflected in the finished assembly automatically. Second, it gives you a model with the finished part number that can be assembled into downstream assemblies that will have the proper bill of material structure.

Using Family Tables

Family tables represent a solution to a situation similar to that described in the previous section. However, in this situation, rather than purchasing an unfinished part and finishing it in house, this part is molded from different color plastics. The use of family tables makes sense in this scenario for a couple of reasons. First, separate drawings will probably not be required for each color. Rather, a single drawing can be used, with a table associated with it that lists the other colors and their part numbers.

✓ **TIP:** *Anytime a table or chart will be used on a drawing, you should consider using a family table. It is a simple matter of placing a family table into a table or chart on the drawing.*

Second, like the part model, downstream assemblies will probably not be required for each color. This allows downstream assemblies to be assembled with the generic model, which can be easily switched out with any of the instances.

The family table is a popular solution because of its versatility. With the ability it gives you to contain parameters, dimensions, features, components, and many more items in assemblies, the family table is being used in thousands of situations.

The Merge Technique

The merge technique is used when the second model needs to contain additional features, such as machining features added to a cast part. The merge process enables you to take one part and use it as a base feature of a second part, or add it to existing geometry by putting the two parts in an assembly and using the Merge command in the ADV UTILS menu to combine them. Because this procedure may not be as familiar to you as the others, the following exercise takes you through an example of how it works.

Exercise 19-1: Using the Merge Feature

In this exercise you will be creating a machined part from an as-cast part using the Merge feature. It will merge the as-cast part into the machined part as the machined part's base feature.

To begin, the as-cast part does not need to be completed, but should have some defining geometry created. The machined part also needs to be created.

1. Create the machined part, but create only the default datums and coordinate system, or whatever base geometry your particular start part includes.

2. Once the parts have been created, start a new assembly by selecting **File** from the pull-down menu at the top of the Pro/ENGINEER window; then select the **New** option. The New dialog box will open. Select **Assembly** and enter a carriage return to accept the default name. The assembly will not be saved, so there is no need to give it a special name.

↦ **NOTE:** *Do not create the assembly using a start part, and do not add any geometry to the assembly, including default datums or coordinate system.*

3. After the assembly has been created, incorporate the machined part in the assembly. Select **Component** ↦ **Assemble** and select the machined part from the parts list.

This part will be placed in the assembly automatically, as if assembled by the default coordinate systems of the part and assembly. The following illustration shows the machined part, MACH_ARM.PRT, assembled to the assembly. The model tree is also shown in the illustration.

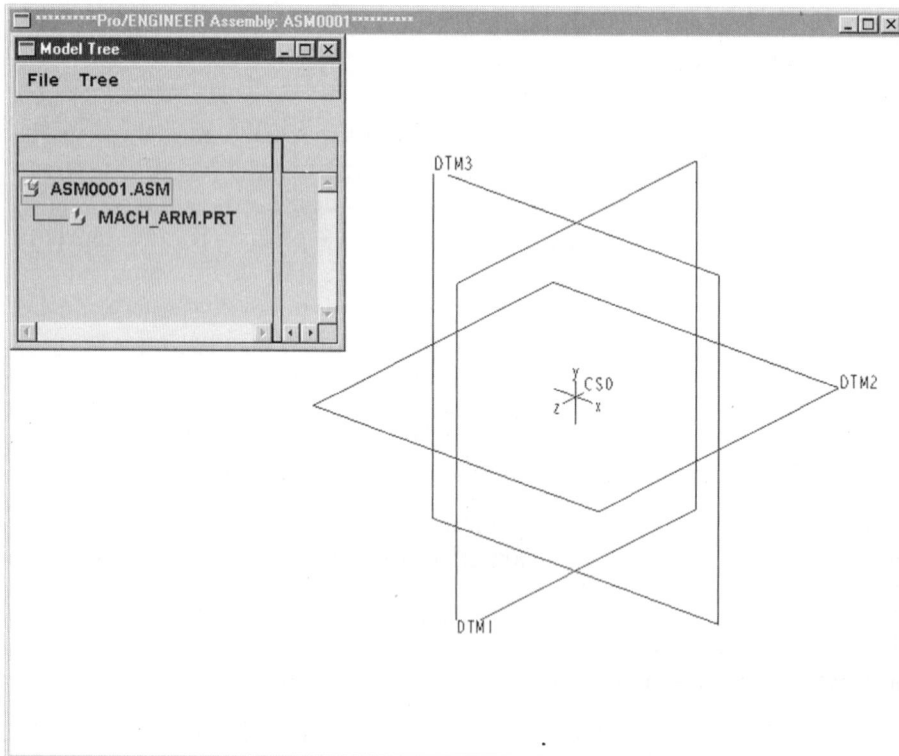

Default assembly with machined part assembled to it.

4. Assemble the as-cast part. Select **Component** ➥ **Assemble** and select the as-cast part from the retrieve list. Incorporate this component in the assembly by selecting the default coordinate systems of the as-cast part and the machined part. The result is shown in the following illustration.

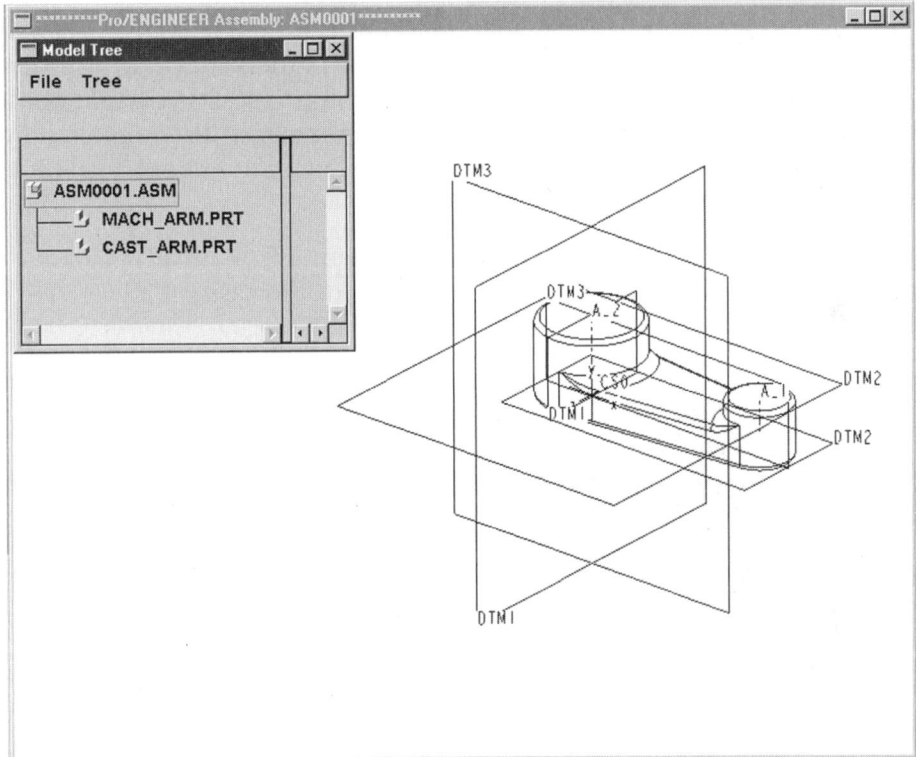

CAST_ARM.PRT assembled to the part MACH_ARM.PRT in assembly ASM0001.ASM.

5. To perform the actual merge, select **Component** ➥ **Adv Utils** ➥ **Merge**. Pro/ENGINEER will prompt you to "Select parts to perform MERGE process to." This is the part you want to add the geometry to. In this exercise, the geometry is added to the machined part MACH_ARM.PRT, because you are adding the as-cast geometry to the machined part. Select the machined part. The next prompt is "Select reference parts for MERGE process." Select the as-cast part CAST_ARM. The as-cast part will now be a reference part of the machined part and will need to be in-session (in the search path) whenever the machined part is called up. Use the model tree as a method of making sure you select the proper model or models for both of these prompts.

Merge Options

To complete the merge, Pro/ENGINEER requires two options to be set. The first is Reference or Copy. Selecting the Copy option will recreate the merged-in model geometry as new individual features in the new part model. In this case, the original model is no longer required for the geometry to be created in the new model. The Reference option means that the geometry being merged in will be retrieved from the original model. This is why the original part needs to be in-session (or in the search path) for the new model to be called up. The second option calls for Datums or No Datums. No Datums recreates the geometry, but does not copy any of the datums from the original part. This helps keep the new part clean of unneccesary datums. The Datums option simply keeps or copies the original datums into the new part.

6. For this exercise, select the **Reference** option, because you want the machined part to always go back and get the latest as-cast geometry. Select either No Datums or Datums. They will work equally well.

Now that the merge options have been set, Pro/ENGINEER will prompt "Detach reference part 'CAST_ARM' from the assembly?" A carriage return will enter the default Yes answer. A Yes answer detaches the assembly from the new part. Therefore, the assembly will not become a required reference of the new part. A No response will make the assembly a required reference.

7. For this exercise, enter a **Y** and a carriage return, or just a carriage return, to accept the default Yes. Now that the merge has been created successfully, the next step is very important. You must quit the assembly window *without saving*. Saving the assembly will make the assembly a reference of the new part. Therefore, select **Close Window** from the FILE menu. Reopen the machined part. The as-cast geometry will now be in the part, as shown in the following illustration.

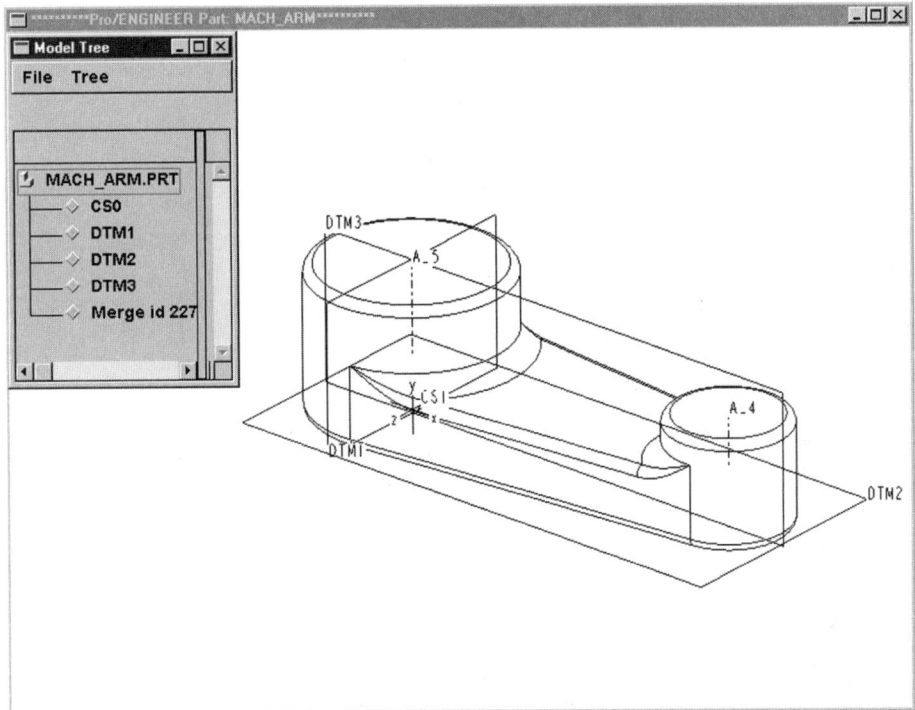

Machined part with as-cast geometry as its base.

8. From the FILE menu, select **Save**, and begin creating the features needed to complete the machined part, which is shown in the following illustration.

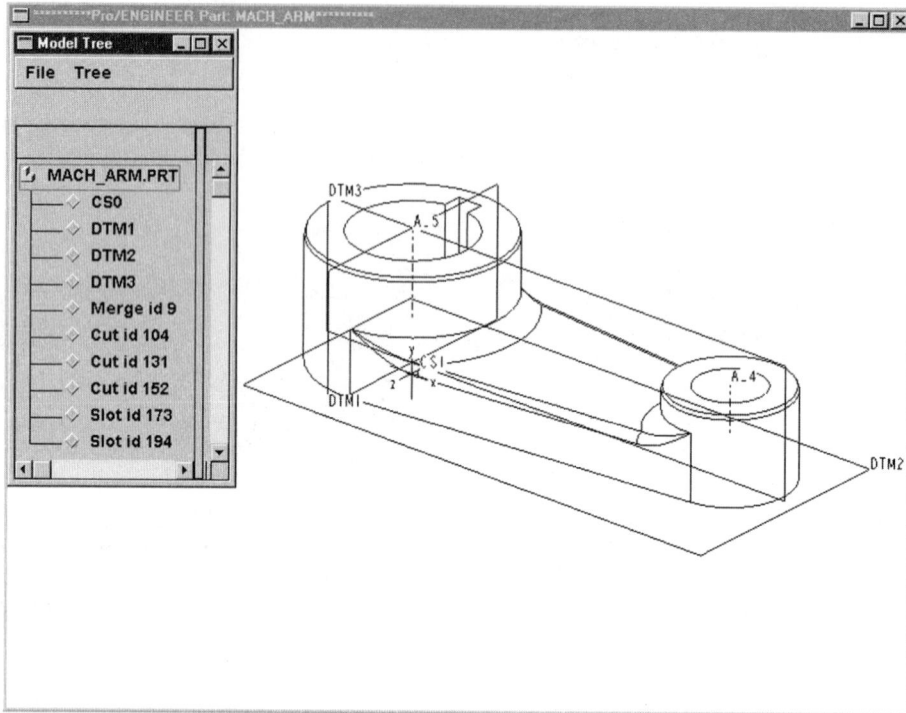

Completed machined part.

Exercise 19-2: Creating the Skin of an Assembly Model

The procedure you will be learning about in this exercise was developed several years ago as workaround to a data exchange problem. The problem was that an automobile manufacturer was requesting that the door lock supplier send data files of the locks to be placed in the full vehicle assembly. However, when surface IGES files were sent to the manufacturer, the files were too large, modem transfer times were way too long, and the files had more detail than was needed.

Simplified Reps were tried as a solution to these problems. However, when the interior features were suppressed, some of the required outside features tied to them by design intent were also suppressed. Simplified Reps also altered the true envelope of the lock, which was deemed unacceptable. What was needed was a way to copy and transfer only the exact outside envelope of the lock, or to send only the "skin" of the lock assembly.

The ultimate solution in this case was a procedure that copies all of the required surfaces in Assembly mode, blanks all of the unwanted components, and then creates an IGES file of the remaining "skin" of the lock. In this case, the smaller IGES file was then sent to the automobile manufacturer. The procedure was easy, and did not take a lot of time on the part of the supplier. More importantly, the customer was satisfied with the results.

The model used in this exercise is similar to the type of lock assemblies for which this procedure was developed. You can see in the following illustration that the vast majority of the detail of the lock is in the interior of the assembly.

Full lock assembly.

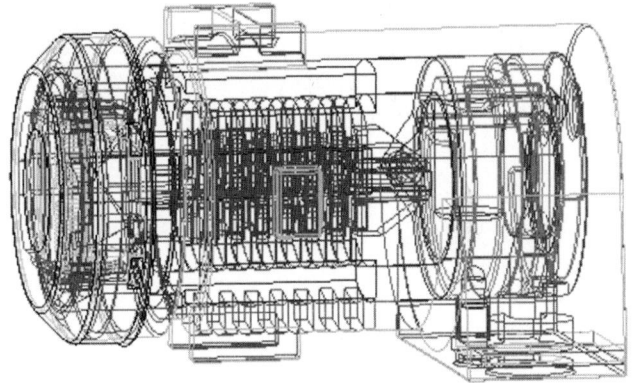

1. Call up the assembly model you want to skin. All of the components need to be on individual layers so that their display can be controlled. Analyze the assembly to determine which surfaces are going to be required to create the envelope and what components these surfaces belong to. Blank all of the components except the one you will be using to copy the surfaces. For this exercise, blank all of the components except the outside case, as shown in the following illustration.

The outside surfaces of this case part will be the first to be copied.

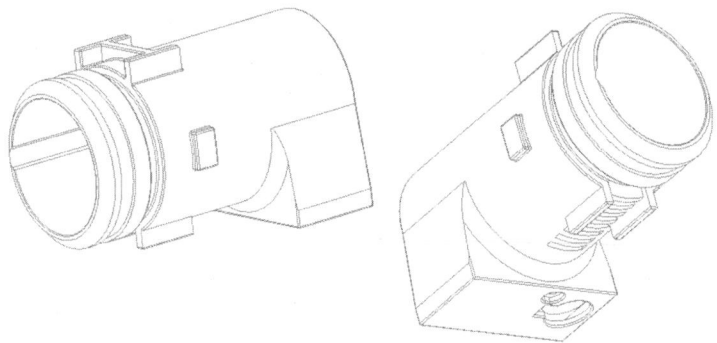

2. To create the outside skin of the case model, select **Feature** ➥ **Create** ➥ **Surface** ➥ **New** (this menu option only appears if previous surface geometry was created) ➥ **Copy** ➥ **Done**. For the part in this exercise, select the **Surf & Bnd** option. To employ the Surf & Bnd selection method, you will select a "seed surface." Pro/ENGINEER will then select all of the surfaces that contact the seed surface, continuing to select surfaces until it contacts a boundary surface. Select the seed surface shown in the following illustration. Once the seed surface is selected, Pro/ENGINEER will switch to the boundary selection menu. You need to select all of the boundary surfaces by adding them individually (the Indiv Surfs option) or by using the Loop option. Select the boundary surfaces shown in the following illustration by selecting them individually.

The results of the surface copied using the Surf & Bnd option.

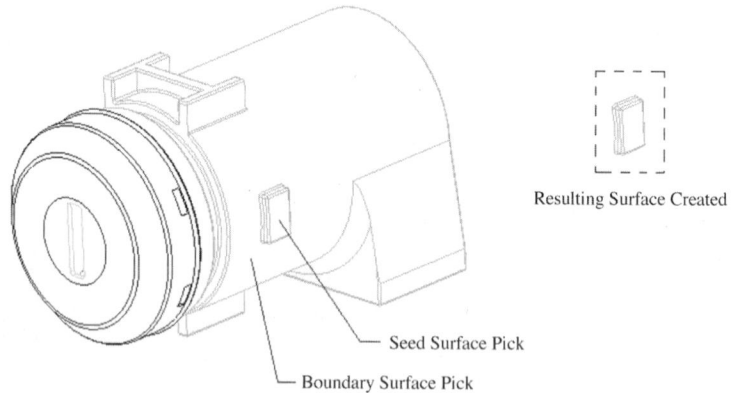

Resulting Surface Created

Seed Surface Pick

Boundary Surface Pick

3. To add additional surfaces to the same surface copy feature, select the surface element in the Surface/Copy dialog box; then click on the **Define** button. All of the surface selection options will be available. Continue selecting surfaces, using whichever method is required, until the entire outside envelope has been selected. After you have selected the seed surface and all of the boundary surfaces, select **Done Sel** ➥ **Done** ➥ **Done** and **OK**. The surface will be created, as shown in the following illustration.

✓ **TIP:** *You can select only surfaces that belong to a single component. A single surface feature cannot contain surfaces from more than one component. This is why it is best to blank all components except the one you are currently working with. This prevents you from selecting surfaces from other components.*

The Surf & Bnd option works well for this procedure because of the flexibility it affords. If the
case geometry changes, the surface will also
update and the surfaces will not need to be re-
created (unless, of course, a hole is added,
requiring that the surface be selected as an additional boundary). The skin surface can be redefined to include additional surfaces or exclude
surfaces that no longer apply.

4. Select **Redefine** and pick the envelope
   surface; then select the surfaces element and click on the **Define** button.
   Under the SURFS SELECT menu,
   select the **Redefine** option. A list of
   available surfaces will appear. As you
   place the cursor over a surface name,
   that surface will highlight in the
   model, which makes it easier to select
   the correct surface for redefinition.

Outer Skin Surface Model

The surface represents the envelope of the case.

Placing the Envelope on a Layer

5. In the top-level assembly, create a
   layer with the name *ENVELOPE*, and
   place the surface that was just created
   on it. Blank the component and the
   ENVELOPE layer. The model should
   be empty. Unblank the next component and repeat the same procedure
   until all of the required surfaces have
   been copied, as shown in the second
   illustration at right.

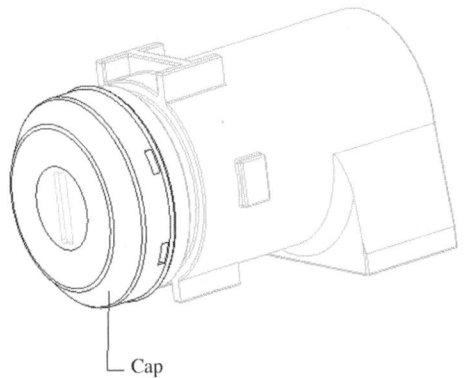

Cap

The cap of the lock assembly is the next component to surface.

6. Continue copying the surfaces that
   are required to create the desired envelope of each component. When all of
   the surfaces have been copied and added to the ENVELOPE layer, the model
   should look as it does in the following illustration.

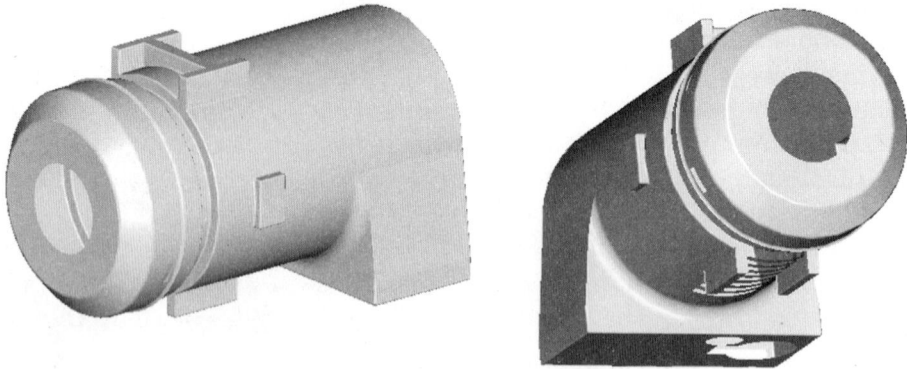

Surfaces are an exact "skin" of the assembly model.

Surface Selection Options

You can use any selection method to pick the surfaces you want to copy. However, you will need to use a combination of selection methods for some components. You will need to use Surf & Bnd to get most of the surfaces and then add additional surfaces by selecting Individual and picking each surface. The other options of the SURF OPTION menu are Loop Surfs, Quilt Surfs, and Solid Surfs. None of these options is used very often in this process on this type of model. However, if they apply to the type of model you are working with, go ahead and use them. If an entire component is exposed and is part of the envelope, you do not need to copy the surfaces. You can simply unblank the component and the surfaces will be created when the file is exported.

Why Create the Surfaces in Assembly?

The skin surfaces need to be created in Assembly mode as assembly features so that they can be separated from the components used to create them. If you create the skin surfaces in Part mode, you cannot select only the surfaces to be placed on the assembly layer ENVELOPE when the components are assembled. This would make it impossible to create the export file.

Exporting the Skin

7. To export the skin to another system, blank all of the assembly layers except the ENVELOPE layer and create the desired export file.

↪ **NOTE:** *Set the config.pro option intf_out_blanked_entities to No to prevent exporting the blanked components.*

Summary

This chapter explored two specialized assembly techniques: the merge technique used with casting and machined parts, and copying component surfaces in assembly to create an envelope. The merge technique, used in exercise 19-1, creates a machining model that always looks back to get the latest as-cast geometry. Therefore, you do not need to worry about updating any casting changes to the machining model. This technique also helps keep the machining model clean. If family tables had been used, all casting and machining data would be in one file, which can be confusing to work on.

Having a separate casting and machining model is especially helpful if geometric dimensioning and tolerancing (GD&T) is being used on the drawings. This is because the targets and callouts will need to have the same names, but may not be the same features. This technique of using the merge function works very well for any situation similar to the casting-and-machining example used in the exercise. Look for situations in which you can apply this technique. It is easy to do, is robust, and, most important, follows your design intent.

The procedure used in exercise 19-2 solved several problems. The amount of information was reduced so that the customer did not have to deal with more than was necessary. The procedure helped the lock supplier in two ways: it cut the modem transfer times considerably, and provided a cost-effective, simple procedure that all of the designers could do.

This procedure has an additional benefit: confidentiality. Because only the outside skin of the model is sent out, the working part of the design stays within your company. The amount of information that lies within a solid model is incredible, and the number of companies willing to work from only a model is increasing all the time. Use this workaround the next time you need to work with the envelope of an assembly or want to keep your design yours.

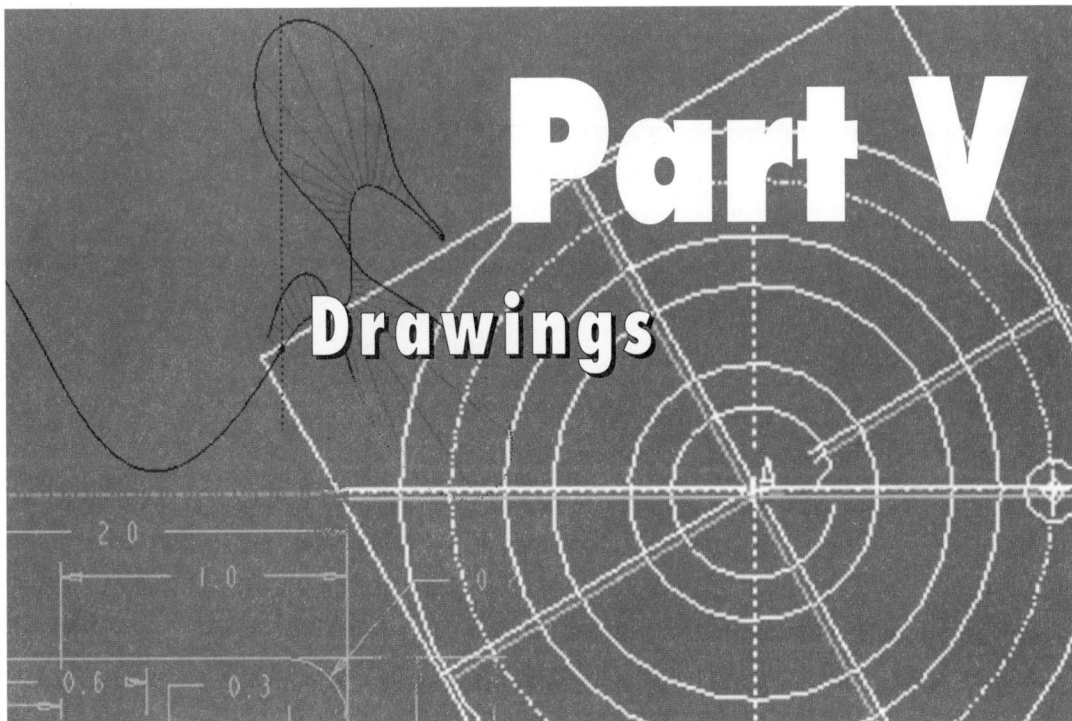

Part V
Drawings

The fifth and final part of this book deals with creating drawings in Pro/ENGINEER. As much as we hear about a "paperless" engineering department or company, the truth of the matter is that most still require that at some point a drawing of a part or assembly be made. The design engineering industry, however, moves more and more toward a "paperless" environment. Many departments within your company and suppliers on the outside are using the model as the preferred medium for part and assembly inspection, manufacture, and cost quotation.

Throughout this book you have been reading about design intent—how to sketch to it, how to dimension to it, and how to select reference features with it in mind. Now it is time for the payoff, which is the fact that by following your design intent throughout the building of a model, the next user of it will be able to create drawings more easily, faster, and more accurately.

The first chapter in Part V looks at drawing formats. The chapter explores some of the techniques that can be used to create drawing formats. The chapter also examines how to make the information contained in a format more useful and automatic, making it more efficient for the detailer to use. The second chapter deals with the drawing itself. The chapter covers drawing views and dimensioning techniques, including some GD&T (geometric dimensioning and tolerancing). You will also learn about using tables to create an automatic BOM (bill of materials) and BOM balloons on assembly drawings, as well as the use of layers in the drawing to help clean up the appearance of drawings.

Drawing Formats

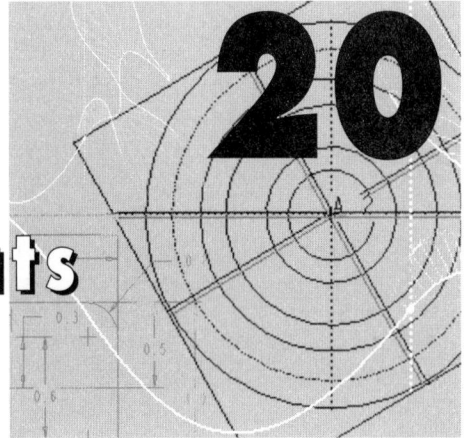

A drawing created for any organization will likely require some type of format (or border, as it is also known). This chapter deals with creating a format that once created and placed in a library can be accessed and used by any user on the system. Formats consist primarily of draft entities and notes, with associated tables created so that information for individual drawings can be added.

You will learn how to add tables to formats, and then how to edit and merge table cells to match the desired title block shape. If a format is constructed correctly, most of the required information will be automatically picked up from the model and printed in the title block. This is accomplished by the use of system and user-defined parameters. The final exercise in this chapter deals with this process.

Creating a Drawing Format

There are two ways to create the geometry that constitutes a drawing format. The first is to sketch the geometry using draft entities; the second is to import existing format geometry from another CAD system. Regardless of which method you use, Pro/ENGINEER requires some general information before you can create the format geometry.

Creating a New Format

A format is a special type of object in Pro/ENGINEER, much like a part, drawing, or assembly. You need to create a format object before you can create format entities. Select the FILE menu from the pull-down menu bar, and select New. The New Format dialog box, as shown in the following illustration, will open, asking for starting information.

New Format dialog box.

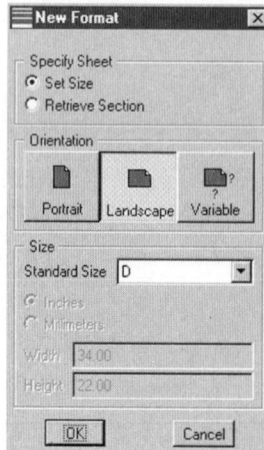

Use Retrieve Section if you have an existing section to be used as the format.

Select the proper sheet orientation or select Variable for non-standard size sheets.

Select the size of the sheet. See text for list of sizes.

The dialog box offers you the format orientation options Portrait, Landscape, and Variable. When you select Portrait or Landscape, the Size area of the box offers you a choice of numerous standard or numerous ISO sizes. The standard sizes are A (8.5 x 11), B (11 x 17), C (17 x 22), D (22 x 34), E (34 x 44), and F (28 x 40). The available ISO sizes are A0 (841 x 1189), A1 (594 x 841), A2 (420 x 594), A3 (297 x 420), and A4 (210 x 297). Once you have chosen one of the standard or ISO size options, you select the OK button.

The Variable option presents you with unit-of-measure (inches or millimeters), width, and height options. For the width and height options, you enter values, which are recorded in the units selected under the unit-of-measure option. You then select the OK button to continue.

Sketching from Scratch

Sketching draft entities in Pro/ENGINEER can be laborious and is never easy. Therefore, having a definite plan of what you want to draw is a must. Sketch the title block on graph paper to determine the exact size and position for each box. This type of sketch, an example of which is shown in the following illustration, will make sketching easier and faster, and will help you figure out the best grid size to use when creating draft entities.

Grid paper sketch.

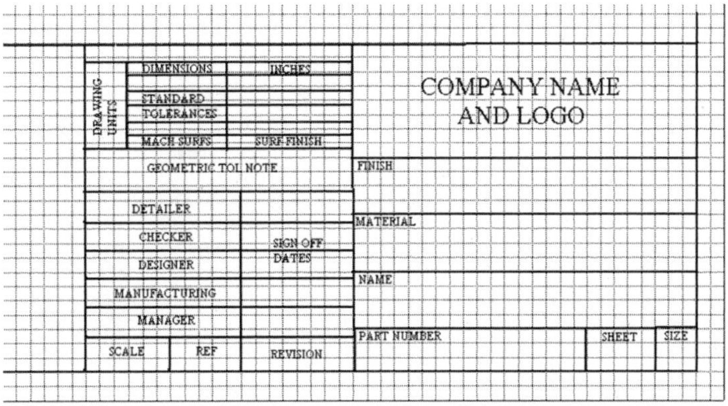

The first step in making a grid sketch is to make sure the Snap to Grid option in the ENVIRONMENT menu is checked. You then modify the grid parameters to a value fine enough to be able to sketch all of the boxes in the title block. To modify the grid parameters, select Detail ➡ Modify ➡ Grid ➡ Grid Params ➡ X & Y Spacing. This will set the spacing in both the X and Y directions to the value you enter. If the value creates a grid too dense to be displayed, set it to a higher value and sketch the border outlines first. Then zoom in on the title block area and modify the grid parameters to the original value.

Because you are zoomed in, the finer grid spacing will display, and you can start to sketch the title block boxes. When you sketch the draft entities, look at the cursor location coordinates in the lower left corner of the graphics window. This will allow you to accurately locate start and end points of the sketched geometry. The following illustration indicates where the cursor location coordinates are found.

Common layout when sketching a format.

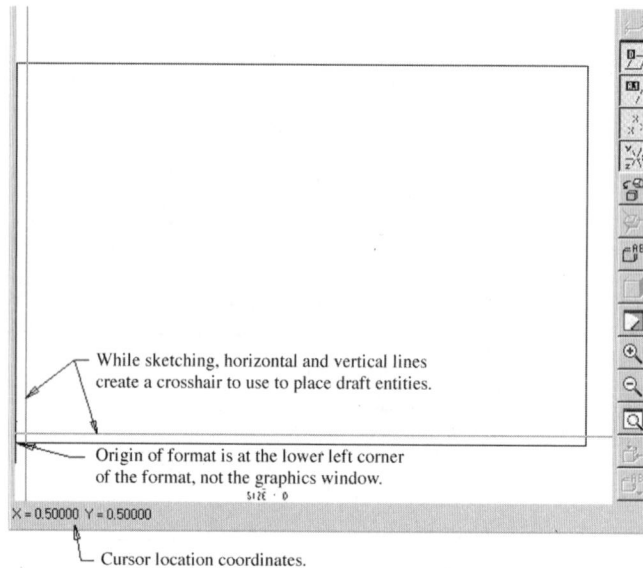

While sketching, horizontal and vertical lines create a crosshair to use to place draft entities.

Origin of format is at the lower left corner of the format, not the graphics window.

SIZE · 0

X = 0.50000 Y = 0.50000

Cursor location coordinates.

Keep in mind that when you create a note, the location of the note will not snap to a grid intersection. The origin of the note will be wherever the cursor is located at the time the note is placed.

Importing Geometry into a Format

Geometry can also be imported into a format. This is a tremendous time saver and can ensure that the geometry of all title blocks is identical. You can import the geometry from an existing set of formats from another CAD system or company. To import the geometry, you need to create a file of the drawing format in one of the following file formats: IGES, DXF, DRW, SET, or CGM. When you create the file to be imported, use the lower left corner of the drawing format as the origin. This will match up with the origin in the Pro/ENGINEER drawing format. Exercise 20-1 takes you through the creation of a format, including a step-by-step import of geometry in addition to the other details of the process.

Tables in Formats

Unfortunately, importing geometry is not a one-step process. Any text in the format that is not in a table cannot be modified in a drawing the format is added to. If you want imported text to be modifiable by the user in Drawing mode, the text must be entered as text in a table. This also holds true for sketched formats. Even if the import file came from another Pro/ENGINEER format when imported, the tables will no longer function, and any text in them needs to be removed. New tables will need to be created and the text reentered.

The title block of your drawing format will most likely involve creating several tables to cover various parts of the title block. In the following illustration, the title block is broken down into several smaller blocks that are grouped by similar size and shape or by similar information. Each of the smaller blocks will be created as a separate table.

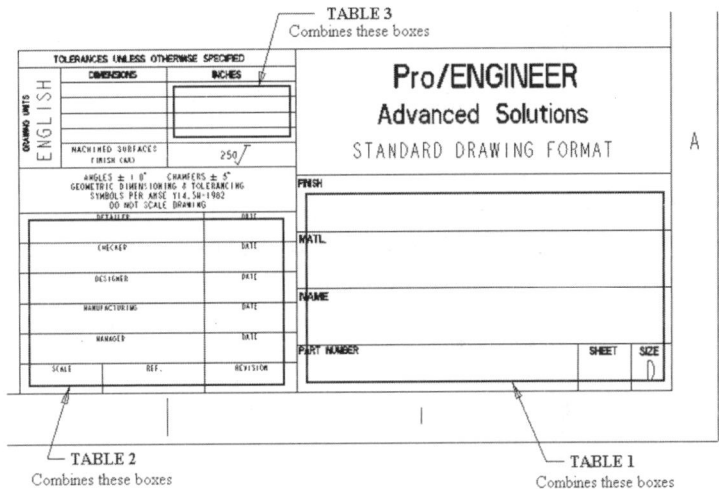

Title block area to be broken down into individual tables.

TABLE 3
Combines these boxes

TABLE 2
Combines these boxes

TABLE 1
Combines these boxes

Table Basics

In exercise 20-2 you will create each of the tables for a format and learn how to modify their appearance to mach the title block. First, however, you need to be aware of your options in creating tables.

When a table is created, it begins at an origin. This origin can be at the vertex of two draft entities or at a selected screen position. Once the origin is selected, you next need to determine the directions in which the table will be created. You must establish either an ascending or a descending order for one direction and a rightward or leftward order for the second direction. An ascending table will have its origin at the bottom row of the table. A descending table has its origin at the top row. A rightward table creates columns to the right of the origin, and a leftward creates them to the left. The following illustration shows the directions in which a table can be created.

Various directions in which a table can be created.

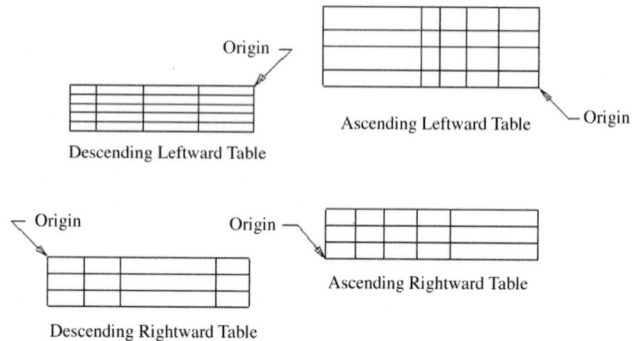

Descending Leftward Table

Origin

Ascending Leftward Table   Origin

Origin   Origin

Ascending Rightward Table

Descending Rightward Table

With the direction of the table set, you can create columns and rows. Columns are created first, by one of two methods: By Num Chars (By the Number of Characters) and By Length. If you choose By Num Chars, a list of 30 digits will appear at the start of the column. Select the number for the number of digits you want to fit inside the column. The same list of digits will appear at the start of the next column until an Enter is selected. At that point, a set of five digits will appear at the start of the row. Select the number for the number of lines you want to be in that row.

This process continues until an Enter is selected and the table is complete. The other method sets the width and height of the table cells by entering an exact value. Because the title block created used exact values, the second method works best for creating tables in formats. No matter how a format is created, imported or sketched, it is created by exact values. Using the Num Char option never makes sense in formats.

Adding Parameters

Drawing formats work well principally because of the parametric note that contains a system or user parameter. These special notes can get information from the model or system and automatically add that information to a drawing without the user having to add or edit anything. The following table lists the Pro/ENGINEER system parameters available for use in a format, with corresponding descriptions.

System Parameters

Parameter	Description
&todays_date	Places the date the note was created, using the dd-mm-yy format. ✓ **TIP:** *The format in which the date is displayed can be modified using the todays_date_note_format" config option. For a complete explanation of format options, see the Drawing Notes chapter in the* **Pro/ENGINEER Drawing User's Guide.**
&model_name	Name of the model used to create the drawing.
&dwg_name	File name of the drawing.
&scale	Current scale of the drawing.
&type	The type of model that creates the drawing.
&format	The format size.
&linear_tol_0_0 through &linear_tol_0_000000	The tolerance value that corresponds to a dimension with the same number of decimal places as the number of zeros at the end of the parameter.
&angular_tol_0_0 through &angular_tol_0_000000	Same as linear tolerance but applies to angular dimensions.
&current_sheet	The sheet number of the drawing currently displayed.
&total_sheets	The total number of sheets the drawing occupies.

The following table lists typical Pro/ENGINEER user parameters, with corresponding descriptions.

User Parameters

Parameter	Description
&name_1 and &name_2	The first and second line of the part name or description.
&material_1 and &material_2	The first and second line of the material description.
&finish_1 and &finish_2	The first and second line of the finish description.
&detailer	Name of the detailer that created the drawing.
&checker	Name of the checker of the drawing.
&designer	Name of the designer or engineer of the part.
&mfg_sign_off	Name of the manufacturing engineer that checked the drawing.
&manager	Name of the manager of the department that created the part.
&drw_rev	The current revision of the drawing.
&detailer_date, &checker_date, &designer_date, &mso_date, &manager_date	The date a person signs off on the drawing.

The following illustration shows a title block with associated tables containing a combination of system and user parameters. These parameters will read the values either from the model associated with the drawing or from Pro/ENGINEER itself. For example, the SHEET box of the title block says "1 OF 1." However, if you were to edit the text of the table, it would say "&current_sheet OF &total_sheets." These two parameters have already picked up their values from the system and displayed them, as shown in the following illustration.

A typical title block using parameters to drive the information.

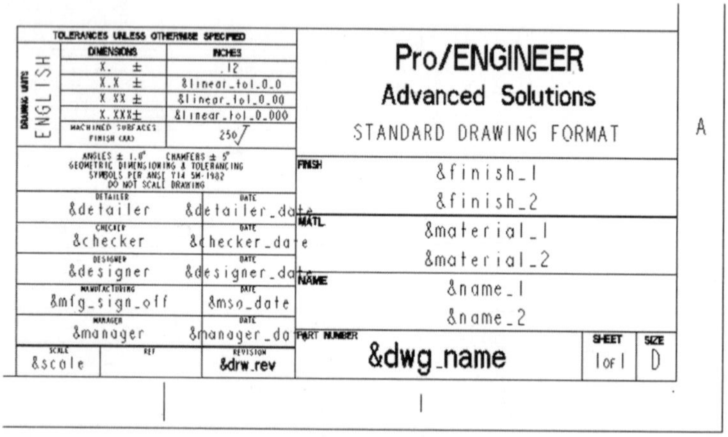

Exercises

In the following series of exercises you will create a drawing format by importing the format geometry (exercise 20-1). You will then create the tables necessary for modifiable text (exercise 20-2). In the last exercise, you will add the text and parameters to the tables to allow Pro/ENGINEER to fill out as much of the title block as possible.

Exercise 20-1: Importing Format Geometry

Before you can start this exercise you will need to create a file to import into Pro/ENGINEER. This file can be in IGES, DXF, or any interface file format you are comfortable working with.

1. Create a file to be imported.

2. You also need a format to export. Create a Pro/ENGINEER format using the title block shown in the previous illustration, or export your company's standard format. If you decide to export your own format, some of the exercise will be lost because the tables and parameters will be different, but there will be enough carryover that the exercise will still prove useful.

Creating the IGES File

Creating the IGES file is a pretty straightforward process.

3. Select **File** from the pull-down menu bar at the top of the screen, and then select **Export**. A menu will expand out to the right. Select the **IGES** option (any of the export options are okay to use). About the only item you need to be aware of is the origin of the export file. The origin must be in the lower left corner of the format itself; it cannot be in the lower left of the screen or the center of the screen.

Creating a New Format and Importing the Geometry

With the export file created, you now need to create a format to import the geometry into.

4. Select the **File** pull-down menu from the top menu bar and select the **New** option. Alternatively, select the New Object icon from the tool bar. The New dialog box will open. Select **Format** and enter the format name *D_SIZE* and click on **OK**. The New Format dialog box, as shown in the following illustration, will open, asking for the proper starting information.

New Format dialog box.

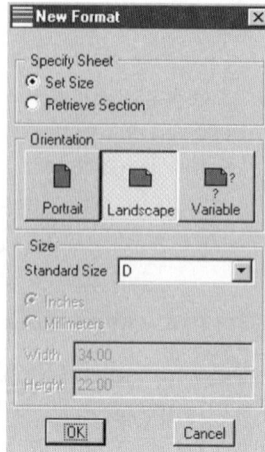

New Format ☒

Specify Sheet
⦿ Set Size
○ Retrieve Section

Use Retrieve Section if you have an existing section to be used as the format.

Orientation

Portrait   Landscape   Variable

Select the proper sheet orientation or select Variable for non-standard size sheets.

Size

Standard Size [D ▾]

○ Inches
○ Millimeters

Width [34.00]
Height [22.00]

Select the size of the sheet. See text for list of sizes.

[OK]   [Cancel]

5. Select the format orientation Portrait, Landscape, or Variable. If you choose Portrait or Landscape, in the Size area of the box, select one of the standard or ISO sizes; then click on the OK button. The standard sizes are A (8.5 x 11), B (11 x 17), C (17 x 22), D (22 x 34), E (34 x 44), and F (28 x 40). The ISO sizes are A0 (841 x 1189), A1 (594 x 841), A2 (420 x 594), A3 (297 x 420), and A4 (210 x 297). If you choose Variable, other options will become available. The first is the unit of measure (check Inches or Millimeters). Next, enter the width and height of the format in the units selected previously; then click on the **OK** button to continue. An outline of the sheet and a coordinate system will be displayed in the lower left portion of the screen. You are now ready to import the format geometry.

6. To import the geometry, select **File** from the pull-down menu and select **Import** and **Append to Model**. This opens the Import Append dialog box, which contains a list of files. Select the file you want to import; then click on **OK**. The format geometry will be placed in the format file and will look like that shown in the following illustration.

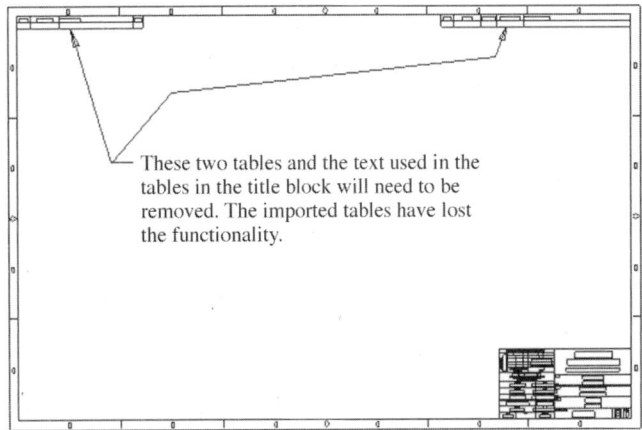

Some cleanup is required when a file is imported in a format.

These two tables and the text used in the tables in the title block will need to be removed. The imported tables have lost the functionality.

7. In this exercise, the size of the format and the size of the import are the same, but if the size of the format is different (larger or smaller), Pro/ENGINEER will tell you that the sizes are different and ask you if you want to scale the import to match the format size. Answering Yes will reduce the scale of the import, causing the title block to be smaller than the title block of the original geometry. If you select No, Pro/ENGINEER will ask you if you want to move the bottom left corner of the import to the screen origin (lower left of the format). Answer **Yes** to this question. The imported geometry will be placed correctly and will maintain the proper size. Use the sketch tools to translate the geometry to the proper location to create the new format.

8. The last step to complete before moving on to creating the tables in the format is to do a little checking and house cleaning. You will want to check the font and text size of the notes imported into the format. Different CAD systems support different text fonts, and when they are exported out and imported back in, the font can change. Make sure the text looks correct before moving on. The house cleaning that needs to be done will only be necessary if you are importing a format from Pro/ENGINEER that contains tables. When Pro/ENGINEER tables are imported, the table geometry is broken down into individual lines. This causes a double line or one line on top of another. Delete the line that used to make up the table. Use query select and the query bin to help identify if there is a double line. Two or more sketched entities will be listed in the query bin.

Exercise 20-2: Creating Tables for Modifiable Text

In this exercise, you will create five tables for the format used. The first will be the BOM table in the upper left corner of the format. The second is the revision block in the upper right corner of the format. The other three tables will be used in the title block itself. You will also be entering text and parameters to complete the tables.

Creating the BOM Table

The first table you will create is the table used to document the BOM.

1. Select **Table ➥ Create ➥ Descending ➥ Rightward ➥ By Length ➥ Vertex**. Select the corner of the format, as shown in the following illustration. The next prompt is to enter the width of the first column. Enter *.75* and enter a carriage return. Pro/ENGINEER now prompts you for the width of the second column. Enter *1.50* and a carriage return, *4.00* and another carriage return, and *.50* and two carriage returns (one to enter the value and a second to stop entering columns and start entering the height of the rows). Create two rows, each with a height of *.31*, and enter a second carriage return after the second row has been created. The table will look like that shown in the following illustration.

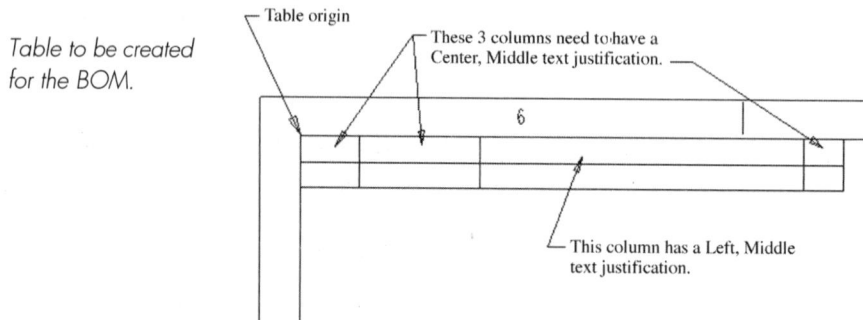

Table to be created for the BOM.

Table origin

These 3 columns need to have a Center, Middle text justification.

This column has a Left, Middle text justification.

After creating the table cells, the next step is to set the justification for the text. The first, second, and last column need to have a center-middle text justification, whereas the third column needs a left-middle justification.

2. To set the justification of a column, select **Table ➥ Modify ➥ Mod Rows/Cols ➥ Justify ➥ Column ➥ Center ➥ Middle**. Pick the three columns that need to

be modified. Select the **Left** option (the Middle option should still be high-lighted) and select the remaining column. Select **Done/Return** from the TABLE menu. The table is now ready for you to enter the text headers of the columns.

3. Select **Table ➡ Enter Text** and pick the top left cell of the table. The cell will highlight and you will be prompted to enter the text string. Enter *ITEM* and two carriage returns. Select the cell to the right of the ITEM cell and enter *PART NO* and two carriage returns. Enter *PART NAME* in the third column and *QTY* in the fourth. The table, shown in the following illustration, is now complete.

Text added to complete the BOM table.

Revision Block Table

The revision block table is created in much the same way as the BOM table.

4. Select **Table ➡ Create ➡ Descending ➡ Leftward ➡ By Length ➡ Vertex**, and pick the corner of the border, as shown in the following illustration. The table will have five columns. Create them in the following order and sizes: *5.75, 1.50, .88, 1.50,* and *.63.* Also create two rows, each *.31* high.

Requirements for the revision block table.

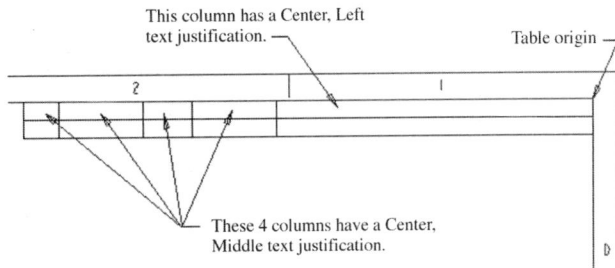

5. Following the instructions from the BOM table exercise, set the text justifica-tion as shown in the previous illustration. Once the text justification is set, add the following text, starting with the top left cell and moving to the right: *REV, DATE, CHG BY, CHANGE NO,* and *DESCRIPTION.* The following illustration shows the completed table.

Completed revision block table.

REV	DATE	CHG BY	CHANGE NO.	DESCRIPTION

Creating Title Block Tables

The first table to be created is the Tolerance Block. This table is where the standard tolerances will be displayed. It is a very simple single-column table with four rows.

6. Start by creating a descending rightward table, with its origin at a vertex. Select the vertex, as shown in the following illustration, and enter a value for the width that is close to the width of the title block area; then do the same with the four rows of the table. The table will look something like that shown in the following illustration.

Table columns and rows need to be modified to fit the format geometry.

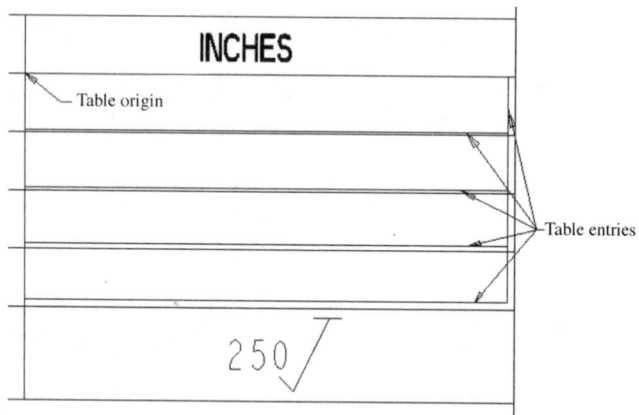

INCHES

Table origin

Table entries

250

You now need to change the sizes of the table cells. You will need to change the column width and row height to make the table match the format geometry.

7. Select **Table ➥ Mod Rows/Cols ➥ Change Size ➥ Column** and pick any cell in the column you want to change. Enter a value for the column width and enter a carriage return. The table will update to the new column width. If the column still does not fit the format, select it again and enter another value. Continue this process until you are happy with the way the columns look in the format. After all of the column widths are correct, select **Table ➥ Mod Rows/Cols ➥ Change Size ➥ Row** and repeat the procedure to modify the height of the rows.

You now need to modify the text justification of the table cells. You have adjusted the size of the rows and columns so that the table exactly matches the geometry of the format. Now it is time to make sure the text will be positioned in the cell correctly. The text for all of the cells in this table needs to be located in the center of the cell horizontally, and in the middle of the cell vertically.

8. Select **Table ➥ Mod Rows/Cols ➥ Justify ➥ Center ➥ Middle** and pick the column. The column will highlight and the justification will change.

The text justification of a table can be set only before any text is entered in the cell. After the text is entered, you must change the text justification through the MODIFY menu.

9. Select **Modify ➥ Text ➥ Text Style** and pick all of the text lines you want to modify; then select **Done Sel**. Select the type of text justification required; then click on the **Apply** button and then the **OK** button.

Now you can enter the text. The text strings you will enter in this table are a combination of simple text and system parameters.

10. In the top row, enter *.12*. In the cell below that, enter *&linear_tol_0_0*. This is a system parameter for the tolerance value attached to a dimension with a single decimal place. The values for these standard tolerances are entered in the *config.pro* file. In the next two lower cells, enter the parameters *&linear_tol_0_00* and *&linear_tol_0_000*, respectively.

To complete the table, modify the text height of the text in the cells. This is done the same way the text height of any type of text is changed.

11. Select **Modify ➥ Text ➥ Text Height** and pick the text strings (make sure you select all of the parts of a string of text); then select **Done Sel** and enter the new height value *.09* and select **OK**. The table is now complete and should look like that shown in the following illustration.

Completed Tolerance Table ⌐

The .12 is simple text; the &linear_tol_0_X notes are system parameters.

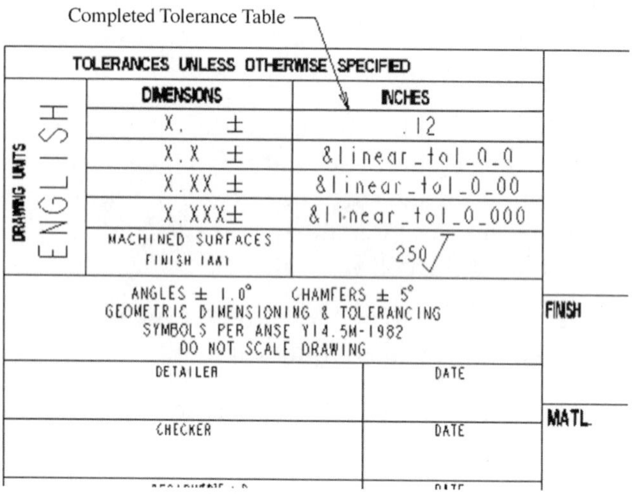

TOLERANCES UNLESS OTHERWISE SPECIFIED		
	DIMENSIONS	INCHES
	X. ±	.12
	X.X ±	&linear_tol_0_0
	X.XX ±	&linear_tol_0_00
	X.XXX±	&linear_tol_0_000
	MACHINED SURFACES FINISH (AA)	250/

DRAWING UNITS / ENGLISH

ANGLES ± 1.0°   CHAMFERS ± 5°
GEOMETRIC DIMENSIONING & TOLERANCING
SYMBOLS PER ANSE Y14.5M-1982
DO NOT SCALE DRAWING

DETAILER	DATE
CHECKER	DATE
DESIGNATE R	DATE

FINISH

MATL.

Sign-off Table

The next table to create is for the names and dates for drawing sign-off. The table will include the scale box and the drawing revision box.

12. Create an ascending, rightward table with the sizes set by length and the origin at a vertex. Select the lower left corner of the scale box as the table origin. The table will consist of three columns and six rows. The following illustration shows a completed table. Notice that two cells occupy the same sign-off box. Later you will merge these cells into a single cell.

Cells from different columns and rows can be merged to form a single cell.

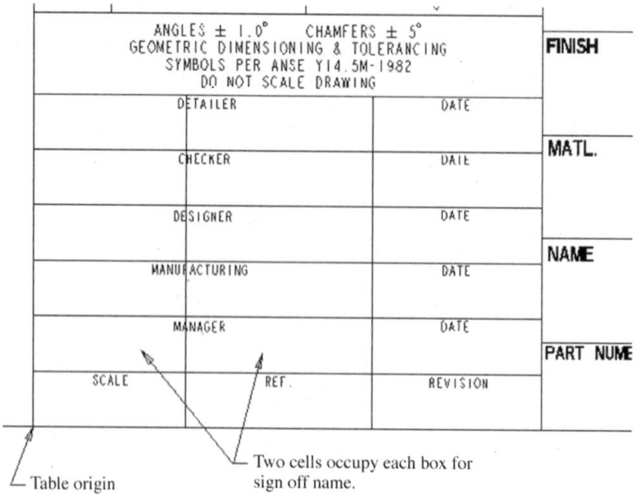

ANGLES ± 1.0°   CHAMFERS ± 5°
GEOMETRIC DIMENSIONING & TOLERANCING
SYMBOLS PER ANSE Y14.5M-1982
DO NOT SCALE DRAWING

FINISH

DETAILER	DATE
CHECKER	DATE
DESIGNER	DATE
MANUFACTURING	DATE
MANAGER	DATE
SCALE	REF.	REVISION

MATL.

NAME

PART NUME

⌐ Table origin

⌐ Two cells occupy each box for sign off name.

13. Before you merge the cells in the sign-off name boxes, set the text justification of the entire table to center-bottom. The two cells that make up each sign-off name box need to be merged so that the cell will function as a single cell. To merge the cells, select **Table ➦ Modify Table ➦ Merge ➦ Columns.** You select the column option because each of the cells being merged belongs to a different column. Select each of the cells in each of the sign-off boxes. The cells will merge and the line that separates them will be removed. The following illustration shows what the table looks like after the cells have been merged.

Table after the cells have been merged.

The two cells that made up each of these boxes have been merged together.

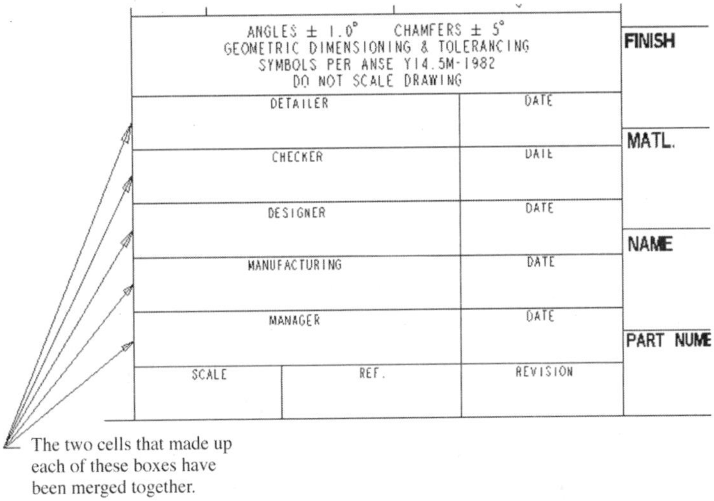

14. Add the following parameters in the cells of the table, as shown in the illustration that follows the table.

Parameters To Be Added

Cell	Parameter
&detailer	&detail_date
&checker	&checker_date
&designer	&designer_date
&mfg_sign_off	&mso_date
&manager	&manager_date
&scale (system parameter)	&drw_rev

The parameters have been added to complete the sign-off table.

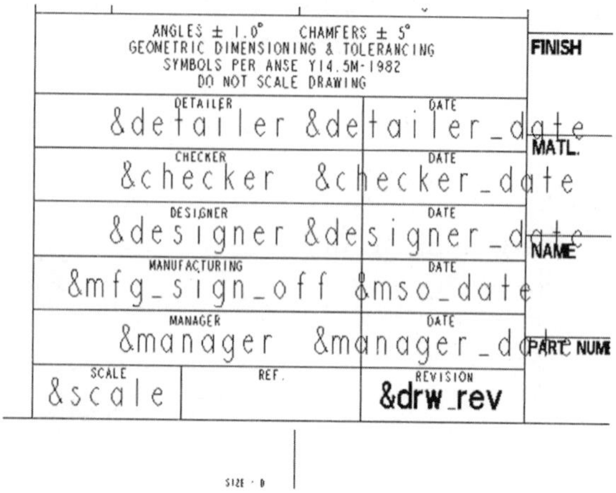

15. In the preceding illustration, you will notice that the font for the drawing revision parameter is in the filled (boldface) font. Modify the font of the parameter the same way you would modify the font of any drawing note. The completed sign-off table is shown in the following illustration. You can now move to the last table, the part information table.

Title block with the tolerance and sign-off tables complete.

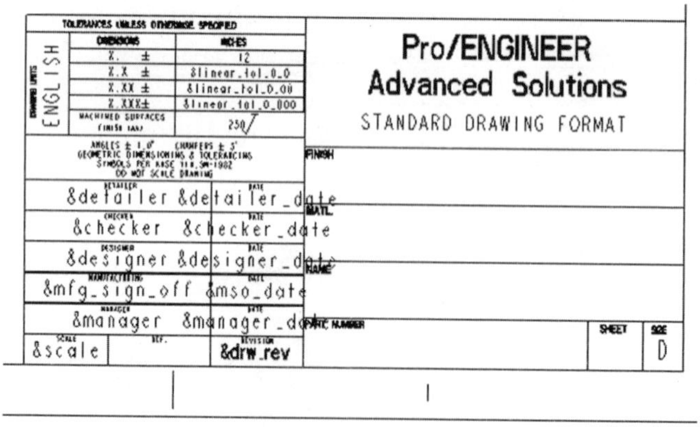

Part Information Table

The last table for the format is created in the same manner as the other tables, with one exception. For the Finish, Material, and Name boxes there is room for two lines of text. To provide for this, a separate parameter is required for each line. You will create a table that will place each parameter in a different cell, and then remove the lines that separate the cells.

16. Create an ascending, rightward table with the cell size set by length and the origin at a vertex. Pick the lower left corner of the Part No. box. The table will have two columns and seven rows. Modify the size of the columns and rows until the table matches the format geometry shown in the following illustration.

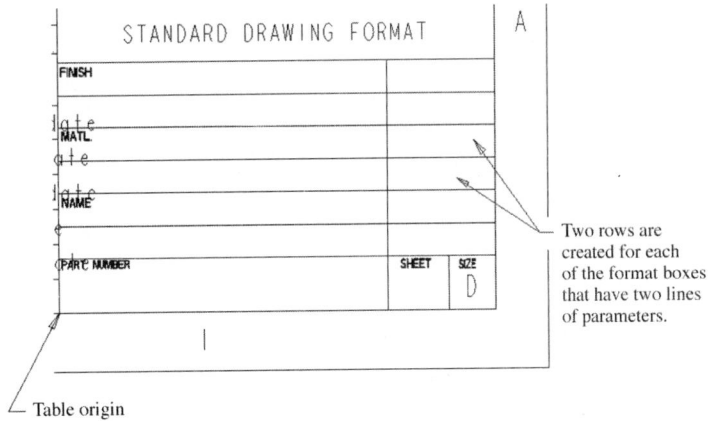

Part information table with two cells in the Finish, Matl, and Name boxes so that they can have two lines of text in them.

STANDARD DRAWING FORMAT

FINISH

MATL.

NAME

PART NUMBER   SHEET   SIZE   D

A

Two rows are created for each of the format boxes that have two lines of parameters.

Table origin

17. Modify the text justification of both columns to center-bottom. The justification of the Sheets box text will have to be modified after the parameters have been added. Merge the column cells of rows that make up the Finish, Material, and Name boxes using the procedure described previously in the sign-off table section. When you finish merging the cells, the table should look like that shown in the following illustration.

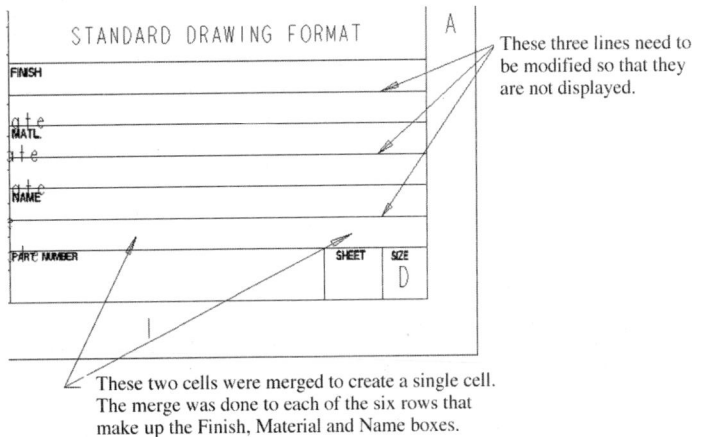

Merged table cells, but lines dividing boxes still displayed.

STANDARD DRAWING FORMAT

FINISH

MATL

NAME

PART NUMBER   SHEET   SIZE   D

A

These three lines need to be modified so that they are not displayed.

These two cells were merged to create a single cell. The merge was done to each of the six rows that make up the Finish, Material and Name boxes.

You now need to modify the table line display. You cannot merge the rows to eliminate the lines in the three boxes because you want the rows to function separately. However, you can modify the line display of a table.

18. Select **Table ➥ Modify Table ➥ Line Display ➥ Blank** and simply pick the three lines you want to remove. The result is shown in the following illustration. The rows still function as if the lines were still there; the lines are simply not displayed. The other options in the LINE DISPLAY menu are Unblank and Unblank All. Unblank will allow you to select individual lines so that they can be redisplayed. Unblank All will unblank all of the lines in a table except for those cells that have been merged.

Lines removed using the Line Display option from the Modify Table menu.

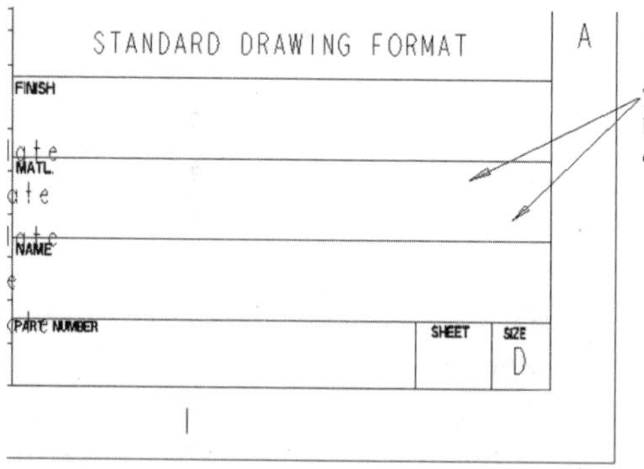

The two cells still remain but the line separating the cells has been blanked.

You can now enter the parameters.

19. Enter the following parameters in the table, as shown in the illustration that follows. Use the same procedure you employed for the previous tables.

Parameters To Be Added

Cell	Parameter
&finish_1	&finish_2
&material_1	&material_2
&name_1	&name_2
&part_name (system parameter)	&current_sheet (system parameter)
&total_sheets (system parameter)	

The only unique item in entering the parameters deals with the parameters in the Sheet box. You want this box to read "1 OF 1," "1 OF 5," or "3 OF 3" (whatever the sheet numbers need to be). To do this, you will use the system parameters "current_sheet" and "total_sheets." These parameters will provide the proper numbers for the Sheet box.

20. To get the Sheet box to read correctly, type what you want to see displayed in the box. However, replace the numbers of the sheet with the proper system parameter. For example, if you want to see "3 OF 5" in the Sheet box, read it as "the current sheet (3) of the total number of sheets (5)." Now replace the numbers with the system parameters and type the following text string in the Sheet box: *&current_sheet OF &total_sheets.* When the system assigns the values to these parameters, they will appear correctly, as shown in the second illustration at right.

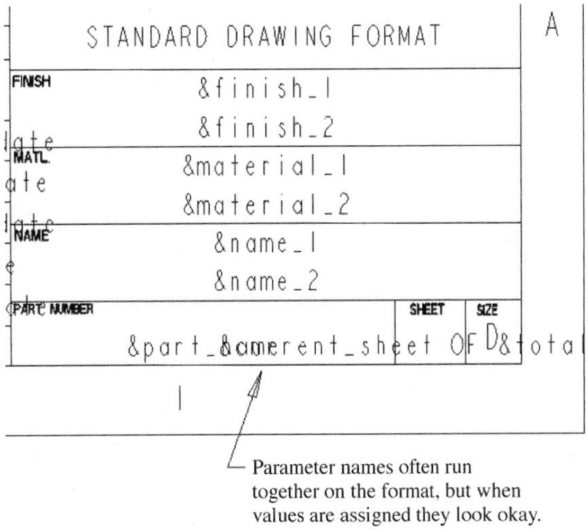

Parameters are added to the table cells.

21. Now that the sheet parameters have been entered you can modify the text justification of the Sheet box to have a Left horizontal justification and a Middle vertical justification. The last adjustment to be made is to the &part_name parameter. You need to change the height and font of the text. Modify the text height to be *.25* and the font to "filled."

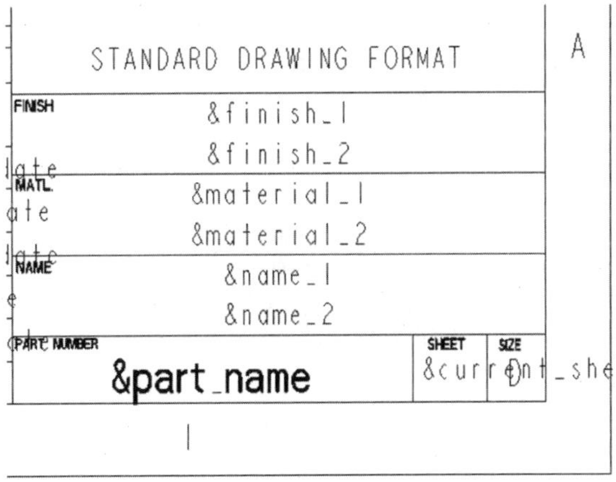

Modified parameter text strings.

The last of the tables is now complete and the format is ready to be saved and placed in a format library to be used by any user on the system. The previous illustration shows parameter text strings modified to the proper text height, font, and justification to complete the table and format.

22. Once the format is complete, save it to a local directory. Create a new drawing using the size of the format you just created. However, do not add the format to the drawing just yet. First, add some views using a model that has all of the user parameters filled out. Now you are ready to add the format to the drawing. The following is a checklist of items to examine and verify before releasing the format to the system.

   • Check all of the tables to make sure the parameters are filled in correctly.

   • Verify that all text fonts are correct. Check all text, not just the tables.

   • Look at text heights. Make sure all text fits in the format boxes.

   • Look for changes you can make to the text justification to make the text fit better in the format boxes.

Summary

This chapter examined the process of creating a drawing format. You learned that the format geometry can be either sketched or imported from another system or from Pro/ENGINEER. Using the grid when sketching a format can make lining up entities easier and more consistent. Much of the chapter was spent covering system and user parameters and tables and how they work together in a format. Parameters can be placed in tables so that the format will automatically update. You also learned that tables are required to be able to modify text in the format when the format is used in a drawing.

The exercises stepped you through the entire process of creating a drawing format. This included importing the geometry, adding a BOM and revision table, and completing the title block by creating three additional tables. Many table techniques were covered, such as merging cells, modifying line display, changing text justification of cells, and modifying the text style of the text itself. All of these techniques were used to complete the drawing format.

Views and Dimensions

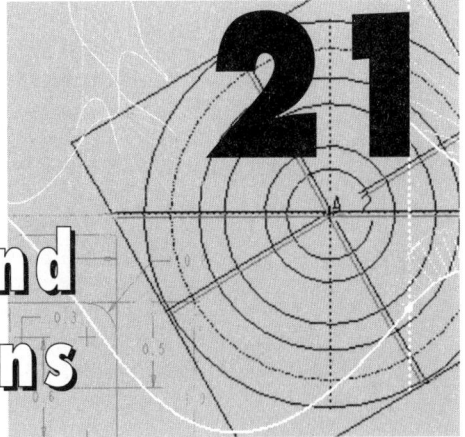

This chapter focuses on drawings. The emphasis is not as much on the "picks and clicks" of how to create views or show dimensions as on how to help your drawings better convey your design intent. You will learn about working with multi-model drawings, and techniques on how to use the Show Dimension option effectively. You will gain a better understanding of geometric dimensioning and tolerancing (GD&T) by exploring datums, basic dimensions, and callouts.

Layers are an effective tool in helping you clear unwanted data from your drawings, making them easier to read and understand. In this chapter's exercises, you will create a datum using datum targets and display them on a drawing. You will also create a Bill of Material (BOM) table and then use the table to automatically display the item balloons. Yet the most important thing to learn from this chapter is how to create a more concise and accurate drawing that relays your design intent to the user of it.

Drawing Views and Multi-sheet and Multi-model Drawings

Your model is complete. You have built the model with design intent in mind from start to finish. The model has been placed in the assembly to check fits and clearances and everything is correct. The toolmaker has even started the tooling, working from a model sent earlier. Yet there is still work to be done. You need to create a drawing, and just as much care and thought needs to go into creating a clear, readable drawing as into creating a sound Pro/ENGINEER model.

Drawing Views

Ideally, you are using a start drawing where much of this work has already been completed, but the following information assumes that this is not the case and starts at the beginning. When a drawing is created, it is not linked to a model until the first view is placed in it. To place the first view, select View ➡ Add View, which will bring up the VIEW TYPE menu. Select Done and Pro/ENGINEER will prompt you to select a model for the drawing.

This first view must be a general view because there is no other drawing information to help place it. Select a point inside the drawing frame to place the center of the view. The view will display in the default orientation and the datum planes will also be displayed. This view needs to be oriented into a drawing view, which can be done with one of two methods. The first is to select the orientation planes in the same way you select them to orient the model in Part mode. If you do decide to select planes to orient the view, you should use the default datum planes. These features are the most stable features in the model, making your drawing views stable in the process.

The second method is to use a saved view. As long as the saved view is based on the default datums, the orientation will be just as stable. After the first view is placed, use projected and auxiliary views as much as possible because these views get their orientation from previously placed views and will stay aligned to them as changes are made to the model.

Multi-sheet Drawings

At times you will need to create drawings that have more than one sheet. Multi-sheet drawings are used for various reasons. Some companies use multi-sheet drawings to document multiple parts of purchased assemblies for which they have design responsibility. Others need multi-sheet drawings just to create the views required to document large or complex parts, such as engine crankcases and cylinder heads.

To create additional sheets, select Sheets ➡ Add. The new sheet will be added to the drawing. All sheets of a drawing will be the same size. If you change the size of one sheet, all of the others will change. The following table lists and describes the other options under the SHEETS menu.

SHEETS Menu Options

Option	Description
Previous	Previous works the same as Next, but goes to the previous sheet. Selecting Previous when the first sheet is displayed displays the last.
Next	Selecting the Next command advances the sheet displayed on the screen to the next sheet. If you are on the last sheet of the drawing, selecting Next will take you to the first sheet.
Set Current	Set Current allows you to select the sheet number of the sheet to be displayed.
Remove	When the Remove option is selected, you are prompted to enter the number of the sheet to be removed. This option also removes any views or data from this sheet.
Reorder	Reorder moves the current sheet to the sheet number you enter. All other sheets are reordered accordingly.
Switch Sheet	Switch Sheet allows you to move items from one drawing sheet to another.

The Switch Sheet option in the SHEETS menu offers you two methods of moving items to another sheet. The Switch Sheet menu is shown in the illustration at left.

The first option in the Switch Sheet menu is Switch Items, which will move the item you select to another sheet and maintain the item's location on the destination sheet. The second option, Switch/Move, will not only move the item to another sheet but allow you to pick a new location for the item. The Switch Sheet option also opens a DWG ITEMS menu. This menu allows you to set which items you want to move. You can choose from Dwg Views to move one or more views. Draft Items will move items such as notes and notes with leaders. However, the leaders will not be carried over to the new sheet. The last item is Dwg Tables, used to move a table to another sheet.

To switch an item from one sheet to another, select Sheets ➡ Switch Sheet ➡ Switch Item and select the type of item you want to move from the DWG ITEMS menu. Then select the item to move and Done Sel ➡ Done. Pro/ENGINEER now prompts you to enter the destination sheet number. This can be a new sheet number. Pro/ENGINEER will create the new sheet for you. After you enter the sheet number, the current sheet is set to the destination sheet with the moved items in place.

To switch an item to a new sheet and change its location on the new sheet, select Sheets ➡ Switch Sheet ➡ Switch/Move. Select the type of item to be

Switch Sheet menu structure.

moved and then select Done Sel ➡ Done and enter the destination sheet number. The destination sheet is now displayed. A GET VECTOR menu opens with options for the nature of the move. The options are Horiz, Vert, Ang/Length, and From-To. Choose the translation method and set the new location.

Multi-model Drawings

When you are working on a drawing, Pro/ENGINEER is getting the information to create the drawing views, dimensions, tolerances, and just about everything else from the model. By adding several models to a single drawing, you can set which model you want Pro/ENGINEER to work with or get information from. The model Pro/ENGINEER is working with is called the current model.

Several models can be added to a single drawing. They can also be various types of models, such as parts and assemblies, or different models from a single family table. Actually, any model type can be added to a drawing to create a multi-model drawing. A very common drawing that makes use of multiple models is that for stamped parts. The complete formed part is one model and the flat pattern (an instance in a family table) is the second.

To create a multi-model drawing, all you need to do is add a model to the drawing. Adding a model gives you the option of creating views and dimensions for the second model as if it were the only model in the drawing. To add a model to a drawing, select Views ➡ Dwg Models ➡ Add Model. A window will open listing the models available in the working directory. You can select one of these models or navigate to other directories (or folders) to select other models.

Adding a model to a drawing does not make any real changes to the drawing; it only gives you the option of selecting which model you want to work with. To change the model Pro/ENGINEER looks at for information, you must "set" a model to be the current model. To do this, select Views ➡ Dwg Models ➡ Set Model. The list of model names that have been added to the drawing will be displayed in the DRAW MODELS menu. Select one of these models from the list and it will become the current model.

When you change the current model to a model that does not have any views of it on the drawing, and then select Views ➡ Add, the only placement option is General, the same as when the first view of the drawing was created. This is due to the fact that this *is* the first view being created from this current model and Pro/ENGINEER has no other references to help place this view.

If a model that has been added to the drawing is not being used for the creation of specific views, it should be removed from the list of drawing models. Select Views ➥ Dwg Models ➥ Del Models and select the model from the list.

Show Dimensions and Design Intent

There are various types of dimensions in Pro/ENGINEER, but they can be broken down into two major types: show dimensions (model dimensions that are "shown" in a drawing) and created dimensions (dimensions that are created in the drawing and are not from the model). The difference between these types is key to creating a drawing that links the model and your design intent.

The reason for paying close attention to design intent while building a model is to be able to use show dimensions exclusively. When you use show dimensions, the person looking at your drawing will be seeing not only your design intent but how your model was constructed. This makes modifying the model for future changes an easy and straightforward process for you and, more importantly, for the user that has never worked on the model. Obviously, using only shown dimensions is not possible all of the time, and some created dimensions will be required, but it is a goal you should strive to achieve on every model you create.

A technique that helps you use as many show dimensions as possible and convey design intent is to show dimensions feature by feature. Even on very large models this technique works quite well. You will be surprised by how much of a large model consists of round and draft features. Often only a few of these features need to have their dimensions shown.

The procedure for this technique is to select Show/Erase from the DETAIL menu. A Show/Erase dialog box will open, shown in the following illustration.

Show/Erase dialog box.

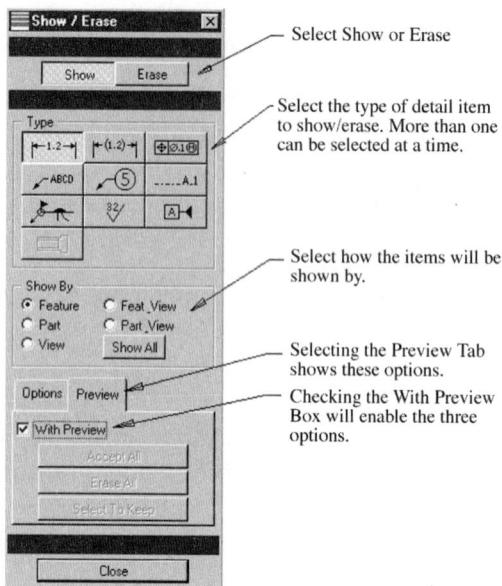

Click on the Show button and select the Dimension symbol from the Type area; then select Feature in the Show By area. Select Sel By in the GET SELECT menu below the dialog box. If the drawing is a multi-model drawing, Pro/ENGINEER will prompt you to select a view port. Select one of the drawing views of the current part. From the SPECIFY BY menu, select the Number option. You will be prompted to enter the number of the feature you want the dimensions to show. You can continue to enter as many feature numbers as you want.

After all of the feature numbers have been entered, enter a carriage return. Once the final carriage return is entered, the dimensions will be displayed. Then select Done Sel. This technique allows you to decide whether or not to display every model dimension on a feature-by-feature basis. Once you have gone through the entire model, you will find that very few, if any, dimensions will need to be created.

If you select the Preview tab in the Show/Erase dialog box and check the With Preview option, you will be presented with three options for handling the dimensions that were just shown. First, you can Keep All, which keeps all of the dimensions displayed on the drawing. The second is Erase All, which erases all of the dimensions that were shown in the last operation. The third option is Select to Keep, which allows you to select all of the dimensions you want to be displayed. After you pick all of the dimensions you want to keep, select Done Sel and the selected dimensions remain and the rest of the dimensions are erased.

> ✓ **TIP:** *When following this technique, use the model tree to help keep track of which feature you are working on and which features you can group. This also helps speed up drawing creation.*

Bi-directional Associativity

The other advantage of using show dimensions over created dimensions is that you can leverage the power of Pro/ENGINEER's bi-directional associativity. Bi-directional associativity means that when a model's dimensions are modified the drawing will automatically update, and when a show dimension (model dimension) is modified in the drawing, the model will also update to the new value. Created dimensions cannot be changed in either the drawing or the model. The value of a created dimension is updated only when the drawing of a modified model is regenerated.

Drawings with a Minimal Number of Dimensions

The goal of most engineering departments is to get to a "paperless" system of engineering data. "Paperless" refers to a condition wherein there are no drawings of parts. That is, all of the data needed to produce, inspect, and assemble parts is contained in the solid models themselves. This may seem like an impossible goal, but many companies are taking great strides toward it. Manufacturing departments and outside suppliers are becoming more and more comfortable working only with a solid model, and coordinate measuring machines (CMM) can be programmed to inspect a part directly from the model.

Because of this preference for working with a model as opposed to a drawing, many companies are finding that creating a fully dimensioned drawing adds little value to the process. These companies are creating drawings with only the key or critical dimensions displayed. These dimensions would show critical fitment areas and key dimensions used for inspection or to maintain control of a manufacturing process. With time to market as critical as it is in today's business world, this type of drawing is becoming more and more popular.

Parametric Dimensions in Notes

Just about every drawing has some form of note on it, and often these notes contain dimensional values that describe geometry. Common examples are "ALL UNSPECIFIED RADII R .06" and "1.750 OD TUBE with .049 WALL THICKNESS." The dimensional values in these notes are just as important as any other dimension on the drawing, and you need to treat them as such.

Typing the dimensional value in the note is the same as creating the dimension in the drawing. The dimensions already exist in the model. It is an easy procedure to place the parametric dimension into the note and gain the same advantages with the note that you get from the show dimensions in the rest of the drawing. The following illustration contains show dimensions as they appear on a drawing.

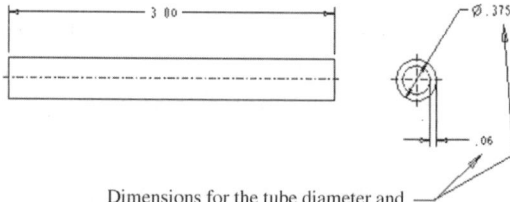

Show dimensions as they appear on a drawing.

Dimensions for the tube diameter and wall thickness to be moved into a standard note.

In the previous illustration, the diameter and wall thickness dimensions will be placed into a drawing note. To create a standard note, select Create ➡ Note ➡ Make Note. The dimensions on the drawing will switch their display from the dimensional value to the dimension name. Type *&D# OD TUBE WITH A &D# WALL THICKNESS*. Typing the ampersand (&) symbol before the dimension name in the note means that the note will replace the dimension name with its dimensional value. Enter two carriage returns to complete the note. The dimensions on the drawing will switch their display back to the numerical value and the note will now read correctly, as shown in the following illustration.

The note is written with the parametric dimensions embedded in the note.

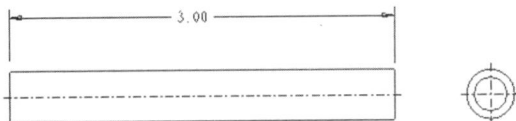

```
|————— 3.00 —————|
```

.375 O.D. TUBE WITH A .06 WALL THICKNESS

Dimensional values are created in the note and removed from the drawing view. The values in the note can be modified and will update in the model.

You will also notice that the dimensions in the note have been removed from the drawing. Having the dimensions in the note and displayed in the drawing is the equivalent of double dimensioning (dimensioning the same feature twice). Pro/ENGINEER will only allow a single dimensional value to be displayed once, and will automatically remove the original from the drawing.

> ✓ **TIP:** *If you are modifying a note, the dimensions will not automatically switch their display from values to names. There are two ways to do this manually. The first is to select Advanced ➥ Relations ➥ Switch Dim. The second method is to select the switch dimensions icon, shown at left. Now you can modify the text of the note to include the dimension name preceded by the ampersand (&).*

Tolerancing and Geometric Dimensions and Tolerancing

The tolerance value of a dimension cannot be displayed on a drawing unless the tol_display option in the *dtl_setup* file is set to Yes. To set the value for this option, select Advanced ➥ Modify Val from the DTL SETUP menu. A window with the *dtl_setup* file will open and you can enter or edit the tol_display option to Yes. Select the File button from the pull-down menu and select Save and then Exit.

Tolerance Display

If you try to change or display the tolerance of a dimension and the tolerance mode area of the MODIFY DIMENSION dialog box is grayed out, which means that the tolerance option is not available to be selected, the tol_display option is set to No and must be changed. The following illustration shows the "grayed out" tolerance mode option.

The tolerance mode option is unavailable in this dialog box.

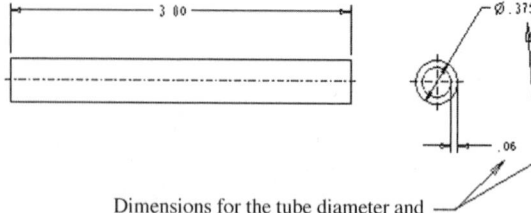

Dimensions for the tube diameter and wall thickness to be moved into a standard note.

Displaying a Tolerance on a Dimension

The following are the four tolerance types available in Pro/ENGINEER's Drawing mode.

Nominal	The Nominal mode displays the dimensional value only.
Limits	The Limits mode displays the tolerance as a maximum and minimum dimensional value.
Plus-Minus	Use the Plus-Minus mode to display a tolerance with an upper value different from the lower value. An example would be a .500 dimension with an upper tolerance value of +.005 and lower tolerance value of -.010.
+− Symmetric	The +− Symmetric mode displays a tolerance that is plus and minus an equal value. For example, a .500 dimension with a tolerance of +.015 and −.015 is displayed as .500±.015.

To change the tolerance mode displayed on a dimension, select Modify ➡ Dimension and select the dimensions you want to change. There are two *config.pro* options you should be aware of that can affect the display of the tolerance or the model geometry itself. The first is tol_mode, which sets the default display of the tolerance. You can enter any of the following tolerance modes: *nominal, limits, plusminus,* and *plusminussym.*

The other option is maintain_limit_tol_nominal, which needs to be understood because of its effect on the model geometry. When the Limits tolerance mode is used and the tolerance values are not symmetrical, Pro/ENGINEER will adjust the model geometry so that the nominal value will be placed at the center of the tolerance range. For example, model geometry with a .500 nominal value and tolerance limits of .505/.490 will be changed to a value of

.4975. This is the center of the upper and lower limits. To be sure that the model geometry will not be modified when using the Limits tolerance mode, set the *config.pro* option maintain_limit_tol_nominal to Yes.

Layers in Drawings

Layers are used in drawings for the same reasons they are used in parts: to group items to control their display. As was discussed in Part I, a set of standard layers needs to be set up in the part and drawing start models so that a consistent naming convention is used throughout the company. Standard layers should include layers for axes, default coordinate systems and datums, other datums, datums used in geometric tolerancing, curves, surfaces, points, and so on. The following are some of these standard layer names.

- Axis
- Curves
- Def_cs
- Datums_all

- Def_datums
- Geomtol
- Pnts
- Surfs

If all of the appropriate items are placed on these layers when all of the layers are blanked in the part, only the part model will be displayed, as shown in the following illustration of the standard layers before and after being blanked.

The same drawing before and after all of the standard layers have been blanked.

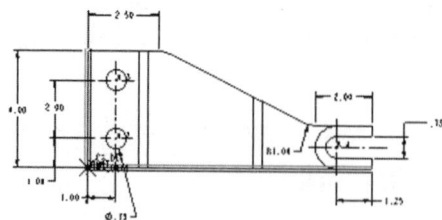

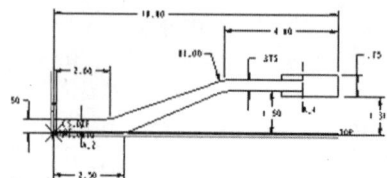

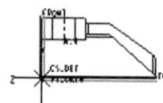

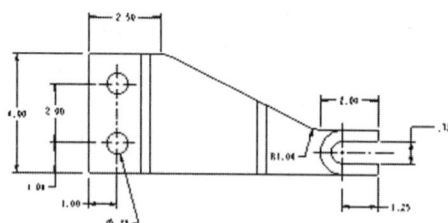

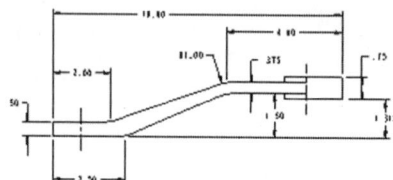

Putting Layers to Work

Now that all of the model items have been assigned to the correct layer in the model, you can put these same layers to work in the drawing to determine how it will be displayed. The best method for working with layers in Drawing mode is to create layers in the drawing with the exact same names as the layers in the part. You can control the display of the items on these layers in the drawing without having to add any items to the drawing layers themselves.

There are two *dtl_setup* file options that need to be set to make this work correctly, which can be set within the Layer Display dialog box. Select Layer ➡ Display or the Layer Display icon, shown at left. Select the Status button in the top menu bar and then Preferences (see the following illustration). A Layer Status Control box will open containing the two options that need to be checked.

Status dialog box to control the two DTL SETUP file options used to control layers in drawings.

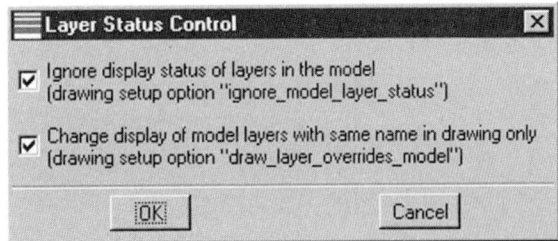

The first layer control option is ignore_model_layer_status, which establishes that the drawing will not look to the model to determine any layer display. In other words, no matter how the layer display is set in the model, the drawing will display the layers as set in the drawing. The second option is draw_layer_overrides_model. This option allows you to use layers in the drawing that have the same names as layers in the model. Changing the display of those layers in the drawing will not affect the display of those layers in the model.

When both of these options are checked, you separate the drawing layer display from the model and the model layer display from the drawing. Changes in the drawing layer display do not affect the layer display in the model, and changes in the model layer display do not affect the layer display in the drawing.

Controlling Text with Layers

There are two items whose display can be difficult to control in drawings: axis names, and view notes such as view names, scale notes, and others. Both of these items are text related and both are easily handled with layers. When an axis is shown on a drawing, the name is also displayed with the axis. You need to create a DRW_TEXT layer in the drawing to put the text on.

To remove the display of the axis name, select Layer ➥ Drawing ➥ Set Items ➥ Add Items and check the DRW_TEXT layer and Done Sel. Select the Text option from the LAYER OBJ menu and select the axes whose text you want to blank. You only need to select the axis once, even if it displayed in several views. After the axes have been highlighted, select Done Sel. Then blank the layer DRW_TEXT and the axis names will be removed. The same method can be used to blank notes created when a view is placed. View names such as

SECTION A-A or DETAIL Z, and the scale note created from scaled views, can also be blanked this way rather than being moved off to the side of the drawing sheet so that they do not show up on the print.

Exercise 21-1: Incorporating GD&T Symbology

This exercise deals with the creation of GD&T symbology. You will start with the bracket part and drawing shown in the following illustration. To this bracket you will add primary, secondary, and tertiary datums, including datum targets. You will then create the necessary basic dimension and some geometric tolerance callouts.

The subject of GD&T on drawings is a large and complicated one, about which many books have been written. This part of the text, and the exercises that follow, cover only the creation of the symbology of GD&T in Pro/ENGINEER and not its application to a design.

Some of the setup work that needs to be done to create the GD&T in a drawing is done in the model. This setup work includes creating datum points to locate the datum targets, and to set the datums, which means that the datum has been defined as a GD&T datum and can be used in a geometric tolerance callout. The following illustration shows the bracket model and drawing you will be creating in this exercise.

The bracket model and drawing required for exercise 21-1.

1. After creating the bracket model, the first step is to create Datum A, the primary datum. This datum will consist of the three datum points shown in the following illustration. Before you create the datum, you need to create the points that will be used to locate the datum targets. To create the datum points, select **Feature** ➡ **Create** ➡ **Datum** ➡ **Point**. Use the **On Surface** option to create the points and select the surface in approximately the same place as that shown in the following illustration. Use the default datums SIDE and FRONT for the placement planes and enter the dimensions in the illustration. Change the names of the points to their datum names *A1*, *A2*, and *A3*. If any of the points were created as a single feature, use the **Other** option when selecting the points to be renamed.

Location of datum points used to create datum targets.

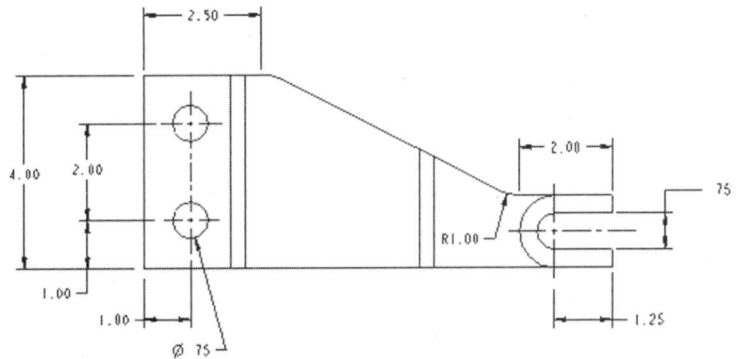

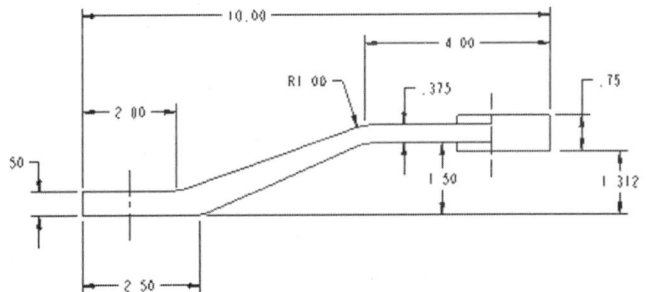

To create the datum targets, the points need to be associated with a datum plane. This can be any model datum plane. Try to use a datum plane that contains as many of the datum points as possible, and that has the same orientation as the plane being defined by the datum points. In this case, you will use the default datum plane TOP because it contains two of the three datum points and has the proper orientation. Before you can set the datum, however, you must change the name of the datum feature to A.

2. Select **Setup ➥ Name ➥ Feature** and pick the default datum TOP. Enter an *A* at the prompt; then enter a carriage return. The default datum plane TOP is now A.

3. Change the names of the secondary and tertiary datums. The secondary datum will be the default datum SIDE. Using the procedure for the previous step, change its name to *B*. The tertiary datum is the default datum FRONT. Change its name to *C*.

With the names of the three datums changed to the geometric datum names, you can now set them so they can be used in geometric callouts.

4. Select **Setup ➥ Geom Tol ➥ Set Datum** and pick datum A. The Datum dialog box (shown in the first illustration at right) will open, allowing you to change the datum name, set the display style, and establish how the datum is to be attached.

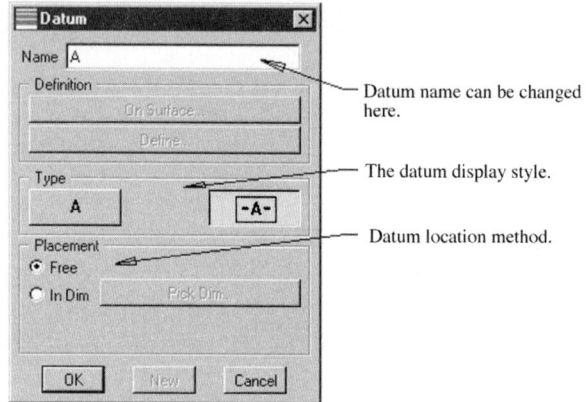

Datum name can be changed here.

The datum display style.

Datum location method.

Datum dialog box.

5. Select **OK**. The datum name "A" now reads "-A-" and is placed in a box. Set the two remaining datums B and C using the same options. The model will look like that shown in the second illustration at right.

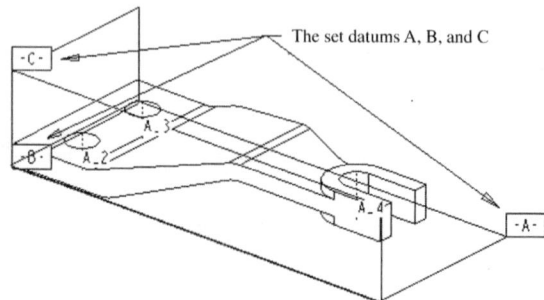

The set datums A, B, and C

Bracket model with default datums set to geometric datums A, B, and C.

When setting the datum in Drawing mode, the menu structure is a little bit different. After selecting the datum, the DATUM LOC menu will open. This menu contains the following options.

DATUM LOC Menu Options

Option	Description
Dimension	The Dimension option will ask you to select a dimension to which the datum symbol will be attached. The only way to display the datum symbol in a drawing is to show the dimension it is attached to.
Geom Tol	Geom Tol places the datum symbol at the bottom of an existing geometric callout.
Default	Default leaves the datum symbol attached to the model datum plane. The symbol is displayed on the drawing wherever the datum plane is perpendicular to the screen.

6. Regardless of which option is chosen, it can be modified later. Select the location and the datum will be set.

An Axis As a Datum

This exercise used only the default datum planes, but they could have been any model datum plane. Another common type of datum is when a hole is used as a datum. When a hole is used as the datum, the datum symbol is attached to the diameter dimension, but the actual datum feature is the center axis of the hole, not the hole itself. You need to change the name of the axis to the geometric datum name and attach it to the diameter dimension using the Dimension option in the Datum dialog box or the DATUM LOC menu.

Because there are two ways an axis can be created (i.e., as a separate feature or as part of a feature), how it is created determines the way the axis is to be selected to change its name. If it was created as a separate feature, use the Feature option in the NAME SETUP menu. If it was created as part of a feature, such as a hole, you need to use the Other option.

7. Exit the bracket model and open the drawing of the bracket. This is where you will be creating the geometric callouts and displaying the datums.

Datum Targets

Although the points and datum plane used to locate the datum targets are created in the model, the actual datum targets are created in the drawing.

8. Show the dimensions used to create the datum points A1, A2, and A3 and then change them into basic dimensions. To convert a dimension into a basic dimension, select **Create** ➜ **Geom Tol** ➜ **Basic Dim** and select all of the dimensions you need to convert. Select **Done Sel** ➜ **Done Sel** and complete the conversion by selecting **Done**. If **Done** is not selected from the EXIT menu, the dimensions will not be converted into basic dimensions.

9. To create the datum targets, select **Create** ➡ **Geom Tol** ➡ **Make Target** ➡ **Simple**. Pro/ENGINEER now prompts you to select the datum plane for the datum targets. Pick the model datum -A-, set earlier. Select the point to locate the first datum target. Pick the point A1. If you have already changed the names of the points to the datum target names, you need to select the points in the correct order. After selecting the point, you will be prompted to select the location of the datum target. Select a spot on the drawing, in the view in which the point was selected. The datum target, A1, is now displayed. Repeat the procedure for the remaining datum points. When all three datum targets have been located, the drawing will look like that shown in the following illustration.

Datum targets for datum A, and the basic dimensions to locate them.

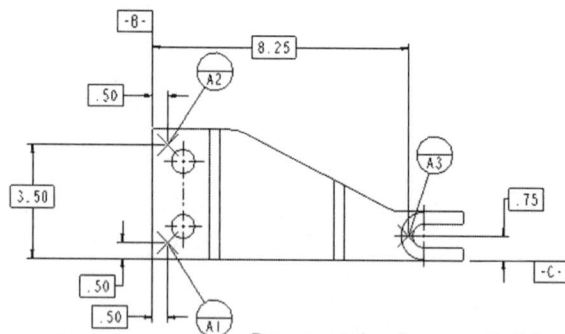

Datum targets have been created and the dimensions used to locate the datum points have been made into Basic dimensions.

✓ **TIP:** *To show the datum targets in other views, select Show/Erase and select the Geometric Tolerance option from the TYPE area and View from the SHOW BY area. Select the view you want the datum targets to be displayed in. This will show all of the geometric tolerance blocks. Therefore, you may have to erase some of the unwanted blocks.*

Creating a Geometric Tolerance Position Callout

Easily the most common geometric tolerance callout is the position callout. You will now create a positional callout for the two holes in the bracket.

10. Show all of the dimensions used to create the two holes, and convert them into basic dimensions except for the diameter dimension, as shown in the following illustration.

Drawing ready for addition of positional callout to the hole diameter dimension.

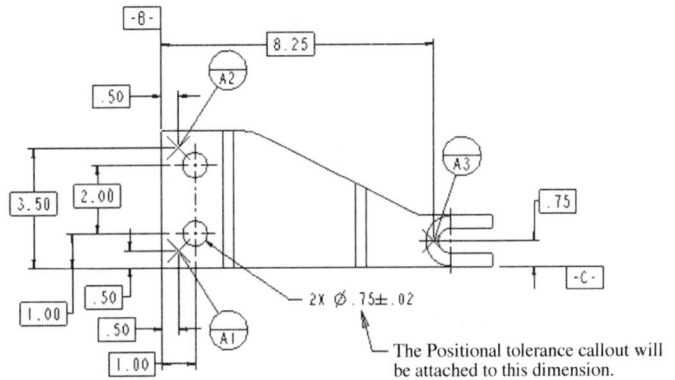

The Positional tolerance callout will be attached to this dimension.

Dimensions for the two holes are shown and converted into Basic dimensions except for the diameter dimension.

11. To create the tolerance callout, select **Create** ➡ **Geom Tol** ➡ **Specify Tol**. At this point, a Geometric Tolerance dialog box will open. This dialog box is shown in the following illustration. Select the type of tolerance you want to use. In this exercise, you will create a position callout. To create the tolerance block, you will step through the tolerance setup tabs at the top of the dialog box, specifying the data required until you are ready to place the tolerance block.

Stepping through the tabs in the Geometric Tolerance dialog box.

Tolerance callout setup tabs.

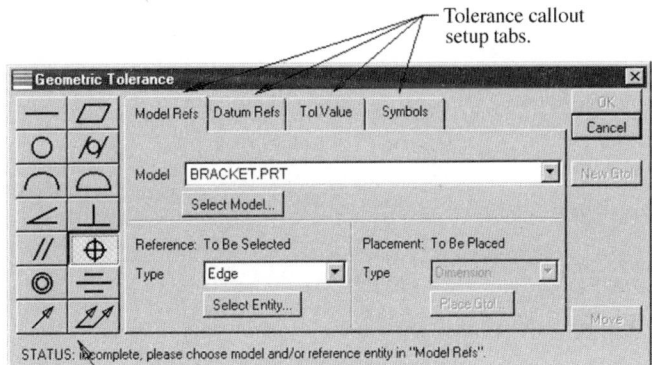

Select the desired tolerance type from this area.

12. Select the **Datum Refs** tab, shown in the following illustration. Select **A** for the primary datum and use the **RFS** (no symbol) option for the datum modifier. Select the **Secondary** tab and select **B** from the list of datums and **RFS** (no symbol) for the modifier. Select the **Tertiary** tab and select **C** for the final datum and again use the **RFS** (no symbol) option.

Datum Refs tab.

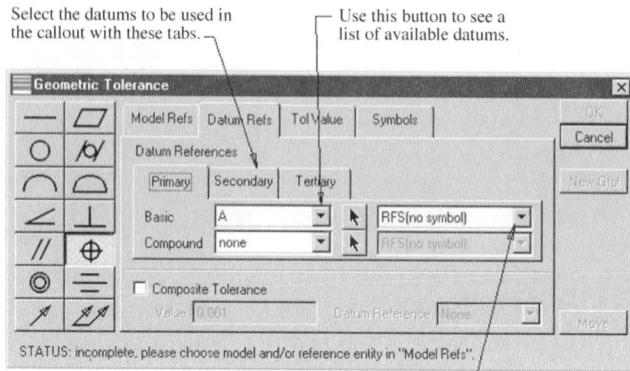

13. Select the **Tol Value** tab to enter the tolerance value and select a material condition for the tolerance value. Enter *.028* for a tolerance value, and select the **MMC** option as a material condition modifier. See the following illustration for an example.

Tol Value tab.

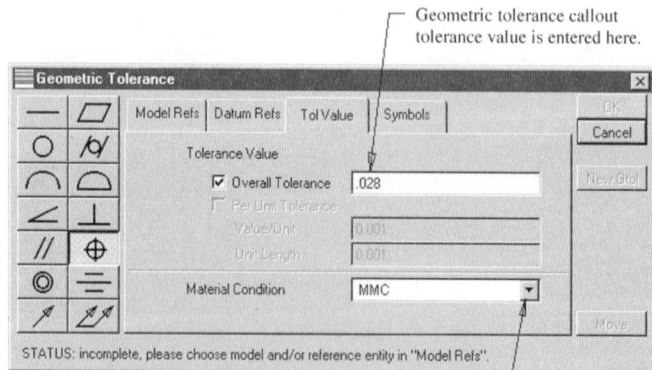

14. The **Symbols** tab, shown in the following illustration, is used to select tolerance value symbols, such as a diameter, as is the case in this exercise. Select the Diameter Symbol box to add a diameter symbol to the tolerance value of the positional callout. It is also where projected tolerance zones are set up.

Symbols tab.

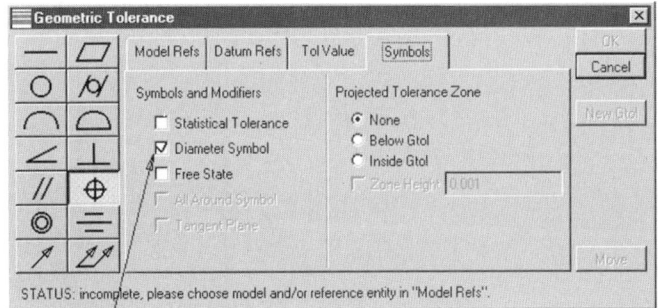

└ Check the Diameter Symbol box.

15. The last step to creating the geometric tolerance callout block is to return to the **Model Refs** tab (shown in the following illustration) so that you can place the callout. See the first of the two following illustrations. A reference entity needs to be selected first. For positional callouts on features such as holes, use the Axis option if an axis is available; otherwise, use the Edge option. For callouts based on surfaces such as perpendicularity or profile of a surface, use the Surface option. You will find that almost all of the GD&T callouts you create are covered by one of these three options.

The reference entity and placement areas of the Model Refs tab.

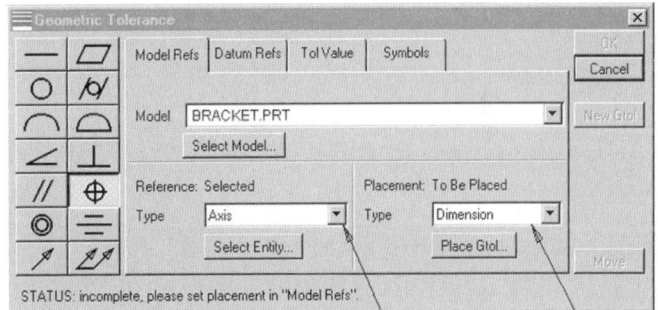

A list of valid entity types can be displayed ⅃ Placement options. ⅃
by using the pull down button.

For this exercise, select the **Axis** option and pick the axis of the hole the dimension is pointing to. After the reference entity has been selected, the callout block can be placed on the drawing. Select the **Dimension** option from the list of options in the Placement area of the dialog box. Select the diameter dimension of the holes. The geometric tolerance callout block will be placed on the drawing, as shown in the following illustration.

Drawing with the positional callout attached to the hole diameter dimension.

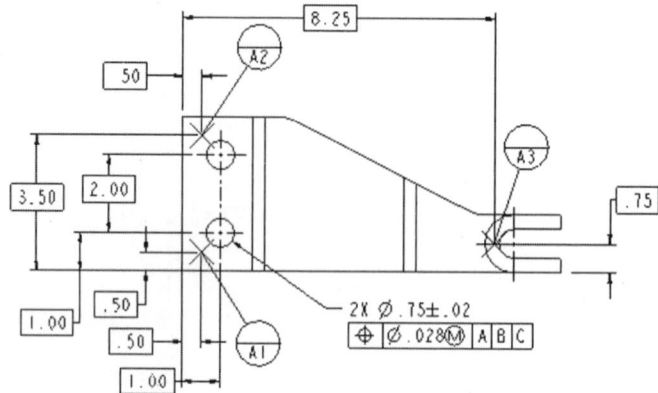

Exercise 21-2: Creating BOM Tables

The type of table most commonly found on an assembly drawing is a Bill of Material table, or BOM. In this exercise, you will create two BOM tables. The first is a simple top-level BOM, and the second is an indented BOM. A top-level BOM table lists only those components used to create that particular assembly. If the top-level assembly contains a subassembly, only the subassembly is listed in the BOM. An indented BOM lists all of the components in the top-level assembly, as well as all components of a subassembly. Both of these BOM tables will then be used to automatically create the item balloons in the drawing.

The first thing you are going to need is an assembly and its drawing. You can create a simple assembly and drawing or select one from the models you work with every day.

1. Call up the drawing (shown in the illustration at right) and create the table shown in the following illustration. This table is constructed in the same manner as the tables you constructed in the format chapter earlier in this book. In fact, if you are using that format the table is already complete and all you need do is set the repeat region.

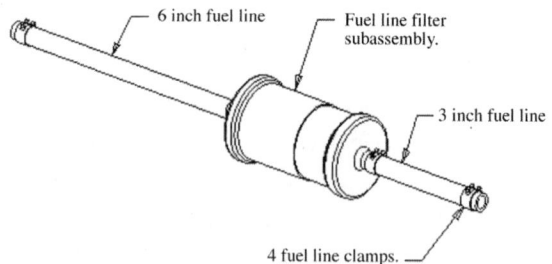

Fuel line filter assembly used in the BOM table exercises.

2. To create the repeat region, select **Table ➡ Repeat Region ➡ Add ➡ Simple**. Pro/ENGINEER now prompts you to select the two cells that define the area of the repeat region. Pick the cells indicated in the second illustration at right. You select the two outermost cells because you want each column to be automatically filled in for each item in the assembly.

Start table for the BOM.

Selection to be made to create the repeat region.

The repeat region is now set, but Pro/ENGINEER needs to know what type of information should be placed in the cells. The text put into the cells is a special type of parameter. These parameters guide Pro/ENGINEER through the model or report information to extract the information you want displayed in the table. This table will use four parameters: ITEM and QTY are report parameters, PART NO is a model parameter, and DESCRIPTION is a user-defined parameter.

➡ **NOTE:** *There are numerous parameters that can be placed in a repeat region. See the "Reports" section of the* Pro/ENGINEER Drawing User's Guide *for a complete list.*

There are two methods of entering the parameters. The first is by keyboard, which means that you type them in as a string of text. The other method is to use the Report Sym option, which presents a system of menus that lets you "build" the parameter into a valid string. Throughout these exercises, you will use the Report Sym option.

3. Select **Enter Text ➡ Report Sym**. Pick the empty cell in the ITEM column and select **rpt... ➡ index**. This prints the index number in the table and is the number

Table with parameters typed in.

of the BOM balloon. Select the cell in the PART NO column; then select **asm... ➡ mbr... ➡ name**. The DESCRIPTION column needs to read the user-defined model parameter *name_1*. Select the cell; then select **asm... ➡ mbr...**

➥ **USER DEFINED** and type in the name of the parameter you want to read *name_1*. The last cell is the QTY column. Select it and select **rpt...** ➥ **qty**. The previous illustration shows the table with the parameters added.

4. Regenerate the drawing by selecting **Regenerate** ➥ **Draft** to update the table, as shown in the first illustration at right.

ITEM	PART NO.	DESCRIPTION	QTY
1	96427	FUEL FILTER ASSEMBLY	
2	65983-3	HOSE, FUEL LINE 3.00 LONG	
3	65983-6	HOSE, FUEL LINE 6.00 LONG	
4	42071	CLAMP, FUEL LINE	
5	42071	CLAMP, FUEL LINE	
6	42071	CLAMP, FUEL LINE	
7	42071	CLAMP, FUEL LINE	

Table as filled out by the repeat regions.

As you can see in this illustration, the table does not display correctly. Each of the four clamps is displayed as a single part, and none of the quantity values appear. You need to change the repeat region's attributes so that it will not duplicate any of the items.

5. To modify the region, select **Table** ➥ **Repeat Region** ➥ **Attributes** and pick anywhere in the repeat region of the table. The table will highlight. Select the **No Duplicates** option and then select

ITEM	PART NO.	DESCRIPTION	QTY
1	42071	CLAMP, FUEL LINE	4
2	96427	FUEL FILTER ASSEMBLY	1
3	65983-3	HOSE, FUEL LINE 3.00 LONG	1
4	65983-6	HOSE, FUEL LINE 6.00 LONG	1

Setting the table attributes to No Duplicates corrected the display.

Done/Return. The table will update to display the items correctly, as shown in the second illustration at right.

Exercise 21-3: Creating a Subassembly BOM Table

In this exercise, you will create a BOM table that will show the information about the components of the subassemblies in addition to the top-level components. This table requires the use of a "nested" repeat region. This means the table will have one repeat region for the top-level assembly and a second, "nested" region for the subcomponents.

Much of the construction of this table is the same as that for the table in the previous exercise. However, one difference is in the shape of the table you start with. The table needs to have two blank rows below the column headings, as shown in the following illustration. The two rows provide the space for creating the nested repeat regions.

The nested repeat regions will be created one after the other, with no menu picks between them. The first will cover the top-level assembly, and the second the subassemblies.

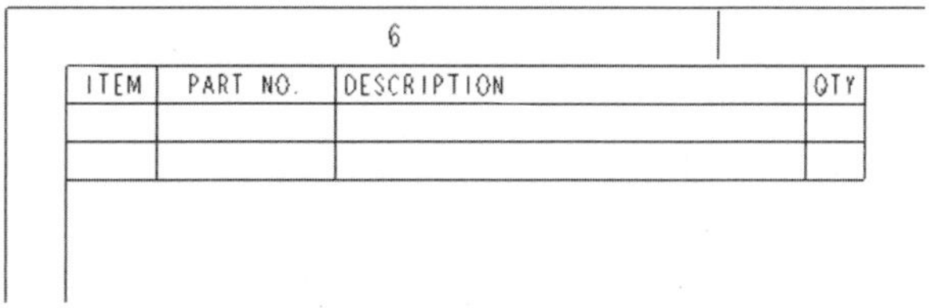

Start table needs "nested" repeat regions.

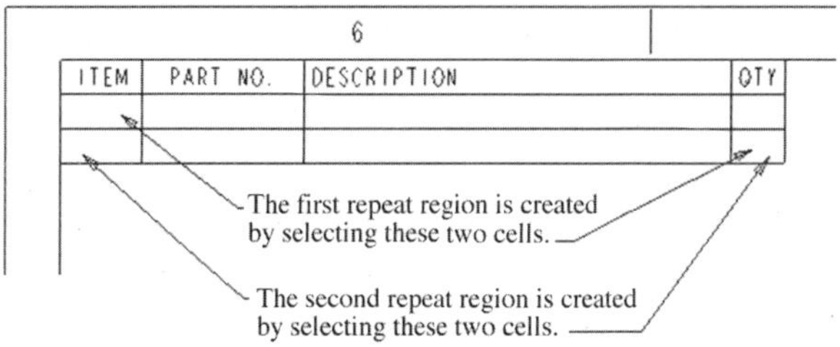

1. To create the repeat regions, select **Table** ➡ **Repeat Region** ➡ **Add** ➡ **Simple** and pick the upper left cell of the two blank rows and then the lower right cell. This creates the first repeat region. Select the lower left cell and the lower right cell to create the second repeat region. The second illustration at right shows exactly which cells need to be selected.

Selections required in creating the two repeat regions.

2. Add the parameters to the cells in the repeat regions. The parameters entered in the top row of cells set the display of the top-level assembly. To enter the parameters, select **Table** ➡ **Enter Text**. Enter the parameters by responding to the entry prompts as follows.

```
ITEM:          rpt… > index
PART NO:       asm… > mbr… > name
DESCRIPTION:   asm… > mbr… > USER DEFINED > name_1
QTY:           rpt… > qty
```

3. The parameters in the second row set the display for the subassembly components. Enter the parameters as follows.

```
PART NO:       asm… > mbr… > name
DESCRIPTION:   asm… > mbr… > USER DEFINED > name_1
QTY:           rpt… > qty
```

Nothing is entered in the cell under the ITEM heading because you do not want additional item numbers being assigned to the subcomponents. The completed table is shown in the first illustration at right.

Table with parameters typed in.

4. Regenerate the drawing to update the table and you will see that the display of the BOM is not correct, as in the second illustration at right. The clamp part is listed several times and there are blank rows between some of the items.

BOM table created by the display is not correct.

Both of these display problems can be fixed by modifying the attributes of both repeat regions. The attributes need to be set to No Duplicates and have a Min Repeats value set to 0. The No Duplicates option removes the listing of the clamp four times and activates the QTY column. A value of 0 for the Min Repeats (minimum repeats) will remove the extra rows in the table. You will change one repeat region at a time to see the effect the change has on the appearance of the table.

5. Select **Table ➥ Repeat Region ➥ Attributes.** To be confident that you select the top-level repeat region, select the top row of the region. The outline of the entire table should highlight. Select the **No Duplicates** option and the **Min Repeats** option and enter a value of *0*. Update the table (as shown in the following illustration) and notice that the clamp is listed only once and all of the top-level information is correct. However, the blank rows are still there and the quantities of the subcomponents are not displayed.

6. To correct the rest of the table, select **Table ➥ Repeat Region ➥ Attributes**. However, this time you need to select the nested repeat region. To select that region, pick the bottom row. Notice how the table highlights. Individual rows highlight, rather than the outside of the table as before. Select the **No Duplicates** option and the **Min Repeats** option, enter a value of *0*, and update the table. The second illustration at right shows the completed BOM table containing information on subcomponents as well as top-level components.

ITEM	PART NO.	DESCRIPTION	QTY
1	42071	CLAMP, FUEL LINE	4
2	96427	FUEL FILTER ASSEMBLY	1
	28317	HOUSING, FUEL FILTER INLET	
	28318	HOUSING, FUEL FILTER OUTLET	
	17220	FILTER, FUEL LINE	
3	65983-3	HOSE, FUEL LINE 3.00 LONG	1
4	65983-6	HOSE, FUEL LINE 6.00 LONG	1

Changing the attributes of the top-level repeat region corrected the top-level display.

ITEM	PART NO.	DESCRIPTION	QTY
1	42071	CLAMP, FUEL LINE	4
2	96427	FUEL FILTER ASSEMBLY	1
	17220	FILTER, FUEL LINE	1
	28317	HOUSING, FUEL FILTER INLET	1
	28318	HOUSING, FUEL FILTER OUTLET	1
3	65983-3	HOSE, FUEL LINE 3.00 LONG	1
4	65983-6	HOSE, FUEL LINE 6.00 LONG	1

Setting the attributes of both tables corrects the display completely.

Exercise 21-4: Showing BOM Balloons on a Drawing

One of the advantages of using a repeat-region-driven table is that you can automatically show the BOM balloons on your drawing. In this exercise, you will use a table from previous exercises to display BOM balloons on a drawing.

1. While in a drawing associated with one of the tables you created in one of the previous exercises, select **Table ➥ BOM Balloons ➥ Set Region**. Select the BOM table and then select **Show**.

2. Select one of the **Show** options to show the balloons. You will probably need to move the attachment points for the leaders and arrange the balloons so that they look right. However, this is faster than creating the balloons individually, and the balloons are linked parametrically to the models. The illustration at right shows

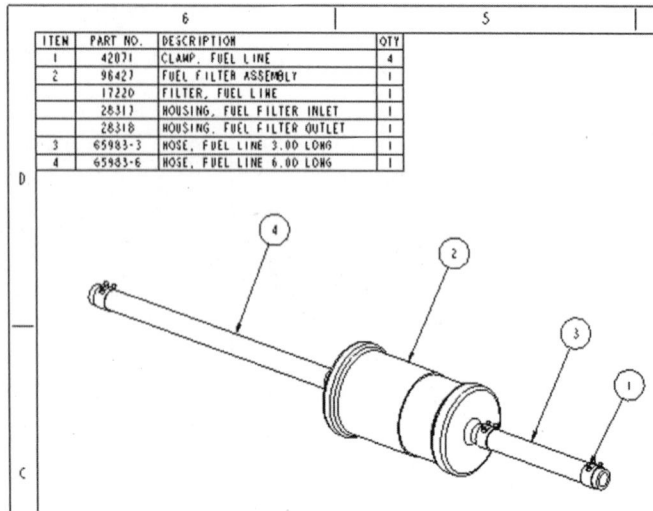

ITEM	PART NO.	DESCRIPTION	QTY
1	42071	CLAMP, FUEL LINE	4
2	96421	FUEL FILTER ASSEMBLY	1
	17220	FILTER, FUEL LINE	1
	28311	HOUSING, FUEL FILTER INLET	1
	28318	HOUSING, FUEL FILTER OUTLET	1
3	65983-3	HOSE, FUEL LINE 3.00 LONG	1
4	65983-6	HOSE, FUEL LINE 6.00 LONG	1

BOM balloons added to the drawing automatically.

the balloons after they have been moved into their proper positions.

Summary

The number of options and techniques used to create a drawing in Pro/ENGINEER is tremendous, and could easily fill a book of its own. This chapter has presented techniques you can apply to everyday situations. You learned to use show dimensions to better convey your design intent to the reader of the drawing.

The chapter also discussed techniques you might not use every day but that are very useful to know when they are needed. These included dealing with multi-model drawings and creating datum targets. In the exercises, you created several BOM tables and automatically created the BOM balloons in the drawing. All of the techniques in this chapter dealt with the idea of using the model information as much as possible in your drawings. The core point of the chapter was that the more information you get from the model, the more automatic the drawing becomes, and the closer it will match your design intent.

Index

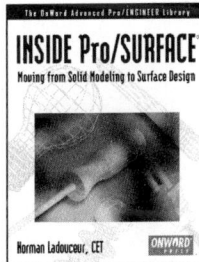

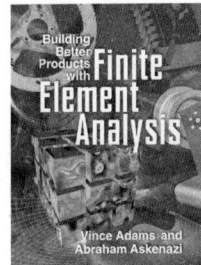